PATRICK GEDDES

One of the great social thinkers of the late nineteenth and early twentieth century, Sir Patrick Geddes (1854–1932) enjoyed a career of astonishing diversity. This new analysis of his life and work reviews his ideas and his philosophy of planning, providing a scholarly yet accessible account for those interested in the history of planning, urban design, social theory, and nineteenth-century British history.

A figure of international importance in the history of modern town planning and environmental studies, Patrick Geddes pioneered a sociological approach to the study of urbanization. He also argued that the city should be studied in the context of the region; predicted that the process of urbanization could be analysed and understood; and believed that the application of knowledge about the city could shape future developments towards life-enhancement for all citizens. Inherent in all these beliefs was the central idea that social processes and spatial form are intimately related. Yet, as an evolutionary biologist, Geddes never underestimated the complexity of that relationship, and his passionate, life-long concern was to understand its nature. Helen Meller believes that the true nature of Geddes' work can be established only by a study of his ideas in the context in which he worked. Her reassessment enables his contribution and his ideas to be viewed as a whole, as it describes his work in Scotland, Ireland, India, Palestine, and France, and traces the way in which Geddes' path crossed those of other pioneers planners and social scientists fired with his enthusiasm for making sense of the modern world.

Increasingly, those worried about the environment and those who contemplate what has been done with dismay are looking again at Geddes' work, and he has become an inspiration to groups varying from Scottish nationalists to ecologists and conservationists. Helen Meller's study shows that his critique of the process of urbanization and modern living is trenchant, relevant, and stimulating in the present day.

The Author

Helen E. Meller is Reader in History at the University of Nottingham. Author of *Leisure and the Changing City 1870-1914* and editor of *The Ideal City*, she has written widely on life in Victorian society, and is an international authority on the city.

Routledge Geography, Environment and Planning series
Series Editor Neil Wrigley

PATRICK GEDDES

Social evolutionist and city planner

HELEN MELLER

London and New York

First published 1990
Paperback edition first published in 1993
by Routledge
11 New Fetter Lane, London EC4P 4EE

Simultaneously published in the USA and Canada
by Routledge
a division of Routledge, Chapman and Hall, Inc.
29 West 35th Street, Nw York, NY 10001

Typeset by Columns of Reading
Printed and bound in Great Britain by
Mackays of Chatham PLC, Chatham, Kent

British Library Cataloguing in Publication Data

Meller, Helen, 1941 –
Patrick Geddes : social evolutionist and city planner.
1. Environment planning. Geddes, Sir Patrick, 1854 –
1932
I. Title
711'.092'4

ISBN 0–415–10393–2

Library of Congress Cataloging in Publication Data

Meller, Helen Elizabeth.
Patrick Geddes : social evolutionist and city planner / Helen Meller.
p. cm. – (Geography, environment, and planning series)
Bibliography: p.
Includes index.

ISBN 0–415–10393–2

1. Geddes, Patrick, Sir, 1854-1932
2. City planners – Great Britain – Biography.
3. Sociologists – Great Britain – Biography.
4. Anthropo-geography – Great Britain – History.
5. Sociology, Urban–Great Britain–History.
I. Title.
II. Series.
HT169.G7M45 1990
301'.092–dc20
[B] 89–10342 CIP

To the memory of
JEANNIE GEDDES

Contents

Contents

Contents

List of illustrations

Headpieces (after the manner of Celtic Ornament) from *The Evergreen: a Northern Seasonal*, by Effie Ramsay, John Duncan, Nellie Baxter, Marion A. Mason, Annie Mackie.

All line drawings and headpieces redrawn for publication by Catherine Houten.

[xi]

Preface

I first came across Geddes' work when, as a postgraduate student, I found a copy of his Dunfermline Report in the stacks of Bristol University library. I was fascinated by it and wanted to find out more about the author. I had no idea then where my quest would take me. Geddes has been like the Holy Grail and I have had to go through fire and water in my pursuit of him! His quicksilver character, his obscure methods of working, and the weight of the devotion of his disciples has sometimes left me in the slough of despond; but the very reverse has been true with much else. His total lack of reverence for convention and authority, his belief in the individual, and his brilliant analysis of cities, citizens and their problems have been full recompense. When his enthusiasm for enjoying himself and giving pleasure to others is added to this, Geddes' humanity shines through. It is really extraordinary how his ideas have grown from, or permeated, British cultural life. It is hard to tell which, since he was so starkly original in his own life and work and yet his ideas have become commonplace for subsequent generations. I hope that this book will give some indication at least of where he found his inspiration and how he worked out his ideas in practice.

Working over a number of years, and in a number of places, has meant that I have incurred a host of debts which it is my deepest pleasure to acknowledge. At their head must come my thanks to the former Social Science Research Council for funding my work in Scotland, India and France, and thus making the whole project possible. In Scotland, I was introduced to the Geddes Papers, University of Strathclyde, by the late Thomas Findlay Lyon. It is with regret that I have to say that many of the people I consulted about Geddes are now deceased. This is because I was fortunate enough to have contacts with many of the older generation who had either known Geddes personally or whose lives had been transformed by contact with his ideas. Tom Findlay Lyon came into the latter category and I am most grateful for his enthusiastic support of my project. I have to thank Professor Nicoll for permission to use the Geddes Papers which were then still uncatalogued; Dr Peter Green for information about the lost Indian Reports by Geddes and bibliographical help; and Jimmy Milligan for heated discussions and liquid refreshment so necessary for anyone working on Geddes' handwritten manuscript material. In Edinburgh, I would like to thank the Keeper of the Manuscripts and his staff at the National Library of Scotland; the archive room there became, for a while, my second home. I would also like to thank Emeritus Professor Percy Johnson-Marshall for help; Professor Christopher Harvie for walking me around Geddes' Edinburgh; Dr R. J. Morris for supplying me with photographs of Geddes' buildings; and Mrs Christine McGegan for finding for me a copy of her father-in-law's unpublished memoir of Geddes.

In London, I should like to thank the Librarian of the Royal Institute of British Architects who sent me copies of two of Geddes' Indian Reports for presentation, together with my collection of the Reports, to the National Library of Scotland. I should also like to thank Mrs Valerie Weston of the India Office Library, and Professor John Harrison of the London School of Oriental and Asian Studies, who guided my first forays into the history of Indian urbanisation. In India, on the Geddes' trail, I was overwhelmed by the kindness and helpfulness of all the people I met. I would especially like to mention the National Library in Calcutta where I was given such service that I was able to do in a couple of weeks what would have taken me months to do elsewhere. The library of the Calcutta Improvement Trust was able to supply me with some rare Geddes Indian Reports, though sadly the complete collection once to be found there no longer exists. My thanks to both Mr Gupta, formerly director of the CIT, and to Dr Narayani Gupta, his daughter-in-law and one of the few experts on the history of Indian cities. Our discussions on Geddes' impact on India still continue. I should like to record my thanks to Dr Gopal who helped me unravel the mysteries of the National Archives in Delhi and the Librarian of the India International Centre who procured a number of books for me. In Bombay, special thanks to Emeritus Professor D. N. Marshall, former Librarian of the University of Bombay, and his daughter, Niloufer, for both guiding me to source material and showing me nineteenth century Bombay. Dr Pheroze Bharucha and Dr J. F. Bulsara, students of Geddes, (the former following him to Montpellier), gave freely of their time to discuss their experiences. Emeritus Professor G. S. Ghurye, another student of Geddes and one of the pioneering professors of sociology in India, kindly consented to give me an interview. In Baroda, I had the great good fortune to spend two days in the company of the late Professor Achewal, then professor of architecture at the University of Baroda, discussing Geddes' Baroda report and his planning ideas in India. In Indore, I would like to thank the Maharaja of Indore for his kind hospitality and for giving me private access to the state archives. I am indebted to the late Mrs Gandhi who most kindly invited me to fly with her from Delhi to Santiniketan where she was going for a degree ceremony. This enabled me to visit Tagore's University which Geddes and his son, Arthur, had helped to build in the early 1920s.

In pursuit of Geddes in Palestine, I would like to thank the late Lord Bentwich who told me about his experiences of Geddes in Jerusalem in 1919, when he had daily contact with him. I should also like to thank Dr Josef Fraenkel of the Jewish Congress in London; and especially, Mr M. Heymann, Director of the Central Zionist Archives in Jerusalem, who went to great lengths to ensure that I had at least seen all the available material on Geddes in the few days I was there. In Montpellier, in the south of France, I was lucky enough to be received by the late Professor Marres, of the University of Montpellier, who had worked with Geddes in the 1920s. I also began a continuing contact with Professor André Schimmerling, who has retired from the School of Architecture at the University but is busier than ever promoting the Patrick Geddes Association, the Collège des Ecossais, and Geddes' message that architecture needs to be related to the historical context of place and people.

Finally, I should like to thank a number of friends and colleagues. First must be Professor Gerald Dix who was responsible for suggesting I should do this work on Geddes when he was the Professor of Planning Studies at the University of Nottingham. He also helped me to contact Mr J. Holliday, who let me have a copy of Geddes' Tel Aviv and Haifa Reports; the late Jacqueline Tyrwhitt; and the late Clough Williams-Ellis. For all this help and encouragement, many thanks. I would like to thank Professor B. T. Robson of the University of Manchester for discussions on Geddes' methodology and also Mark Dale, Adult Education, University of Nottingham. Professor Anthony Sutcliffe, University of Leicester, has given help on the international town-planning movement and much general encouragement as well. The late Emeritus Professor K. C. Edwards, University of Nottingham, gave me his personal recollections of the LePlay Society in the 1920s; and Professors R. Osborne and J. P. Cole, University of Nottingham, have given me guidance about the early history of academic geography. Mr Michael King, Dr Danny Lawrence and Dr Ken Levine, all of the Department of Sociology, University of Nottingham, have been very helpful. To those who have read part or all of the manuscript, I am deeply grateful. Dr David Massey, University of Liverpool, has read the chapter on regionalism and town planning; Professor J. E. Thomas, University of Nottingham, has commented on the sections devoted to Geddes' views on education; and above all, Professor William Ashworth has given most generously of his time and read the whole manuscript very carefully, saving me from innumerable errors. For any errors that remain, I have to accept full responsibiliy.

My warmest thanks to Mrs Jenny Chambers who joined my struggle with the ageing word-processor and helped me with the burden of typing, and corrections and yet more corrections. I would also like to thank Ms Catherine Houten for all the excellent artwork she produced which quite transforms the appearance of the book; Mr Peter Whitehouse who drew the diagram of the proposed dome for the Hebrew University of Jerusalem; and Mrs Anne Shalit who supplied the photographs of Palestine in the 1920s. I would like also to acknowledge the original artists for *The Evergreen*, 1896, Helen May, Nellie Baxter, Annie Mackie and Effie Ramsay, whose work is included here.

I cannot conclude this list of debts incurred without a special mention of the late Mrs Jeannie Geddes. She looked after me whenever I was in Edinburgh and became a close and dear friend. This book is dedicated to her memory as a measure of my deep affection and appreciation of her friendship; and also as a tribute to the warm welcome she gave to the many Geddes scholars who turned up on her doorstep in search of information about Patrick Geddes. I would also like to include her children, Anne, Marion, Claire and Colin, who have made me feel part of their family. Last but not least, I want to express my deep thanks to my own family and friends who have put up with me while I have been engaged on this project. My daughter, Meesha, has lived with the making of this book all her short life, and her irreverent attitude and irrepressible good humour on the subject of Geddes has sustained me more than she knows.

Frontispiece   Pencil drawing of Patrick Geddes, November 1916

CHAPTER 1

Introduction

THERE HAS BEEN A RENEWAL OF INTEREST IN THE IDEAS and planning philosophy of Sir Patrick Geddes. Who was this man who pioneered a sociological approach to the study of urbanisation; who discovered that the city should be studied in the context of the region; who confidently predicted that the process of urbanisation could be analysed and understood; who believed that the application of such knowledge could shape future developments towards life-enhancement for all citizens? Inherent in all these beliefs was the central idea that social processes and spatial form are intimately related. Yet, as an evolutionary biologist, Geddes never under-estimated the complexity of that relationship. His passionate, life-long concern was to unravel, as far as was humanly possible, the nature of these complexities, moving into areas outside the conventional limitations of the social sciences.

Only recently have modern scholars returned to these preoccupations and few of them even refer back to the work of Geddes.[1] The prediction of Lewis Mumford, one of the most famous of those influenced directly by Geddes, has remained unfulfilled. Mumford wrote in 1950:

Patrick Geddes is fast becoming a rallying centre for the best minds of this generation; his thought, like that of his old associate and friend, Kropotkin, will probably guide the future, since the mechanists and Marxists in the present hour of their triumph, demonstrate the failure of their philosophies to do justice to either life or the human spirit.[2]

Perhaps Mumford's outburst had as much to do with his feelings about the 1950s as his championship of Geddes. Yet the fact remains that Geddes' ideas are not accessible, like those of his friend Kropotkin, even for those of the present generation who might be interested in them.[3]

The reasons for this lie partly in the way he chose to work, and partly in the character of the man himself. Working outside any conventional framework either institutional or academic, Geddes never laid claim to any particular body of knowledge. He would parry criticism of his various expositions by producing new ideas rather than defending an established position. He had a brilliant facility for demolishing the ideas of others from which he gained much pleasure. He was a restless 'entrepreneur' in the newly-developing social sciences, who preferred to test his own ideas in personal debate which tended to give him the advantage. He reached out to as many individuals as he could by constant travelling, fleeting exhibitions, and lecturing, and was happy when he met people receptive to his ideas. Believing that achieving his objective: social and environmental improvement and regional self-determination, could not be done by book learning alone, he was too impatient to spend his time developing his ideas in a major treatise.

Yet he wrote and published prolifically. Apart from his commissioned town planning reports (between forty and fifty in total) the bulk of his output was short articles, lectures, reviews, pamphlets, and books with collaborators.[4] *Cities in Evolution* was the only full-length work he produced on his own. He tried, characteristically, to illustrate his ideas in the writing of his planning reports where theoretical shortcomings could be excused, or in some instances even illuminated by practical considerations. In the late 1920s he made his last great effort to communicate his ideas by adding some chapters to a textbook on biology written by his old friend and former pupil, J. Arthur Thomson.[5] Thomson, who became Professor of Natural Sciences at Aberdeen University, was the earliest and best of Geddes's collaborators, writing with him the esteemed but controversial study *The Evolution of Sex*, which, in 1889, first established Geddes' reputation, and a number of other biological works. During the First World War Geddes became involved in a new series with another collaborator, V.V. Branford, entitled *Making of the Future* which was designed especially as a vehicle for his ideas on environmental planning and social change. Victor Branford, a former student of Edinburgh University and an amateur sociologist of some influence in Britain, did most of the editorial work as well as contributing his own volumes.[6] However, it was an inauspicious time for Geddes since he was fully occupied with his planning and exhibition work, and the books produced in his name were of very inferior quality. His published work therefore tends to belie Mumford's famous claim.

The problem with Geddes' publications is not only their varying quality. Much of what he wrote is difficult to follow because he developed a totally idiosyncratic approach to the very concept of knowledge and created his own unique methodology. Most of his published work was aimed at initiating readers in his special outlook and approach. Thus, in every short article he began again at the beginning, outlining and repeating his basic ideas endlessly. This is what makes his publications unhelpful for a full understanding of his contribution. Mumford was ready to recognise that 'his books, even when supplemented by his manuscripts and notes, were only a small part of his total productivity'.[7] Geddes used his vast, private correspondence as a vehicle for developing his ideas, seeking and demanding the specific response of an individual. In this way he was able to communicate what he was trying to do rather than what he had achieved. He constantly craved a sympathetic indulgence towards the approach he adopted, which was outside the confines of any known discipline. What then, was the true nature of his quest and the special contribution he was to make?

Geddes was a child of his generation in that his mission was to foretell the future and ensure that it would bring improvements in the quality of life for all. Like his slightly younger contemporary, H.G. Wells, who shared with him the experience of being a student of T.H. Huxley, what fascinated him most was the potential brought by modern knowledge to transform society.[8] He saw the most fundamental question challenging the present and future generations as the relationship of man with his natural environment, whether that relationship was defined in global terms, or on the purely local scale of countryside and town. Knowledge in the natural and physical sciences had the potential to change completely the traditional equilibrium between human society and the environment. The problem was to motivate people to make the right choices in using their new-found power and this was both a matter of cultural conditioning and a moral challenge. Geddes wanted to transform the nineteenth century ideal of progress: 'from an individual Race for Wealth into a Social Crusade of Culture'.[9]

The keyword here was 'culture'. Geddes stretched the term to encompass his entire philosophical approach. What determined the quality of life in his view was the interaction of spatial form with the culture of the people. The future depended on cultural evolution. With this belief Geddes was able to view the political debate about the future as largely irrelevant. He chose to consider that his position was above the fierce discussions about capitalism and its social consequences. As he wrote in the 1880s, as far as the capitalist and the labourer were concerned: 'For both life is equally blank at present: the capitalist, in his big ugly house, is no happier than the labourer, in his little ugly one'.[10] During and after the First World War his disciple, V.V. Branford, was to help him launch his crusade for the 'Third Alternative'.[11] It was to be a crusade with the humanitarian object of cultural evolution which would be produced by an interaction of environment, modern knowledge and the historically determined values of the people. Such an objective cut across party lines and could appeal to people of all political persuasions, or so Geddes and his followers believed. This may have been a pipe-dream. Yet working for his cause he was to give voice to many concerns which have subsequently been recognised as vital challenges to the survival of mankind in the twentieth century.[12] His role as a pioneer has thus been amply confirmed.

By the time that Geddes was beginning his life's work in the mid-1880s, technological advances and urbanisation had already altered profoundly the relationship between man and his natural environment. Great Britain, by the third quarter of the nineteenth century, had become the prototype of a modern, industrial, urbanised civilisation, which was to be experienced in differing degrees by many other countries in the world. Geddes set himself the task of acquiring an understanding particularly of the process of urbanisation, which he called 'city-development', in the hope that, through such knowledge, there was a chance of directing change away from trends which were destructive to the individual, to the community, and to the human spirit. He sensed the unrest amongst the youth of his generation, disturbed by the social consequences of industrialisation and urbanisation. The desire for change, revolutionary movements, the growth of nationalist agitation amongst subject nations, were all, he believed, the result of changes which if left undirected could lead to catastrophe. It was in the period of social turbulence and unrest after the First World War that he was to write in his Town Planning Report for Indore in 1918:

> It is from the section of youth least contented with the present, most determined to advance upon it, and thus more or less in 'unrest', that revolutionaries are at present drawn; yet these are but so many strayed pioneers. . . Let us educate such restless spirits in the main

aspects of life, in appreciating the corresponding departments of its activity, and sharing in them too — Industrial and Aesthetic, Hygienic and Agricultural, Educational, Economic and Social. Yet also Ethical; with faith and effort in the possibility of these, in their community, their city, and its betterment around them.'[13]

Comprehensive in approach and lacking in compromise, Geddes led his life from the earliest days dedicated to 'the cause'.

He was born in October 1854, at Ballater in west Aberdeenshire, Scotland, the youngest son of a one-time sergeant-major of the Highland Regiment, the Black Watch. The family subsequently moved and he was brought up and educated in Perth. The circumstances of his family and early life were important to him as he, and his subsequent biographers, have laid great stress on them to explain the extraordinary nature of his genius and his career.[14] He was ready to claim that he was descended from peasant stock and had frequent recourse to a vision of himself as a sturdy independent Highlander and Scotsman. In cultural terms he saw himself as an outsider in the mainstream of English life, the dominant world culture of the time. But in the broadest sense of the term 'culture' he believed he was a countryman with a close understanding and empathy for nature. Such a feeling for the natural world, he was prepared to argue, was in fact an ideal starting point for studying change in modern civilisation. His claims were a little romantic. The nearest relative he possessed who might be claimed as a peasant was his grandmother. She came from farming stock in the Highlands though by no means the poorest. It was a sad but perhaps appropriate stimulus for her grandson's career that she, and her merchant husband, went to Glasgow to seek a living and both soon died, victims of an outbreak of cholera in the unhealthy city.

Their orphaned son, Geddes' father, followed the path of many an impecunious Scotsman by enlisting in a Highland Regiment. He made a success of this career rising to the highest non-commissioned rank, the limit to his promotion being determined by his lack of a private income. By the time his last son had been born (he was a late, unexpected child), his father had been put on the reserve and was now permanently based near Perth. Patrick therefore grew up living in a small cottage in the hills outside Perth which his religiously inclined parents called Mount Tabor, with his family of ageing parents and the older children, two boys and a girl. His elder brothers left home while he was still a child, both to seek their fortunes overseas. The nearest in age to Patrick was his sister Jessie who was thirteen years his senior and who was closely concerned with the welfare of her young brother in his childhood. He was much loved and his father, who had been an extremely strict disciplinarian with his older sons, was able to spend

[5]

more time with him, and to be more indulgent to him than he had been with the elder children. Geddes was often to claim that his father was his first and best teacher, and to eulogise on how he had given him the finest education for life by teaching him especially how to care for a garden. The skills, discipline and understanding that this involved were, so Geddes believed, just those which needed to be applied on a broader more advanced level to the problems of man's control over his environment.

Geddes' subsequent career was as idiosyncratic as his approach to modern problems. After periods of study in the natural sciences in London and Paris, much travelling in Europe, and a visit to Central America, Geddes went back to Edinburgh in 1880 at the age of 26. He was to make the city his home base for most of his life, although he was unable to make a living there for much of that time. His first paid employment was at the University of Edinburgh where he acted as an Assistant to the Professor of Botany and as a part-time lecturer in the natural sciences at the School of Medicine. He supplemented this with some extramural lecturing, though most of his spare time was occupied with unpaid voluntary work. He applied for four Chairs in the 1880s, the last being the Regius Chair of Botany at Edinburgh, but with no success. His reputation for originality and his idiosyncratic approach to learning did not help him when disciplines, especially the natural sciences, were seeking to define and extend their limits and to win academic respectability.[15]

He finally achieved his first permanent post in 1889 at the age of 35. His friend and benefactor, James Martin White, heir to a large estate near Dundee, endowed a Chair of Botany at Dundee College, affiliated to the University of St Andrews, on special terms. These were that Geddes should occupy the chair, but that it should be clearly understood that his major energies were to be left free for activities outside the university. He was only required to teach and be in residence in Dundee for three months a year coinciding with the summer term. These terms were agreed upon because Dundee College badly needed funds to expand its academic activities and such an agreement suited Geddes. He remained Professor of Botany at Dundee for the next 30 years, though for much of that time his work there remained insignificant in comparison with the range of educational, philanthropic and town planning activities which he undertook over the same period. He was able to build for himself a career in the social sciences on an *ad hoc* basis while retaining his title as Professor of Botany. It was an arrangement which suited him well.

Geddes was supported throughout his life in his work by his wife, whom he met in Edinburgh, his three children, and a small group of friends, many of them former students at Edinburgh University. To this small nucleus was added a number of Edinburgh friends from artistic and literary circles whom Geddes met through his cultural activities, and a few philanthropic individuals, both men and women, whom Geddes met in the course of his voluntary social work. The centre of Geddes' circle considered themselves as radicals and members of the 'avant-garde' in contemporary society. But the radicalism was literary and artistic rather than political. Socio-cultural questions such as man's relationship with nature, neo-Romantic poetry, and the 'woman question' were the kind of issues which concerned them. The 'woman question', indeed, was much debated by them, and the outcome of these discussions provided the context for Geddes' first and most influential monograph, which he wrote with J.A.Thomson, *The Evolution of Sex* (1889).[16]

Women played an important part in promoting Geddes' work. The 'New Woman' of Geddes' small circle was, however, no suffragette or Fabian lady lecturer. According to the Geddesian canon she was liberated:

> to fulfil her true biological destiny as wife, romantic companion and ideal mother. Her role was to inspire her man; to know intuitively what needed to be done; to nurture cultural and spiritual values which were the potent elements for generating the highest quality of social evolution.[17]

Geddes' willingness to commit himself fully on controversial subjects such as women's role in society and the relationship between the sexes added an aura of daring and lack of respectability to the Geddes' circle. They counteracted this by adopting a high moral tone in their aspirations and personal relationships. Much of Geddes' success in leading his little group successfully in such socially difficult terrain was due to his wife. An Englishwoman, daughter of a Liverpool merchant, nurtured in the traditions of middle-class social mores, she lent credibility and authority to her husband's social crusade of culture.

Anna Morton's support for her husband, whom she married in 1886, was the keystone of his life and career. Their marriage appears to have been successful despite the enormous strains put upon it by Geddes' total devotion to his work, financial difficulties, and long separations.[18] Anna complemented her husband in many respects. To his fierce energy and wild enthusiasm, she brought a calm level-headedness and a strong common sense. While he indulged in grandiose schemes, she undertook responsibility for the essential details on which any successful outcome depended. She wholeheartedly supported his mission to the world, believing his work to be of the utmost importance, but she did not share his neo-Romantic, somewhat self-indulgent approach at all. Her one outlet, in an often busy and harassed life, was her music. She was a gifted piano player whose music played a part in the many social gatherings centred on the Geddes home. This was something she did not share with her husband. Geddes himself was tone deaf, gaining his greatest aesthetic pleasure through the eye rather than the ear.

Many friends paid tribute to the support she offered him. Geddes, after her premature death from enteric fever in India in 1917, admitted that he had 'subjected poor Anna to overstrain'.[19] He wrote to his daughter of the difficulties, mostly financial, that they had faced together, saying:

> I trust too you will have less of the anxieties which could not but weigh on us, in circumstances you do not yet fully know, and of which I must tell you — the more since I knew you have sometimes felt the atmosphere (and though we did our best to keep it from you all, it could not but be felt).[20]

Geddes and his wife had very little private time together as 'home' for the family was usually Edinburgh, while Geddes, after his short spell in Dundee each year, spent much time either travelling or, when actually in Edinburgh, out working on his schemes around the Old City. On a typical occasion Anna rushed home to meet him on his return from Dundee (she had been out on business for him) only to find that he had returned and gone out again immediately.[21] Geddes tried to sustain the 'romantic' element in his marriage by occasionally writing his wife special love letters but while he started these professing his undying affection, he had always moved to more general discussion of environmental problems before he reached the end.[22]

What Anna, and Geddes' closest disciples such as J. Arthur Thomson and Victor Branford, were all prepared to do, was to sacrifice their immediate interests for the sake of the cause. What Geddes brought to them, apart from an uninhibited capacity to give and receive love, was excitement, interest and sparkle. His most fruitful collaborations were with men such as J. Arthur Thomson, and his future son-in-law Frank Mears, who had naturally dour, introverted personalities from which they longed to escape. As Thomson wrote: 'I was born stiff', and Geddes was the release, the stimulus which his friends found intoxicating.[23] The price that they and his family had to pay though was that all other interests had to be subservient to those of Geddes himself. His elder son Alasdair, who coped with this better than his siblings, his older sister Norah and his younger brother Arthur, could still be reduced almost to tears by his father's autocratic waywardness. One such occasion is recounted by Geddes' first biographer. It concerned the hanging of the Cities and Town Planning

Exhibition at Ghent in 1913 when Alasdair was 22 years old. Geddes had sent him on in advance to put up the exhibition, then arrived late and insisted it should all be re-hung on lines that had just occurred to him on the journey there.[24] His youngest son, Arthur, was treated by Britain's first psychoanalyst, Dr David Eder, who diagnosed him (in his mid-twenties) as suffering from a 'father-complex' which reduced him to periods of acute depression.[25]

Notwithstanding these faults, Geddes was deeply loved. Despite the high moral tone and lofty objectives of his social mission, those belonging to the inner circle of Geddes' friends had enormous fun. The romanticism, the nature worship, the forays into a (somewhat fictitious) Celtic past, were delightfully unconventional in comparison with the norms of social behaviour of Edinburgh society.[26] The social activities of Geddes' Summer Schools included day-long rambles in the hills, followed by evenings singing highland songs with Mrs Kennedy-Fraser.[27] Geddes' friends and supporters were convinced that this was a preferable way to be radical and avant-garde rather than other alternatives of the time. Rachel Annand Taylor, a friend of the Geddes', and a poetess with a sharp wit, gave her response to the Brave New World of the artistic avant-garde in the England of 1910 after a visit to Letchworth:

> It was really worth while going, for the relief of coming back was so great that it sent me absolutely into high spirits — I would not live in the thin cold air among conscious little 'arty' houses and gardens of Letchworth for untold gold. The hard white light withered me, the formal Fabian 'intelligence' of Mrs. Ratcliffe (a most estimable lady, a perfect wife and mother in her way) drove into my dark, quivering heathen soul.

She had to go to Edinburgh to get 'recharged'.[28]

Geddes' two most important disciples, J. Arthur Thomson and Victor V. Branford, were both Edinburgh students in the early 1880s. Both were deeply concerned about moral issues and personally fought their way from feeling the need for a religious faith to acquiring a non-religious moral code no less demanding but more in tune with the challenges of modern society. J. Arthur Thomson had actually got as far as being on the point of entering the ministry of the church when he met Geddes. Geddes pointed him in the direction of the Comtean post-theocratic state and convinced him of the importance for mankind of working in the natural sciences. In 1886, aged 26, Thomson wrote to Geddes in deep gratitude for his advice on this. His language is redolent of the crisis of conscience which had beset earnest and eager students since the wide dissemination of the concept of evolution

[9]

and the conflict it appeared to engender between religion and science. Thomson wrote to Geddes:

> By your help I was slowly led, not without pain, to a wider synthesis and surer knowledge. I was born again of hope, and under my impenetrable hide a new enthusiasm burns; all things become a second time new in the light of scientific synthesis. For this I thank you.[29]

Geddes' circle was augmented by the generations of students whom Geddes met through his student halls in Edinburgh, his summer meetings, or the very select few whom he took as his assistants to Dundee.[30] They formed a faithful band of supporters who worked for him not only in Edinburgh but also at the Paris World Exposition Special Summer School of 1900, and in London, when Geddes moved his main base there after the turn of the century. A faithful one or two such as T.R. Marr were still around to help him with his Collège des Ecossais in Montpellier at the end of his life.[31] Beyond this charmed circle Geddes' warmest contacts were with individuals working in disciplines in which he had a particular interest, such as the geographer, H.J. Fleure, the Bengali natural scientist, Sir Jagadis Bose, and the architect planners, H.V. Lanchester and Patrick Abercrombie.[32] Of these, perhaps, Bose became closest to Geddes on a personal level. He was a particularly warm-hearted man, and he showed Geddes much loving concern after the deaths of Alasdair and Anna in 1917. Soon after the opening of his own research institute in Calcutta for which he had fought long and hard, Bose wrote to Geddes: 'Your letter warmed my heart. Write me always like this. Let me speak to you as my other self and you do the same . . . You have very few even in your own country who understand your aims and I am in a worse predicament'.[33] Bose's view of Geddes, coloured by his own experience, as the lone outsider misunderstood by the world, was shared by Geddes himself and all his intimate friends. He was aware that people most often regarded him with his self-imposed mission as a crank. He himself summed up rather sadly the common reaction to the kind of social crusade he wanted to promote in his Indore Report:

> But though in our duller everyday moods, we mostly incline to be more or less preponderatingly fossil in the past, or philistine in the present, there is also another class of minds, thinking and dreaming in the future; and whom we therefore call Utopists, when they seem simply dreamers, or else 'Cranks' when they seek to accomplish something towards the future, and so obtrude it upon our present.[34]

Geddes' views of himself as an outsider did nothing to help him develop a style of writing which was readily comprehensible. He was almost incapable of writing simple prose. He shared with many pioneer sociologists an unsureness of touch when it came to expressing his ideas. C. Wright Mills has made some interesting comments on language style and the pioneer sociologists. He suggests,

> lack of ready intelligibility, has little or nothing to do with the complexity of subject matter and nothing at all to do with the profundity of thought. It has to do almost entirely with certain confusions of the academic writer about his own status.[35]

Those that write in readable prose recognise themselves as 'a voice', and assume that they are speaking to an educated and wide-ranging public. Those who recognise themselves as a voice, but are less sure of their audience, develop tendencies towards a lack of intelligibility in their prose style. If they feel they are less 'a voice', and more the agent of some impersonal sound, then the style becomes a formula, and the public, if one is found, will be disciples of the cult.[36] Geddes' writing would appear to fall in all three categories with instances of the second and third increasing over the course of his life. It was part of the price that he paid for continuing his activities outside the institutional framework of academic life.

In the very last major statement of his ideas that he made in *Life, Outlines of General Biology*, (published in 1931), Geddes wrote specifically about his feelings on this matter. He was aware that, in spite of all his efforts, he had never found an audience. Characteristically, he recounted this experience in the form of a dream. It is worth quoting this passage in full because it gives not only a flavour of Geddes' prose style but also an insight into how he saw his life's work as a mission to the world. In his dream, heavy with symbolism, 'Life' is represented as a great organ:

> AN ACTUAL DREAM − After long and perplexed thinking, of how it has come to be that Life, and its evolutionary development and expression, still so generally fails to interest either the experts of the physical world or the scholars of the humanities − came sleep, and then dream. In the vast hall of a great building its organ is taking shape. Its main pipes, stops and swells are already in place: the musician is at his keyboard, and the audience encouragingly streams in. But they are gesticulating in active debate with each other; and as they come nearer, he finds with astonishment they are deaf, and so wonders what interests them here. For most, he finds the interest of the organ is as a great and complex machine: or, for a few others, an unusual form of architectural façade: but neither discern its real nature, much less its possibilities. A scholarly group open and scrutinise the music-books; but when he hopefully turns to them,

he finds it is their strange notation and printing which interests them; and that they conceive of such characters, such books in libraries as all that music has to offer; for they too are deaf.

A new group enters by the Eastern door; and he is relieved to find them free from these various material views of the West, indeed frankly contemptuous or pitying of them. They know that music is in the soul; but, alas, there they leave it: they have no use for his organ, his scores, or his active voicing of them; for them all alike are but material and external: so they are deaf to him also. And though many bear upon their breast or brow an outward and visible symbol of the movement of life, they have too often turned it the wrong way.

Wellnigh in despair he makes sign for silence: he begins to play, but none listen; he can reach no ear! Indeed, each group shows dissatisfaction, and that increasingly; till at length, and from all sides, they pounce on him to tear him in pieces. So thus − as nightmare − ended this actual dream. Yet first it was a dream of hope, indeed of full confidence in the Muse-world of Life; and that in its unified material and psychic expression it could not but reach all minds − the esthetic [*sic*], the mechanical, and the learned alike: or surely at least those of the Orient, with their inner life oftener awake, and more deeply cultivated. But neither the minds confined to external outlook, nor those of strictly interior meditation and discipline, suffice for this. Vital Synthesis is not reached by either: since Life, from its simplest to its highest manifestations, evolves through increasing interaction of the inward with the outward world, and conversely. With this conception hope returns, with faith in the possibilities of this full Muse-Organ of Life, even to the awakening and arousal of its spectators as auditors − sometimes even to organists in their turn.[37]

Geddes belonged to a whole generation of writers, thinkers, and philanthropists who were piecing together a critique of the Industrial Revolution and its social consequences.[38] This was the first vital step towards coming to terms with the vast social changes which were coming in its wake; changes which challenged society to adjust ever more speedily to new styles of living. Pioneer social scientists like Geddes often turned to the natural sciences for guidance in their work.[39] The concept of evolution seemed so well suited to this particular task. Where Geddes was different was his conviction that this concept was so revolutionary itself that it required an entire break with all known approaches to knowledge and its application. What was needed, Geddes suggested, following John Ruskin's lead, was to create a new way of thinking centred on the production and development, not of goods, but of people.[40] For Geddes this was the lesson of the discovery of the theory of evolution, and it was also a matter of common sense once society had achieved means of generating wealth, so that no one need go hungry.

This was his 'humanistic philosophy' which so inspired Mumford. Benton Mackaye, another member of the Regional Planning Association of America, recounts how the meeting he had with Geddes helped to confirm him in the direction his work was taking. He had been talking to him about his work in conservation and planning. Geddes told him that his subject was 'not conservation, not planning, not even geography. Your subject is geotechnics'.[41] Mackaye discovered geotechnics was defined as the applied science of making the earth more habitable. In the broadest possible sense it was what Geddes himself had always had as his goal, and what he believed was possible in the circumstances of active mass democracy. Capitalism, socialism, or state intervention were not the direct means to these ends. Survival of the human species depended on achieving a new equilibrium between a natural and man-made world. That went beyond physical environmental planning to cultural evolution, and that was the challenge of modern civilisation.

What follows in subsequent chapters is first a discussion of the origin and context of Geddes' theories and approach to knowledge. The most formative years of his life in this respect were those in which he was pursuing his own education, roughly between the ages of 17 and 27. In the 1880s he developed his views on the role of the social sciences in society, and worked out his biological approach to economics. He also undertook his first practical work renovating property in Old Edinburgh. His wide-ranging work in pioneering new forms of educational activity, particularly his summer schools, and the contribution of his special museum, the Outlook Tower, are the subjects of the next chapter. This is followed by discussions of his involvement in the evolution of the social sciences in Britain, especially geography and sociology. His increasing commitment to the town planning movement after the turn of the century, culminating in the Cities and Town Planning Exhibition at the Royal Institute of British Architects Conference in 1910, the Ghent International Exposition in 1913, and his work in Dublin before the First World War, provides subject matter for another chapter. During the war Geddes moved away from the main stream of the conflict in Europe, to take up the challenge of social reconstruction in the mainly rural Indian subcontinent. This was a challenge which took up most of his extraordinary energy in the sixth decade of his life, though he still found time to work in Palestine during the same years making a considerable impact wherever he went. Finally there was his retirement project in Montpellier, and in planning terms, the vicissitudes in the use of Geddes' concepts of regionalism, one of the most critically important aspects of his legacy to the planning profession. Geddes' full and varied life is only with difficulty contained, even within this wide range of subjects. It is with reason that Mumford, in his book *The Condition of Man*, likened Geddes to the modern equivalent of Renaissance man, man full of energy, ideas and creativity, the most fully alive person that Mumford had ever met.[42]

Notes

1. See, for example, D. Gregory and J. Urry (eds) (1985) *Social Relations and Spatial Structures*, London: Macmillan; R.D. Sack (1980) *Conceptions of Space in Social Thought*, London: Macmillan; Peter Saunders (1986) *Social Theory and the Urban Question*, London: Hutchinson, 2nd edn; David Harvey (1973) *Social Justice and the City*, London: Edward Arnold.
2. Lewis Mumford (1950) 'Mumford on Geddes', *The Architectural Review*, August, p.82.
3. There have been a number of books, theses, pamphlets and articles written about Geddes. These are cited in the **Select Bibliography**.
4. See **Select Bibliography**.
5. John Arthur Thomson, born 8 July 1861, died 12 February 1933. Brought up in East Lothian, son of the Reverend Arthur Thomson, he was educated locally and at Edinburgh University; he also attended the universities of Jena and Berlin. He was first appointed lecturer in the School of Medicine at Edinburgh University. He gained the Chair of Natural History at Aberdeen University in 1900 and stayed there until he resigned in 1930 when he was knighted for his services to education. He published extensively mostly popular science books and textbooks, but his most important publication was *The System of Animate Nature* (1920). He was particularly interested in studying the compatibility of Darwinism with religion.
6. Victor V. Branford, born 1864, died 22 June 1930. Brought up in East Anglia and educated at Oundle and Edinburgh University where he first encountered Geddes. He had a career in business with special interests in South America. He became chairman of the Paraguay Railway. He always wanted, though, a career in academic life, and kept in touch with Geddes, attending the Edinburgh Summer Schools and being the moving force behind the setting up of the British Sociological Society in 1904, which he hoped would become a vehicle for publishing Geddes' ideas. He edited the *Sociological Review* from 1912 to his death. In 1920, with his second wife, Sybella Gurney, he founded Le Play House. His main publications were *Interpretations and Forecasts* (1914) and *Science and Sanctity* (1923).
7. Mumford, op.cit., p.82.
8. H.G. Wells (1905) *A Modern Utopia*. See also P. Alexander and R. Gill (1984) *Utopias*, London: Duckworth.
9. P. Geddes (1886) 'On the Conditions of Progress of the Capitalist and the

Labourer', *'Claims of Labour' Lectures*, no.3, Edinburgh Co-operative Printing Co., p.34.
10. Ibid., p.36.
11. Branford invented the term the 'Third Alternative' as a propaganda slogan.
12. For example, the ecological balance between man and his environment, the conservation and preservation of historic cities, the role of women in modern society, the preservation of peace, and so on.
13. P. Geddes (1918) *Town Planning towards City-Development: report to the Durbar of Indore*, Indore: Holkar State Printing Press, Part II, p.37.
14. P. Boardman (1978) *The Worlds of Patrick Geddes: biologist, town planner, re-educator, peace warrior*, London, Routledge & Kegan Paul, pp.8−26; P. Mairet (1957) *A Pioneer of Sociology: life and letters of Patrick Geddes*, London: Lund Humphries, pp.1−13; P. Geddes (1925) 'The Education of Two Boys', *Survey Graphic* (New York) 54, (September), reprinted in M.Stalley (ed.) *Patrick Geddes: Spokesman for Man and the Environment* New Brunswick: Rutgers University Press, 1972. pp.365−80. See also chapter 2.
15. See chapter 2, pp.25–31.
16. See chapter 3, pp.79–84.
17. J. Conway (1972) 'Stereotypes of Femininity in a Theory of Sexual Evolution', in M. Vicinus (ed.) *Suffer and Be Still: women in the Victorian Age*, Bloomington: Indiana University Press, pp.140–54.
18. She is sympathetically drawn in all the biographies of Geddes, particularly in Boardman, op.cit., *passim*, and P. Kitchen (1975) *A Most Unsettling Person: An introduction to the ideas and life of Patrick Geddes*, London: Gollancz, pp.102−111. See also correspondence between Geddes and his wife, Geddes Papers MS10503, National Library of Scotland, Edinburgh.
19. Patrick Geddes to Norah, 11 March 1921, Geddes Papers MS10502, NLS.
20. Patrick Geddes to Norah, no date, Geddes Papers MS10502, NLS.
21. Anna Geddes to Norah, 2 July 1911, Geddes Papers MS10501, NLS.
22. Patrick Geddes to Anna, letter from New York 1899, Geddes Papers MS10503, NLS.
23. J.A. Thomson to Geddes 1886, Geddes Papers MS10555, NLS.
24. A. Defries (1927) *The Interpreter: Geddes, the Man and his Gospel*, London: Routledge & Kegan Paul, p.69.
25. Patrick Geddes to Norah 17 January 1921, Geddes Papers, MS10502, NLS. Geddes' three children gave him lifelong support and loyalty: 1.His eldest daughter, Norah (1887−1967) was able to benefit from the unusual education she received and from her father's belief in the cultural mission of women. She became a landscape gardener able to collaborate with her father on planning schemes, but much of her work was directed towards social and philanthropic ends. She worked in open spaces around Old Edinburgh and organised an Outlook Tower team to help her. She went to Dublin with Geddes to plan the little parks and open spaces designed to bring light into the lives of the slum dwellers and their children. She married Frank Mears, a young architect, who was working with Geddes before the first World War and who went with him to Palestine in 1920. Norah's relationship with her father was sometimes stormy when she found the burdens he laid on them all unbearable. 2.The elder son, Alasdair (1891−1917) provided Geddes with the example which justified his educational ideas. (V.V.Branford 'A Citizen Soldier: his education for

war and peace being a memoir of Alasdair Geddes' *Papers for the Present* No.5 London: Headley, 1917.) Geddes' believed in the need to cultivate practical as well as academic skills. Alisdair was taught craft skills from farmers, shepherds, and fishermen. He was an expert gardener. He worked as a laboratory assistant in a Zoological marine station and he was a qualified coastal lifeguard. He spent a summer in the Antarctic on board the *Scotia* on a scientific expedition with Dr W.S.Bruce, mapping unidentified areas. He was fluent in French and German, had some Flemish and Gaelic and was encouraged to travel. He bicycled around much of Europe on his own. He was an expert in Highland dances and could play the pipes. His academic training, apart from the Home School, was given to him by Geddes' contacts in different universities. He graduated in 1914 (aged 23) and won the Vans Dunlop scholarship which was to be the first step towards his doctoral thesis. But his career was always secondary to his role as his father's collaborator. He was very successful in this and helped Geddes in the most successful years of the Cities and Town Planning Exhibition in the UK and Belgium from 1911–1913. He went with Geddes to India in 1914, but decided after one season to enlist in the armed forces. He went to France and became an observer in the Army Balloon Corps and was promoted to Major in 1916. He was killed in May 1917. 3.The youngest of the family, Arthur (1895–1968) was the least able to adapt to Geddes' unconvential ways. A nervous, sensitive child, he suffered in comparison with his elder brother who was both versatile and phlegmatic. Arthur was 18 when the First World War broke out and he became a pacifist. He eventually volunteered for the Army Field Ambulance Unit but was turned down on medical grounds. He was to find a place instead with the Society of Friends War Victims Relief Organisation and was sent to France. After Alasdair's death, he endeavoured to fill his place as his father's main collaborator, going out to India in 1920. Here he worked alongside C.F.Andrews and Leonard Elmhirst helping to build and set up Rabindranath Tagore's new national college and ashram at Santiniketan in West Bengal. He was temperamentally unsuited to work with Geddes but never gave up trying. In the 1920s, he presented a doctoral thesis at the University of Montpellier, the subject matter being a geographical study of a Bengali Indian village. At Montpellier, he met and married Jeannie Colin, the granddaughter of Elisée Reclus, the great French geographer and anarchist, and an old friend of his father. He became a lecturer in geography at the University of Edinburgh.

26. A description of Geddes' Edinburgh group is given by a contemporary, E.A. Sharp (1912) *William Sharp (Fiona Macleod): A Memoir*, London: Heinemann, 2nd edn., p.50.
27. See chapter 4, p.97.
28. R. Annand Taylor to Geddes, 20 June 1910, Geddes Papers MS10572, NLS.
29. J.A. Thomson to Geddes 1886, Geddes Papers MS10555, NLS.
30. Edinburgh students, sympathetic to Geddes and his ideas in the 1880s and 1890s, included J. Arthur Thomson, V.V. Branford, E. McGegan, T.R. Marr, John Ross, A.J. Herbertson, Cecil Reddie, W.S. Bruce, and many others. Assistants at Dundee included T.R. Marr, Robert Smith, Dr Marcel Hardy, R.N. Rudnose Brown and Rachel Annand Taylor.
31. T.R.Marr (born in Scotland in the 1870s and died in France in the early

1940s) became Geddes' assistant at Dundee in October 1894. From 1895–1901, he acted as the bursar of the Outlook Tower. From 1901–1909, he was co-warden of the newly amalgamated Art Museum (set up by T.G.Horsfall) and the Manchester University Settlement. The period of his so-called 'captainship' at Manchester was described as a 'golden age' for the settlement. He was very active in promoting civics, acting as secretary to the Manchester Citizens Association. Subsequently, he became a successful entrepreneur in the French construction industry. He renewed his contact with Geddes in France in the late 1920s, advising and investing in the Collège des Ecossais. He was business manager of the College from 1930–1939.

32. For further discussion of these relationships see chapters 5, 7 and 9.
33. Bose to Geddes, 11 February 1918, Geddes papers MS10576, NLS.
34. P. Geddes, *The Indore Report* op.cit., p.37.
35. C. Wright Mills (1970) *The Sociological Imagination*, Harmondsworth: Penguin Books, p.239.
36. Ibid., p.243.
37. J.A. Thomson and P. Geddes (1931) *Life: Outlines of General Biology*, London: Williams & Norgate, 2 vols, pp.1,417–18.
38. A. Toynbee published his famous historical analysis of the Industrial Revolution which publicised the term in 1883.
39. Talcott Parsons (1968) *The Structure of Social Action: A study in social theory with special reference to a group of recent European writers*, New York: The Free Press, pp.1–5.
40. P. Geddes (1884) *John Ruskin: Economist*, Edinburgh (pamphlet).
41. B. Mackaye (1950) 'Growth of a New Science', *The Survey* LXXXVI, 10, pp.439–42.
42. L. Mumford (1944) *The Condition of Man*, London: Secker & Warburg, pp.382–90.

CHAPTER 2

The formative years:
the biologist's viewpoint

GEDDES, IN HIS STUDENT DAYS, WAS TO DEVOTE himself full time to the study of biology at possibly one of the most exciting periods in the history of the discipline. The concept of evolution had penetrated scientific and social thinking in every direction.[1] Geddes, as an enthusiastic student, was aware that every question about the universe, its inception and evolution, had to be asked afresh and new answers found. He was determined that in his own studies he would try and find a new cosmology. In good evolutionary fashion, he made his starting point an exploration of former great eras in cosmological thought, the periods of the cultural flowering of ancient Greece, Rome, and the European Renaissance. He was steeped in the idea, common to natural scientists of the time, that the crucial dilemma of modern science, preventing the formulation of a satisfactory new cosmology, was that the sciences of mind and matter had become separated.[2] This was the legacy of the scientific discoveries of the Renaissance period which had ushered in a 'mechanical' or 'chemical' approach to the universe. Modern biology, with its understanding of life and evolution was the antidote and the way forward.

However, the concept of evolution, which had seemed so pregnant with potential for the future in the 1860s, began to prove, with the passage of time, to hold even more problems than possibilities. The concept, applied to cosmic phenomena, covering the history of man as well as plant and animal life, offered no hard or fast guide-lines to the would-be interpreter of a modern, scientific, cosmology. In 1880 T.H. Huxley wrote, with his usual perspicacity, in his essay 'On the Coming of Age of *The Origin of the Species*': 'History warns us that it is the customary fate of new truths to begin as heresies and to end as superstitions'.[3] One of the greatest dangers in the new cosmology was the difficulty of defining and studying the evolutionary life-process if it transcended both mind and matter. There was a great temptation to create the possibility of a life-force, the 'élan vital', of mystical origins, beyond the bounds of reason.

Geddes was caught in the dilemma that he did not wish to rely on the mysticism of the 'vital' biologists and yet he wanted to believe in a life-force.[4] It was an intellectual conflict which he was never to resolve. Instead he took up the idea that a resolution to this problem could not be made within the confines of conventional knowledge and scientific methodology. The new insight necessary to direct work along more fruitful paths could only be produced by going back to fundamentals, and questioning the nature and theory of knowledge itself. Geddes, in the company of other optimistic students of his generation, was prepared to argue that the total knowledge and understanding of creation and change had been transmitted through the generations by book knowledge, supplemented by religious belief.[5] Now they had to apply the new understanding that the concept of evolution had given them to question, not only the validity of traditional knowledge, but also the validity of the very methods by which knowledge was created.

Geddes began, independently, to follow a path being pursued at that time by Henri Bergson, one of the major prophets of the 'vital' biologists.[6] Bergson and Geddes were to meet in later life at the Paris World Exposition of 1900 and to correspond, and Geddes was always warm in his appreciation of Bergson's work. However, although they both started from the same point, the questioning of the nature of knowledge, their paths diverged. Bergson continued to try and build up a new philosophical viewpoint and an understanding of the 'élan vital'. Geddes tried to apply his evolutionary insight in practical ways, especially in his work for educational reform. It seemed essential to him, if there was to be hope for progress in the future, to oppose the continued transmission of established knowledge in the conventional way, and to nurture instead the elements of a 'real' education. He had thus at a very early stage in his career begun on the idiosyncratic path which was to take him outside the mainstream of academic life, and eventually from the natural to the social sciences.

His theory was that the new cosmology would only be found by people trained in new and evolutionary ways of thinking. This demanded, in effect, a revolution in education. Even what constituted knowledge was something which was open to question. As an

[19]

evolutionist, Geddes was sympathetic to the idea that knowledge could only come directly through intuition and not by the reasoned use of the intellect. This was especially the case for a new cosmology since it had to supersede the bounds of all current knowledge. To oversimplify grossly the view of the young evolutionists, it was believed that the creative element of the human mind was the instinct. This was not just a matter of a simple response but something which could be developed with self-awareness. A creative instinct was actually intuition, an instinct developed by self awareness. The intellect was of a lower order and was used merely as a means of interpreting and classifying what was already known. Fellow students with Geddes in London, also studying with Huxley, such as C. Lloyd Morgan, were receptive to these ideas. They became interested in the study of the processes by which we learn and what we learn.[7]

Geddes' search for a new cosmology thus became sidetracked by the absorbing problem of how to refine instinct into intuition. He became a life-long devotee of the idea that individual progress could not be made except by a process of interrelated thought and action. Intuition required physical experience as much as anything else and he resolved to go personally as soon as he could to all the great centres of scientific endeavour in Europe. He thought that he would gain much more by working alongside great scientists than by just reading their papers. But he was already confident that his past stood him in good stead. The best refining influence on natural instinct had to be nature. As a rustic youth from the backwoods in Scotland, he had spent his childhood in close communication with nature, and had observed at first hand the life-force of creation in the hills, woods, fields, and garden near his home. He became convinced that he had to rely on his own, thus refined, intuition, in his search for a new cosmology. To follow Geddes' idiosyncratic path to knowledge, therefore, it is necessary to highlight some of the formative influences of his childhood and youth since these were Geddes' own guide to developing his unique intuition.

Experiences of childhood and youth

On the surface it would appear that he had had a most uneventful, even dull, childhood, the adored, youngest child of elderly parents. There were no distractions except the kirk on Sundays (of which his father was an elder), and the freedom to ramble in the fields, woods, and hillside near his home. His mother, sadly going blind, partly from overstraining her eyes by undertaking fine embroideries and needle-work, wanted him to become a minister of religion. In her simple

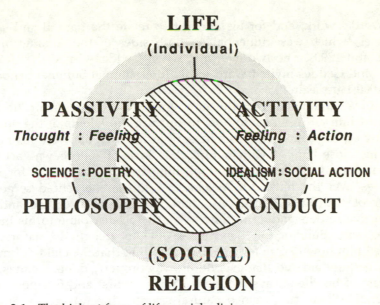

LIFE
(Individual)

PASSIVITY   ACTIVITY
Thought : Feeling   *Feeling : Action*

SCIENCE : POETRY   IDEALISM : SOCIAL ACTION

PHILOSOPHY   CONDUCT

(SOCIAL)
RELIGION

Figure 2.1   The highest form of life: social religion

Source: P. Geddes, 'A Needed Research Institute: geographical and social', reprinted in *East and West* (1903), Madras National Press (1904), p. 31.

piety, she believed that such a step would be a fitting vocation for her youngest son.[8] For the young Geddes himself, such an objective could not have been further from his desires. A highly nervous and energetic child, the long services every week in the kirk were a torture which, in itself, was enough to turn him away from religion. Yet two characteristics of his later life are thrown into relief by this simple, devout, and disciplined upbringing. Geddes consciously felt the need for a religious dimension even when he had ceased to believe in orthodox doctrines; and throughout his life he retained a desire to serve humanity, the same kind of idealistic and non-self-seeking desire as would well befit a minister of religion.

These were traits shared by many pioneer sociologists in the nineteenth century. The founding father of sociology, Auguste Comte, had prescribed a substitute for god-inspired religion in his *Religion of Humanity*. More on a par with Geddes, the American, Albion W. Small (born in 1854, the same year as Geddes), was to become the first professor of sociology at the University of Chicago in 1892. In his approach to the subject, Professor Small shared a similar sense of personal commitment to social service and had an idealism based on mystical experiences. Like Geddes, he was to define his mysticism as part of the reality of human experience, with its roots firmly in this world. His inspiration, too, came from the natural sciences, through his guide and mentor, L.F. Ward.[9] Geddes was to gain a great respect

[21]

for Ward's work, and for his attempt to relate the natural and social sciences, which was influential on both sides of the Atlantic in the 1880s and 1890s. The two men met when Geddes visited the USA in 1899, and Geddes invited Ward to his International Summer School at the 1900 Paris Fair.

Geddes' childhood brought him emotional commitment. It also brought him an insight into how to nurture the talents of the young. The two great educative forces he saw in his own life were his father and his close contact with nature. His love of nature was actively encouraged by his father, who took him, from an early age, for long rambles over the hills. But Alexander Geddes, the retired sergeant-major of the Black Watch, also had a belief in discipline and an instinctive understanding of how to teach self-discipline to his bright young son. Building on the child's obvious delight in nature, he trained him how to care for a garden, insisting that the child did much work without aid and that everything was properly done. Geddes, for the rest of his life always thought of himself, first and foremost, as a gardener. When, as a trained natural scientist, he was given, at last, his own special Martin White Chair of Botany at Dundee, the first thing he did was to create a botanical garden.[10] He delighted in telling students at his lectures (often held in the garden) how all the great natural scientists were gardeners; and he aimed at a threefold objective for his garden: as a scientific laboratory; as living material for a study of the concept of evolution (he even used plants representing elements of Greek mythology, as he was to do again in his garden at the Collège des Ecossais at Montpellier); and finally as an object of beauty, a meeting point between the munificence of nature and the aesthetic appreciation of man. Gardens always figured prominently in all his later town-planning activities, because he believed they had a vital role to play in the production of a good environment. They brought pleasure and delight; they were of educative value as simple illustrations of the evolutionary process; and they had a typical Geddesian bonus, the practical value of the produce that could be grown in them.

But gardens, for all their beneficial influences, were too cramped to be a totally satisfying environment for the growth and development of young people. Ultimately, the most important experience was freedom: to ramble, experiment, and investigate in the liberating atmosphere of the countryside. Geddes was to become particularly concerned about 'town' children as he felt that an urban upbringing must stultify and perhaps permanently blight the growing processes of the young. In the city, social behaviour and a concern for law and order repressed the natural curiosities of the young. In *Cities in Evolution* he wrote:

But — though there is obviously nothing more important either for the future of industry or the preservation of the state, than vigorous health and activity, guided by vivid intelligence — we have been stamping out the very germs of these by our policeman-like repression, both in school and out of it, of these natural boyish instincts of vital self-education, which are always constructed in impulse and in essence, however clumsy and awkward, or even mischievous and destructive when merely restrained, as they commonly have been, and still too much are.[11]

He was a lifelong supporter of the national and international youth movements of the early twentieth century, aimed particularly at bringing town children in touch with the countryside. He often praised the Boy Scout movement in articles and in his planning reports, and the American Woodcraft Folk also received his approval.[12] Through these organisations, Geddes believed, 'townlings' could have first hand knowledge of the countryside and thus gain in reverse, some of the qualities the 'rustic' child could bring to the city. He wrote:

The rural upbringing gives more than health and vigour; for its immigrant youth awaken to the stimulus of the city, with a freshness beyond that of town youth to which it is so habitual; whence often doubled. And as every society has much of coherence, newcomers find it hard to enter its average mass: hence have often to begin at lowest; yet with fresh mind and powers, of active efficiency, they even rise to the top.[13]

Yet in that 'rise to the top' the 'rustic' child could easily eschew the average 'townling's' measure of success, the acquisition of money. In the course of his own lifetime Geddes raised and spent large amounts of money, and he and his family generally lived in beautiful surroundings. But he never aspired to the conventional comforts of affluence. Money was merely a tool; real wealth was altogether different, it was a harmonious balance of organism and environment, mutually supportive and mutually satisfying. Geddes found that one of his most revered masters, Frédéric Le Play, had had childhood experiences which had led him to take up an attitude towards money very similar to Geddes' own. Dorothy Herbertson, who produced her biography of Le Play whilst under the influence of Geddes, placed a special emphasis on this point. Describing the poverty of Le Play's childhood she wrote:

The hard lessons of these childish years were never forgotten. The child learned, and the man remembered, that the wealth of a family is measured not by money, but by that body of collective resources which later was called 'real wages' . . . Not least among the misfortunes of the city poor is the fact that they are obliged to buy everything out of their earnings.[14]

This was to be Geddes' own starting point in his approach to poverty. In Edinburgh, Dublin, and the Indian cities he worked in, he always tried to ensure that his planning activities, mostly through the provision of gardens, would bring more 'real wealth' within the range of the poor. With their own physical effort, they could then work towards improving conditions for themselves and their families.

This was the core of his 'biological' approach to the social problems of cities. His concern, though, was not only with the material condition of the poor, but also the cultural interaction of city and countryside as a way of leading to the permanent elimination of poverty. He believed that future favourable evolution in human society would only take place with a combination of the best the town and the country could offer, and in this better society, poverty of mind or body would cease to exist. He wrote:

> How to combine this fundamental vividness of rustic life, with the subtler, yet it may be even more strenuous life of productive urban culture is perhaps the main problem before the evolutionist. In modern everyday phrase, this task is, in fact, already opening before us; already we are seeking to advance rural development here and town-planning there; we have next to co-ordinate both in regional renewal.[15]

His concern for rustic and urban interaction, and his understanding of the geographical concept of the region, was something else that Geddes claimed he had first experienced emotionally as a child.[16] The insight came to him, when, as a child, he climbed the hill behind his home and looked down on his favourite view, the valley of the River Tay. There, the town of Perth nestled on the valley floor, framed by the mountains behind it. Geddes was to insist that this gave him the idea of the city and its region which his subsequent evolutionary studies were to extend and deepen. A sense of geography, and the concept of the region as a suitable unit for evolutionary study, were ideas currently common amongst natural scientists in the third quarter of the nineteenth century. On a national scale there was a growth of interest, on an amateur basis, in the study of the natural sciences in different areas, which even reached the town of Perth. An indication of this can be gauged from the records kept by the British Association for the Advancement of Science and Art, which show the decade of the 1870s to be a peak period for the founding of new amateur scientific societies.[17]

In Perth a local naturalist, Dr Buchanan White, founded just such a society of amateur enthusiasts to study, collect, and classify the flora and fauna of the Tay Valley. In time, Dr White's society flourished and became the Perthshire Society of Natural Science with its own museum. Thirty years later, when Geddes was trying to inspire the founding of a Naturalists' Society in Dunfermline, he was to describe the Perthshire museum as 'admittedly the best regional museum in the country, or probably anywhere'.[18] His view of regionalism was thus not unusual amongst keen amateur natural scientists of the time. What was unusual was his determination to pursue his amateur interests on a full-time basis without recourse to the conventional ways of studying natural science on an approved degree or diploma course. His desire to work in this way was encouraged by the special circumstances of his home background and education. He had left the Perth Academy at the age of 16, without a clear idea of a future occupation. His father suggested a broadening of his talents and experience through self-education (he built his son a 'shanty' by the cottage to house his museum collection and to be a small workplace and laboratory); and through the acquisition of skills. Geddes worked alongside a local chemist, attended lectures at the local art school, and learnt the skills of carpentering from a local joiner and craftsman.

However, after two years his father was anxious for him to settle down in a safe and respectable career and, acting within the limitations of Perth's local economy, chose the career of banking for his son. Patrick's much older brothers had already been successful in this field. The young Geddes thus spent his nineteenth year as a bank clerk, though only on a special condition he had agreed with his father. If he did not like it after one year, he could leave and pursue his interests in the natural sciences. Inevitably he did not like it. His nature and personality made him an alien in such an environment. Yet he worked conscientiously for his year of apprenticeship, and the knowledge he was to acquire about financial transactions was to aid him greatly in later life, in all his fund-raising activities.

London and Huxley

At the end of the year his father kept his promise, and Geddes enrolled at the University of Edinburgh for a course in botany and the natural sciences, starting in October 1874. However, he suffered a great emotional disappointment on reaching Edinburgh when he discovered the kind of work he was going to do. He wanted to study living nature in evolution. Instead he found himself cutting up and classifying dead specimens. In a dramatic gesture he left the university after one week, determined instead to seek out the most famous natural science teacher of the day, T.H. Huxley, who was working at the Royal School of Mines in London. Huxley's job was to give the student engineers a concentrated five-month course in the natural sciences as part of their training in their second year of study. He was not able to offer a degree course in the natural sciences. However, Geddes was supremely uninterested in degree courses and qualifications. He wanted only to sit at the feet of the master.

Ironically, if he had stayed in Edinburgh he would have had the opportunity to work with Huxley there. Sir Charles Wyville Thompson, who held the Chair of Natural History at Edinburgh, was an old friend of Huxley's. When Wyville Thompson went on survey work in the summers of 1875 and 1876 in HMS *Challenger*, he asked Huxley to come and give a course on natural history in his place. Huxley came and was a great success. Six hundred people attended his first lecture in May 1875 and 353 enrolled for the full course, which was a record for any Edinburgh class. Huxley's reputation was well known in Scotland. He had shaken Edinburgh society with his lectures in 1862 and 1869 on the 'Evidences as to Man's Place in Nature' and 'On the Physical Basis of Life'; and his popularity amongst students was enormous. In 1872 he had almost been elected as Rector of St Andrews University by students who had not even asked his permission to put his name forward.[19]

Geddes' desire to go to Huxley in London, therefore, was understandable. But it probably cost him the chance of a successful career in the natural sciences. Wyville Thompson's work in marine biology was to make Edinburgh 'one of the chief centres of zoological activities in the world' and in 1879 he was succeeded by John Murray, who became the founder of the modern science of oceanography.[20] In 1884 Murray established the Scottish marine station at Granton for biological study, and he was to call on the expertise of Geddes for help and advice. Geddes had, by this time, worked in the marine station of Roscoff in France and gained much useful experience. But in the 1880s, when Geddes was applying for chairs in the natural sciences, one of the factors which told against him was the unconventional nature of his education. A five month 'sandwich' course at the Royal School of Mines was not the equivalent of a degree course at Edinburgh; and Wyville Thompson had been one of the last of the old school of scientists who were appointed to chairs without ever having passed a scientific examination.[21]

However, in 1874 the 20-year old Geddes was hungry for knowledge and excitement, and he had a contempt for examinations which never left him. On arrival in London he suffered a minor setback, as Huxley would not accept him on the course until he passed the same preliminary examinations in chemistry, physics, and geology, taken by all students. His lack of respect for examinations was increased when he was able to pass this hurdle with little further effort, other than reading a textbook just before the examination. Yet he had to wait for a year to get on to Huxley's course. Thus for a year he roamed London, and absorbed all there was to see and hear in the greatest capital city of the world at that time, with a passion that was all his own, but a capacity that indicated his 'rustic' upbringing.[22]

At last, however, the winter session of 1875 arrived and Geddes found himself at the feet of his master. The impact Huxley had on his students has been well documented. His lucid and brilliant lectures, liberally illustrated with diagrams and drawings, and his insistence on daily practical work as the essential training for a scientist, made an enormous impression on his students. One of them, H.G. Wells, describes the sensation of arriving in Huxley's class for the first time:

> All my science hitherto had been second-hand – or third- or fourth-hand; I had read with a sense of being a long way off from the concrete facts and still further off from the living observations, thoughts, qualifications and first-hand theorising that constitute the scientific reality . . . Now here were microscopes, dissections, models, diagrams close to the objects they elucidated, specimens, museums, ready answers to questions, explanations, discussions. Here I was under the shadow of Huxley, the acutest observer, the ablest generaliser, the great teacher, the most lucid and valiant of controversialists.[23]

The effect on Geddes was immediate and profound. Huxley demanded the total commitment of his students for five months and they undertook no other course work during that period. His argument was that such an initiation in the natural sciences would have such a deep impact on their young minds that the knowledge that they gained from this experience would last throughout their lives. He adopted this method, not only because he thought it was educationally sound, but because nearly all his students would never again need to return to the study of the natural sciences in the pursuit of their careers. For the impressionable and enthusiastic Geddes, such a course was ideal, and in later life he warmly recommended short periods of intensive study as the best way of getting to understand a subject. However, after this initiation Geddes had some problems with his education. He did not wish to qualify as an engineer at the Royal School of Mines and yet there was no provision for him to continue his studies in the natural sciences. Huxley, however, had been favourably impressed by Geddes' response to the subject and his application, and suggested that he might like to return the following year to act as a demonstrator on the course. Huxley also then hoped he could undertake some research and thus acquire the necessary status for a future academic career. Geddes was very happy to seize these opportunities and his closer relationship with Huxley proved to be one of the most formative on his development and thinking.

Huxley, the teacher of science, could not be separated from Huxley, the philosopher and controversialist, the defender of Darwin and the concept of evolution, and the spokesman of scientific materialism.[24] Huxley's agnosticism was of no concern to the young Geddes, but his theory of life was another matter. Huxley himself never put forward any dogmatic claims about his interpretation of the concept of evolution. In fact, with his students he simply presented the known data, usually of a palaeontological nature, and left them to make their own conclusions about it. Huxley was always ready to confess that his own talents were more those of the engineer than the naturalist. What really excited him was the study of 'the mechanical engineering of living machines' and he had, he wrote, 'very little of the genuine naturalist in me'.[25]

For Geddes, this meant that gradually he became aware of the Achilles' heel of his master. Central to the concept of evolution was the theory of natural selection, and this could not be analysed or developed conclusively from fossil data. The two key aspects of the theory, the ordinary reproductive discontinuity of the species, and the nature of variations, demanded a botanical and zoological study of living specimens. This is what Geddes had always wanted to do, and his contact with Huxley merely confirmed him in his direction. Having absorbed all the knowledge that he could from his beloved master, he began to react intellectually against him. He began to examine critically the outlook and philosophies of those whom Huxley attacked more vigorously.

One of Huxley's public targets was, in private, his old friend Herbert Spencer. In the 1870s Spencer's reputation was at its height. His books were widely read, not only in Britain and Europe, but also in America. The work he published in these years included his *Classification of the Species*, the second volume of the *Principles of Psychology*, *The Study of Sociology*, *The Comparative Physiology of Man*, and in 1876, *The Principles of Sociology*, vol.I, to be followed, in 1879, by *Principles of Sociology*, vol.II. What attracted Geddes to his work was his development of a general theory of evolution. Huxley warned his young student, however, that it was not wise to take Spencer's work too literally, especially his attempts to relate the natural and social sciences. But Geddes found his appetite grew the more he read, and he identified strongly with Spencer's stated objective: to seek an 'order among those structural and functional changes which societies pass through'.[26] To apply the concept of evolution to society and to use a knowledge of how the process took place in nature as a guide, opened up a new vista of possibilities to Geddes. He accepted gladly Spencer's view of society as an organism of functionally independent parts, and he was warmly appreciative of Spencer's informed attempt to trace the evolutionary forces working towards changing society.[27] Spencer, however, in the last resort, did not depend on natural sciences for his theories. He sought a 'law of progress' which had to be based on some universal principle and he turned thus, to metaphysics. He created the concept of the 'unknowable', a mystic force, ultimately responsible for generating change.

The concept of the 'unknowable' was to be the cause of a fierce public debate in the early 1880s between Spencer and a leading positivist, Frederic Harrison.[28] Harrison insisted that for the social scientist, the 'unknowable' was not some mystic force but the religion of humanity. Just what this was the young Geddes had begun to find out, again reacting against Huxley. Huxley had written a disparaging article about 'The Scientific Aspects of Positivism'. Geddes, feeling that the materialistic explanation of life espoused by Huxley was unsatisfactory in some respects, decided to study positivism for himself. He began to attend the Positivist Church in Chapel Street, London, run by Dr Richard Congreve, who then introduced him to the Positivist Society.[29] Through these activities he discovered the work of Auguste Comte. Furthermore, he found, in the religion of humanity, a passionate religious commitment outside the framework of orthodox religion, and ostensibly based on scientific principles. He did not continue his attendances at Positivist Society meetings. But Comte remained a key inspirational influence on his thinking for the rest of his life.

Geddes was open at every pore to the ideas that were circulating at this time, and he could hardly fail not to come under the influence of another of the great social prophets, John Ruskin. Geddes, the 'rustic' philosopher, was prepared to argue that, just as the natural sciences were the inspiration of the 'rustic' student, so the arts were the province of the 'townling'. To complete his own evolutionary development to the greatest extent, he needed thus to learn and absorb what the most gifted men of letters had to offer. Ruskin had become a controversial figure in the 1870s, moving from his art criticism to confronting the social problems of contemporary society.[30] In 1884 Geddes wrote a paper, *John Ruskin, Economist*, in which he outlined what he had found valuable in Ruskin's work. His central argument was that for all Ruskin's literary and artistic background, the approach to society that he adopted was that of a natural scientist. Ruskin's tirades against the modern city, against mechanised industry, against the market economy, were not the 'incoherent, hyperaesthetic, and even hysterical' outpourings of an art critic.[31] Ruskin wanted an end to the control of the economy by market forces and, in its place, the creation of a system designed to serve the biological and aesthetic needs of humanity.

Ruskin was not only a critic of modern society, he also initiated practical experiments aimed at creating the desirable social order. Long before he founded the Guild of St George he had given the initial financial support to Miss Octavia Hill, which enabled her to embark on her career as social reformer and improver of the housing of the working classes. The social environment of cities, and especially the nature of education, deeply interested him. On education, Geddes discovered that Ruskin spoke 'lightly of the three Rs and . . . [threatened] to make even the first of them optional'.[32] In their place should be scientific observation and reasoning, and artistic co-ordination of hand and eye, developed through the practice of the arts. Geddes' views on Ruskin's practical activities present, in embryo, the ideas he was to develop as his 'civic reconstruction' doctrine, the key to his approach to town planning. Geddes wrote of the activities inspired by Ruskin:

> The so-called 'aesthetic revival' with its outcomes like the Kyrle and other 'Environment Societies', represent in fact the small beginnings of the Industrial Reformation, of that re-organisation of production — of products and processes, of environment and function, which is the nearest task of the united art and science of the immediate future.[33]

But in the late 1870s he still hoped to make his career in the natural sciences. In the summer of 1876 Huxley had sent him to Cambridge to undertake some vacation work in embryology, under the leading British embryologists of the day, Michael Foster and Francis Balfour. However, although he was working in a vital area for evolutionary studies, Geddes found the atmosphere of Cambridge, and the prospect of studying there, thoroughly uncongenial. He was restless. He wanted to travel and explore before settling down to a career. Huxley was deeply concerned that he would have no career if he did not get further training first, and he put him up for the Sharpey Physiological Scholarship at University College, London. Geddes spent the winter of 1877–8 thus in London, as a student and a demonstrator, working for Professor Schafer.

However, in the early spring of 1878 Geddes fell seriously ill, and by Easter, though recovering, he was ordered, on medical grounds, to take some kind of change. Again, Huxley showed great kindness and concern for his welfare and arranged for him to have his heart's desire, a foreign trip to the marine station at Roscoff in Brittany, run by Huxley's friend Professor Lacaze-Duthiers of the Sorbonne. This convalescent period proved to be a turning point in Geddes' life. It was his first visit to France, a country for which he was to develop the deepest attachment. It was his introduction to marine biology which he found the most satisfying of all his studies in his quest to understand the concept of evolution. Finally, later that summer in Paris, he was to pick up ideas on the study of society, emanating from the work of French social scientists who had already gained a far more important place in academic life than their British counterparts. It was from these ideas that Geddes was later to try and build his theory of social evolution.

France, the University of Paris and the study of marine biology

When however, Geddes set foot in France for the first time during the Easter vacation 1878, he went, not to absorb developments in the social sciences in France, but to convalesce from his severe illness contracted in the early months of 1878, and to pursue his scientific investigations into biological evolution at the marine station at Roscoff. For Geddes it was a revelation. He had never before worked at marine biology, and he found the work, and his environment and fellow scientists, completely congenial. Marine biology at Roscoff brought him near to the exciting prospect of exploring the evolution of life. He made a special study of the different types of protozoa, trying to determine how 'the greatest of all steps in morphological progress, that from the Protozoa to the Metazoa' came about. The paper in which he wrote up his discoveries 'Sur la fonction de la chlorophylle chez les planaires vertes' was published by the Académie des Sciences, Paris, in the autumn of that year.[34]

Geddes found the working conditions at Roscoff so ideal that he was, later, to model his ideas on the best kind of educational institutions on the marine station and its activities. The pattern of work and social life that he found there were to be the basis for his own attempts to develop new paths in education at the early Edinburgh Summer Meetings in the late 1880s when he used the Granton Marine Station. The work elements that impressed him were the combinations of outdoor practical activity, followed by the study of collected specimens in the laboratory. Each day began early with outdoor expeditions, followed by indoor work in the afternoons. In the evenings, the students, who all shared communal accommodation, were free for social activities and discussions. Geddes was able to join in fully, with both the work and the social life, since he quickly developed his grasp of school-learned French, to a complete mastery of the language. Geddes had to return to London for the summer term. But, as soon as he could, he went back to Roscoff. As the autumn approached he decided not to return to London where, in any case, his scholarship had now come to an end, and instead he went to Paris with his Roscoff friends, who were students at the Sorbonne. His introduction to Paris coincided with the 1878 Paris Exhibition, a triumphant celebration of all that had been done to restore the damage done by the Franco-Prussian War of 1870 and the Paris Commune, and to reiterate faith in the new Republic. It was to have the greatest impact on the susceptible and emotional Geddes. Here he found a capital city which was everything that London was not, a city in which the university had a leading role, both in educating students and in advising governments – a city where culture and city life had reached new levels of integration and achievement.

When Geddes later developed his theory of Social Reconstruction, he had in mind an idealised version of what he found in Paris in 1878. In London, working under Huxley, Geddes had witnessed at first hand Huxley's efforts to induce the City of London and the ancient Guilds to undertake responsibility for the development of technical and scientific education in the city.[35] He had been made aware of the cultural gap between the sources of wealth in the city and the sources of scientific knowledge. London University in the 1870s was still largely an examining body with restrictions that were so cumbersome that University College, London, seriously considered affiliation to the Victoria University, established in Manchester in 1880, as a way of gaining greater academic freedom and a chance for development. The cultural life of London was fragmented, both socially and intellectually, and the application of scientifically progressive ideas usually met with a hostile reception. In Paris everything was different. Here the university had an influence and importance for the cultural life of city and nation which was greater than was found, probably, anywhere else. T.N. Clark describes how there was a liberal consensus, a cartesianism, upheld by the bourgeoisie, and 'identified with order, hierarchy, authority and the bureaucratic institutions exemplifying the *esprit de géométrie*: the state, the military and the university'.[36] All were united by a common objective, an enlightened pursuit of reason, which was conducive to encouraging a positivist outlook and scientific mentality. Clark suggests that 'The shared ideology of many academics and political leaders during the Third Republic led to university-generated efforts to further national and cultural goals'.[37]

This political involvement of the university only occurred at certain times, usually of crisis, and the decade of the 1870s, in the aftermath of the events of 1870–71, was one of these. Geddes was thus introduced to the intellectual life of Paris at a special moment. He was so impressed with what he found that he was prepared to overlook the domination of examinations over the university system, the lack of flexibility caused by centralised control, and the political intrigues surrounding the appointments of professors. He was impressed by the integration of the state with the university; with the university's attempts to reform its teaching activities; and above all, by the development of regional universities with a strengthening of the Faculties at Montpellier, and newer foundations from Lille and Nancy, to Aix and Marseilles, from Bordeaux and Toulouse, to Grenoble. The new university at Clermont-Ferrand with its emphasis on regional studies and the development of the rural hinterland especially interested him.

It was Paris itself, however, where he felt most stimulated by all the social and intellectual activities going on around him.[38] In the 1870s almost 50 per cent of all French students still enrolled at Paris and Geddes was able to feel an immediate sense of identity with his student companions, many from rural backgrounds. Many years later he was to write of what this experience had meant to him, still in deeply emotional terms:

In that vivid time . . . from 1878 onwards, when the French regional Universities were reappearing, Paris also was rising with them, not falling, as the senile among her authorities had feared. Among the many active little papers which rise so frequently in Paris, and run their course with their new, active group − . . . one bright little weekly boldly called itself *L'Université de Paris*. Why so? It explained broadly to this effect: 'Because it is time to see, and to say to all, that this University of ours is not merely the Sorbonne renewed from the Middle Ages, the Collège de France, continued from the Renaissance, the Schools of Medicine and Law, the Ecole Polytechnique and so on; in short the established centres of higher teaching. It includes, of course, the great institutions of science − the Observatory, the Jardin des Plantes, the Pasteur Institute and the like; but it is more than all these. Spiritually and educationally it has also a focus in the National Library, and another in the yet vaster treasure-house of the Louvre. Its school of literature and language is not merely of lectures: it is above all the House of Molière, the Comédie Française; and with this many a minor theatre of living art. It is the French Academy too; and the free writers as well, the young poets even more. Music, too, is part of this true modern University, as of old: and not only in the Conservatoire, but in the Grand Opera, the concerts and more, wherever a young composer can express his dream. The visible arts, too, are part of this University proper; and

[33]

not merely in their great school; but even in the Salon, the rival and minor exhibitions; but above all in the studios. And there not merely collective, but individual; witness the Impressionists; witness the sculptors; witness too the architects, who are striving to raise our city beyond the meretriciousness and monotony of its Second Empire style . . . Our true University is thus in the City; nay more, it is the City, great Paris herself. She is ever stretching out for us her fresh ideas, in the bright conversations of the salon and of the cafe; and so she diffuses them into the intellectual atmosphere and at every social level.'

Geddes gets so carried away by his version of cultural evolution in which all citizens join together that he has no time for current analyses of the divisions within society especially class structure. He continues:

The University is thus no specialised caste of culture: it is *solidaire* with all true citizens in all their occupations, all their classes, above all, then, with the People, whose sons we largely are, from whom all classes rise, and whom we students are in training to serve . . . So in this faith let us work on, to develop City and University together, augmenting, multiplying all their culture-institutions, farther and without end, yet never forgetting in our academic life their larger civic and social purpose; so that every citizen, that is every worker, woman and child among us, shall increasingly enter into their manifold inheritance, and continue it for themselves in their own day and way.[39]

The discovery of Le Play and the Le Playist school

His excitement and admiration of Paris and her intellectual life was further increased when curiosity led him one day, during that winter session of 1878−9, to attend a lecture by a M. Edmond Demolins, expounding the social theories of Frederic Le Play. It seemed to him that at last he had found what he had been seeking unsuccessfully in London, a serious attempt to provide a model for the practical, scientific study of society. Le Play's work was being developed by the Le Playist School in Paris[40] and Geddes was eventually to become, as he turned more to the social sciences, a leading exponent of their ideal, which he tried to promote in Britain. His influence in this respect was particularly marked from the 1890s to the 1920s when, in the development of sociological theory, Le Play's ideas had already become eclipsed by those of Durkheim, Weber, and others.[41] In many ways Le Play's ideas had had their most crucial influence in Britain before Geddes took up the cause.[42] It was Le Play's work which contributed to an understanding of the importance of environmental factors and industrial change in causing the destitution and degeneration of the poor, which was such a feature of the intellectual response to poverty in the 1880s.

Le Play's ideas began to go out of fashion after his death in the early 1880s, both for political and theoretical reasons. Le Play's support for Napolean III's government between 1855 and 1870 branded him as a political reactionary which his work for social reconstruction did nothing to repel. Le Play, although he largely eschewed politics, was conservative, though his attitude grew from a hatred of violent change rather than from a desire to maintain the status quo. He had grasped that social change was related to economic change, and that the crucial relationship between men was determined by the means of production. However, this did not lead him towards a sympathy with the work of Marx. Le Play's outlook was broad enough for him to encompass his catholicism with elements of Positivist philosophy. But it would not stretch as far as accepting dialectical materialism, class conflict and an economic interpretation of history. For Le Play and his followers, Marx's analysis was too abstract. Their concern was with actual communities, existing industries and the variations between them, and how such variations were enhanced by differences in geographical location.[43] Geographical and environmental factors, particularly obvious in the case of industries such as mining, were accepted as a vital determinant of social structure, which could best be studied by starting with a basic social unit, the family, and studying it in the context of its environment. Le Play established the key units for study as Lieu, Travail, et Famille: Place, Work, and Family.

In order to study these fully, Le Play was to develop the technique of the social survey to a new level. The social survey, as a method of providing an accurate description of a situation as a basis for future reform, had been already well established both in England and France.[44] Le Play's contribution was to suggest that this method could be used as a basis for a scientific study of society. He saw the survey as a practical means for uncovering the facts and also as a method for the selection of problems to be investigated. Le Play's great monograph, *Les Ouvriers Européens* (1855), illustrated how this technique could be used to reveal insights about working-class life. A second edition was brought out between 1877 and 1879 whilst Geddes was in Paris. This proved to be by far the most influential aspect of Le Play's work. He tried to show that income and the standard of living were not always closely related, and the cultural and environmental context of work and family life could, to some extent, offset low wages.

In the late 1870s, when Geddes first reached Paris, the promotion and dissemination of Le Play's ideas and work was in the hands of two leading disciples, the Abbé Henri de Tourville and M. Edmond

Demolins. De Tourville was an aristocrat, who, having taken holy orders, became vicar of St Augustine's church in Paris in 1873, the year he met Le Play. He became deeply involved in the intellectual potential of Le Play's ideas, and tried to work out a means of overcoming one of the more obvious weaknesses in his methodological structure. This was the impossibility of systematically linking the individual family to the rest of society. If the challenge was to explain change by relating family structure to environment, then this could not be done on the basis of the analysis of different individual families. These needed to be fitted into an all-embracing context which would relate them in all directions, geographical, economic, and cultural, with society at large, and environmentally, by specific reference to the neighbourhood, the local parish, the city, the state, even the state in relationship to other foreign countries. The Abbé was no fieldworker or collector of data himself. He saw his contribution as the provision of a method of classification which would then become a tool for the analysis of social change. His great system of classification was called *La Nomenclature Sociale d'après Le Play* (Paris 1887).[45] He had already taken the Le Play school down the environmental determinist's path.

M. Edmond Demolins was to go one stage further. Demolins, in the 1870s, was building a reputation for himself as a social scientist, whilst earning a living as secretary of La Société d'Economie Sociale, founded in 1856, and subsequently much dominated by Le Play and his ideas. The perspective provided by men such as Demolins differentiated this society from the British version, the National Association for the Promotion of Social Science, though in many respects the two institutions shared similar objectives. Demolins was an urbane figure with a developing reputation as a good lecturer. When he and Geddes met for the first time in 1878 Demolins was 26 and Geddes 24. Inspired by similar enthusiasms, they found an instant sympathy for each other, and Geddes learnt much of his knowledge of Le Play and his work through Demolins. Demolins was dedicated to taking up and developing the scientific basis of Le Play's work which he said the master had damaged by his tendencies to draw premature conclusions and generalisations from the available evidence.[46] Demolins, particularly, was involved with de Tourville in his attempt to develop a more accurate system of classification.

In his enthusiasm, however, Demolins was to fall head first into the trap of environmental determinism which was in vogue in the 1870s amongst European social scientists. The arch environmental determinist, the German, Friedrich Ratzel, was in the process of working out his deterministic theories, where even cultural phenomena were regarded as products of their environment. In the 1870s Ratzel had been travelling in Europe and the Americas and he brought out his first book, on North America, in 1878. He had fully developed his ideas, though, four years later, when he published the first volume of *Anthropogéographie* (1882), to be followed by the second volume in 1891. In these volumes the whole life of men, all their multiple activities, communities, and societies, were studied methodically and scientifically in relation to their geographical environment. However, the techniques and assumptions upon which this approach was based soon became heavily criticised as too mechanistic and unsubstantiated.

One of the fiercest attacks on it was made by Lucien Febvre in his monograph *Geographical Introduction to History* which, though not published until 1925, was mostly written before the First World War. The major butt of his attack was borne by the luckless Ellen Semple, disciple of Ratzel, whose book *Influences of Geographic Environment* (1911) displayed the weaknesses of this approach to the full. However, Febvre was concerned with the kind of influence that the concept of evolution was having on geographical and historical studies, and the activities of M. Demolins also came in for some scathing criticism. Febvre wrote:

A well-meaning populariser, with a very confident belief in his own capacity shuts himself up in his own closet to reflect, as so many others have done, on the whole history of nations, and to discover the principle, the bond, and the explanation. By the side of M. de Tourville's *Nomenclature des faits sociaux* (we are dealing with an adept in social science), we imagine him putting on the table (presumably in order to support and at times excite the springs of his imagination) several good historical dictionaries, two or three recognised text-books, and the *Géographie Universelle* of Elisée Reclus, that Providence so often unacknowledged . . . Then, starting with a brilliant idea, an ingenious hypothesis worthy of romantic fiction, he sets himself with a kind of mechanical fury to extract from it universal consequences, and we have in twice five hundred pages *Comment la route crée le type social* by Edmond Demolins.[47]

The problem that Demolins and many others, including Geddes, faced, was the old and thorny one. How does one relate Man: 'the physical and moral, individual and social, "natural" and "political" being, with Environment: the Earth; or if it be decomposed (and what an effort of analysis!) the ground and the climate'. Febvre suggests that Demolins' contribution partook of the nature of an 'Apocalypse revealed to the Elect'.[48] Demolins and de Tourville were never lacking in confidence in their own methods, and from the late 1870s were full of optimism that the *Nomenclature des faits sociaux* would provide the bridge necessary to relate the two. Their work was published in a Le Playist journal *La Réforme Sociale*, and by the mid 1880s Demolins had 120 students enrolling in his classes for study of the Le Play method and its developments. By 1886 the old school of Le Playists had become sufficiently alarmed by the direction in which Demolins and de Tourville were taking their studies, that they forced Demolins to resign his post as editor of *La Réforme Sociale*. However, de Tourville put up new funds and Demolins, confident in the support of his students, continued his work and started a new journal, *La Science Sociale suivant la méthode de Le Play*. This became shortened to *La Science Sociale*, an abbreviation which emphasised the stronger concern in this journal for science rather than reform. Through its pages Geddes was to remain in touch with the work of his teacher and friend, Demolins, who thus continued to exercise a profound influence on his sociological thinking.

Elisée Reclus and the study of the region

The 1880s and 1890s marked a watershed in the social sciences in France when enthusiasm was great and activities intense.[49] The battle was on to incorporate evolutionary ideas of social change stretching over periods of historical time. Much effort was expended on reformulating the long-held universal sequence of man's development from hunter to herder to ploughman, from ploughman to citizen. The progressive Le Playists were by no means alone in this attempt, or in the development of the idea that great civilisations were socially differentiated from each other by the consequences deriving from their different staple diets of rice, maize, or wheat. Anthropologists and geographers, with a fascination for cultural evolution, found themselves moving in the same direction. However, Geddes was able to bring to this study the knowledge of a technique and approach that he had learnt from T.H. Huxley, when he had been studying evolution in the natural sciences. One of Huxley's outstanding qualities was the lucidity with which he was able to describe and reveal the complex factors leading to natural evolution, which was most in evidence in the popular textbooks he wrote for the general reader. In two of his more famous popular textbooks, *Practical Elementary Biology* (1875) and *Physiography: an Introduction to the Study of Nature* (1877), Huxley had solved the difficult problem of relating organisms to their environment without gross over-simplification, by drawing his examples from one particular region, the Thames Basin.

The idea of the region, as a representative section of the universe, and thus a suitable subject for study, had a considerable potential appeal. But Geddes was not unaware of the crucial difference between the study of the natural, and the social, evolution of the region. In nature, especially Huxley's nature, which he presented as an almost mechanical system, evolution took place as a result of the ruthless struggle for survival – nature red in tooth and claw. In human society, change was not necessarily totally predetermined. It could take place as a result of human decision. Besides, Geddes was dedicated to the idea of the life-force, the *élan vital*, as a creative, not a destructive force. In his view, natural selection was not the prime moving force in evolution, the outcome of the survival of the fittest. Natural selection was instead, a curb on evolutionary tendencies, the pruning tool which enabled the better development of the plant/organism. He was to find a great deal of sympathy, later, for the ideas of the Russian émigré, the naturalist geographer, Prince Peter Kropotkin, who was to argue that even in natural evolution, there was evidence of co-operation amongst species for their mutual support and development; and that groups of men, uncorrupted by modern ideas of political economy, would naturally co-operate with each other and help each other if they lived in small, anarchic communities.[50]

Geddes was particularly receptive to the concept of regionalism which was to be France's major contribution to the development of geographical studies. From the 1870s, in the light of a better understanding of the military, economic, and political importance of geographical studies, the French made a determined effort to catch up with the Germans in this respect. The German geographers, particularly Karl Ritter and Friedrich Ratzel, had contributed much to the development of human geography and to the conception of climatic and economic regions. But the leading German-trained French geographer, Elisée Reclus, was not able to develop this tradition in France in the 1870s, since as a Communard he had been expelled from France after 1871. Exiled in Switzerland, however, he became an anarchist, under the influence of Proudhon, and he embarked upon his life's work, the *Géographie Universelle: la Terre et les Hommes*, the first volume of which was to appear in 1875 and the last in 1894, when the London Geographical Society presented him with a medal to commemorate the event.

Geddes was to become a profound admirer of Reclus and was able to persuade him to come to the Edinburgh Summer Schools in 1893 and 1895. However, he was not altogether impressed by Reclus's attempt to place man in his environment. As he was to write in an obituary for Reclus in 1905:

Comte had subordinated the individual to humanity, as now in our day Tolstoi would do yet more; but to Reclus, as later to Nietzsche, it is in the individual that humanity must find its expression. In this vivid idealism of the species and its type he, as indeed this whole school of thinkers, no doubt too much lost sight of the intermediate categories of the city and state, of the nation and empire, of the unity of language, of occidental and oriental civilisations – in short, of the whole graduated social framework in which we find support, albeit too often also our limitations.[51]

Reclus's work was, however, the source for Geddes' attempt to create a general unit for the study of a region which Geddes was to call the Valley Section. This was a diagrammatic method of depicting the subsoil, the natural environment, and the economic life of the region at the same time. The diagram was a cross-section of a river valley starting from its source in the hills to the estuary on the plains (see Figure 2.2). Placed along the valley at appropriate intervals were the working implements representative of the 'natural' occupations as defined by Demolins: the pickaxe for miner, axe for forester, bow and arrow for hunter, crook for shepherd, scythe and plough for peasant, spade for gardener and at the coast, nets for fishermen. On the plains there was also a city representing the more complex and sophisticated aspects of civilisation.

Geddes was to use the Valley Section in lectures and exhibitions at the Outlook Tower and as a means for teaching Demolins's sociological ideas to enthusiastic amateurs for whom sociology was a strange new subject. To the basic diagram Geddes was to add industrial occupations in which primary occupations such as mining, forestry, hunters/furriers, shepherds, and so on, became translated directly to their secondary stage: that is, miners became iron and steel workers; foresters, workers in timber and paper making; hunters/furriers, shepherds, and so on, woollen-mill workers; and so on, with the retailing and provision of food and drink being related to peasant, gardener, and fisherman. The Valley Section proved an excellent educational tool and worked well in exhibitions. But Geddes was always willing to suggest it had potential for developing new ideas which proved to be rather optimistic.[52] However, as the individual volumes of Reclus's great work appeared, each devoted to a major country or region, the emphasis on regional and climatic differences on their physical variations provided a point of contact between Reclus's work and the mainstream of French geographical studies being developed under the influence of Vidal de la Blache. De la Blache was a pioneer of one of the most fruitful developments in the evolution of the social sciences in France. His work as a 'human geographer' was important both intrinsically and as a contribution to the range of studies in sociology, ethnology, history, anthropology, and economics, which made France in the 1880s and 1890s a leading centre for the study of 'human ecology'.[53]

De la Blache and his followers were particularly concerned with developing techniques for analysing the human geography of the region. By the turn of the century an impressive series of monographs

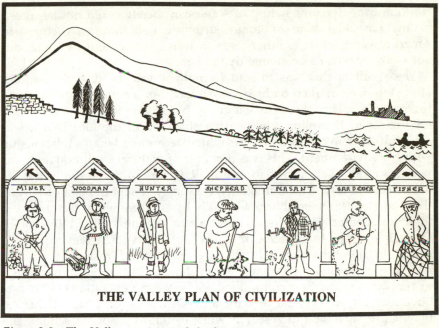

THE VALLEY PLAN OF CIVILIZATION

Figure 2.2   The Valley section and the basic occupations

'By descending from source to sea we follow the development of civilisation from its simple origins to its complex resultants; nor can any element of this be omitted. . . In short, then, it takes the whole region to make the city. As the river carries down contributions from its whole course, so each complex community, as we descend, is modified by its predecessors.'

Source: P. Geddes (1905), 'Civics: as applied sociology', Part I, *Sociological papers*, (ed.) V. V. Branford London: Macmillan, pp. 105–6.

were to emerge as a result of their work. Knowledge of these developments encouraged the Le Playists to pursue their own ideas on using the region as an appropriate area for study. But many of the problems of relating man and his environment had begun to appear much more difficult to solve than they had done in the optimistic days of the late 1870s. Even Edmond Demolins and the progressive Le Playists were running into trouble. A major setback occurred in 1892 with the publication of Paul de Rousiers's study of the United States, *American Life*. De Rousiers was one of the most successful Le Playists, who built up for himself an influential career as a politician, as a teacher at the leading Paris business school, the Institut des Hautes Etudes Commerciales, and finally as the holder of the coveted Chair in Social Science at the Ecole Libre des Sciences Politiques from 1908–32. However, the result of his monograph on *American Life* was effectively to undermine the whole Le Playist attempt to relate particular family structures to specific kinds of economic and geographical environments.

He found the dynamic nature of American society could not be traced to any particular kind of family structure and family structures in America were similar to other areas of European settlement which did not enjoy America's economic dynamism.[54]

The result of this was to lead Demolins, and his disciple Geddes, along less-well-marked paths in their search for a comprehensive social theory to explain the interaction of particular societies with their environment. Demolins now became absorbed in the idea that it was social rather than economic factors which determined man's relationship with his environment. This was a matter of cultural evolution and was perhaps more vital even than economic factors in determining the life of the community. The Demolins-led Le Playist school began to argue that value patterns, established initially by environmental factors, were then transmitted from one generation to the next. From this point Demolins began to perceive that the possibility of achieving 'social peace' in the future, the Le Playist goal, could rest, not on just adapting to changing economic circumstances, but, more importantly, to transmitting the best cultural values between generations. This was obviously an educational activity and Demolins gave up all his Le Playist propaganda and research work to found a special school, the Ecole des Roches, dedicated to serving this new objective.[55] The children were to be inculcated with the best values to be found in each type of custom or tradition nurtured by different families over the generations.

For guidance, Demolins turned to the social history of Britain to find out how the educational system had produced a nation capable of attaining and ruling the greatest empire the world had ever seen. He visited Britain in the 1890s to collect material for a monograph on the subject. Geddes invited him to the Summer Meetings at Edinburgh in 1893 and 1895, and through these contacts Demolins learnt about developments in the British public school system. In the socially 'radical' environment of Geddes' circle, Demolins heard about the new 'progressive' public schools of the 1880s, particularly Cecil Reddie's Abbotsholme in Derbyshire.[56] This school, inspired by Edward Carpenter and the ideals of the Fellowship of the New Life was quite atypical of the English public school. Geddes had been a friend of Reddie's since the latter's student days at Edinburgh University in the early 1880s and greatly encouraged him in his work.[57] Demolins wrote much about Abbotsholme without even actually visiting it and, because his work on English education was popular, a curious view of English public schools became widely current in France. Demolins's reputation as an educational reformer was not one whit diminished by the inaccuracies of his scholarship.[58]

Geddes felt a powerful attraction for the ideas behind Demolins's École des Roches and he organised a Home school for his own children in the 1890s. They learnt skills and values working with local craftsmen, and Geddes tried to provide them with the same kind of nurturing that he had received from his father. He taught them all how to garden, and he remained for all their lives as a 'father-friend', seeking from them their innermost thoughts and attempting to cultivate their talents. Since he was away from home a great deal, many of these intentions did not work out as planned, and only Geddes' eldest son, Alasdair, seemed to have coped well with this treatment, possibly due to the fact that he was by nature already equipped with a cheerful and equable temperament.[59]

August Comte and the Comtean definition of sociology

However, as far as the social sciences were concerned, Geddes did not find himself in the same kind of impasse as the French Le Playists. This was because he always continued his allegiance to his other great master, the pioneer French sociologist, Auguste Comte. His introduction to Positivism and the works of Auguste Comte while in London had seemed to hold out much promise that here was a sociology with an evolutionary perspective altogether more sophisticated than that of the Le Playist school. In Paris the Comtists were well represented in the first sociological society founded in 1872. They were led by Pierre Lafitte who was nearly deposed in 1877 as he underestimated the strength of the religious fervour of the group. A move was made to invite Dr Congreve from London to come as a new leader, but Lafitte managed to stave off this challenge.[60] However, in the Paris of 1878, where the impressionable Geddes was collecting his ideas, the work and message of Comte was being enthusiastically propagated. Geddes' early interest became confirmed into discipleship and he avowed a life-long debt to Comte.

Geddes was confident that with the help of the Le Playist school and the Comtists, he was on the right track for developing his own evolutionary approach to the social sciences. Confidence, however, was not enough, and the two key problems that Geddes addressed himself to were: how to determine the right path for the conscious direction of evolutionary forces and, secondly, how to find the means to ensure that such self-direction took place. Comte provided Geddes with some starting points.[61] The prerequisite for the first objective was a synthesis of all knowledge. Comte constructed a hierarchy of those sciences which he believed were the bases of knowledge, and then placed at the summit of this the science of sociology. Comtean sociology thus encompassed all knowledge. The solution to the second objective was the education of an elite group in an understanding of sociology, that is all knowledge, who, thus armed, would direct effort towards higher evolutionary goals. Geddes took up this latter idea enthusiastically and made it the mainspring of his criticism of the modern university system. He raged against increasing specialisation in modern studies, suggesting instead that students should be encouraged to have an encyclopaedic comprehensiveness in their approach to knowledge. Only then could discoveries of importance in one field be understood, not only in narrow academic terms, but also in other disciplines and in the future possible applications of this new knowledge for the benefit of mankind.[62]

Lecturing as an old man in India in 1919, Geddes reiterated his faith in his interpretation of the Comtean approach to knowledge. Comte, he suggested:

realised with unprecedented clearness, that the preliminary sciences are not so purely abstract or externally phenomenal as their students have mostly supposed, and thus socially detached; but are each and all of them a development of the social process itself. That is to say, logic and mathematics, physics and chemistry, biology and psychology, are none of them independent of social life from which their cultivators may seem isolated, even to themselves: they are, on the contrary, direct products of the social conditions and changes of their times, and so have advanced or stagnated with these. Comte traced too, with the same French lucidity, still too rare in other countries, the rise and progress of all these sciences in interaction with the associated arts; and so he insisted on our handling and grasping all these together, towards our adequate sociology and economics truly abstract, because substantially and connectedly concrete . . .

The main conceptions of the sciences are not only ordering their sub-specialisms but are themselves the unifying specialisms of the cosmodrama. Yet these are not only products of human development; but each and all necessary organs of its activity and intelligence, like the very limbs, the very eyes, and no one of which thinking man can dispense with, without injury even greater than physical mutilation or blinding. For he is mathematician, physicist, biologist, psychologist, sociologist, moralist — i.e. with all these elements of the thought and activity of men, and not merely in serial order, but in organic unity also like the fingers of the hand. Whereas our existing curricula — still so often merely mathematical, even distinct from physical, or too often these in insufficient clearness for any of the above remaining three — are thus comparable to old rituals of mutilation, rather than to education proper . . . So for dealing with the social and economic world in its complexity we need the complete and unmutilated grasp.[63]

For Geddes, the 'complete and unmutilated grasp' was the justification of his position as a generalist and synthesiser of knowledge and he worked hard to be effective. His knowledge may, in some instances, have been less than profound. He was far too impatient to become a specialist in any area, though his friends and disciples attested to their admiration for the breadth of his knowledge. His confidence that he had something original to offer as a generalist was based on his belief that he had invented a new and potentially powerful methodology with which to study the connections between all disciplines. This was the last and possibly, for Geddes himself, the greatest legacy of his youthful search for a new cosmology. He invented a series of diagrams which he described as 'thinking machines': graphic methods for encouraging different ways of thinking. By using these thinking machines Geddes hoped to initiate and educate others to accept the need for educational reform and the need to pursue a synthesis of all knowledge. In the early twentieth century in Britain, Victor Branford tried to promote Geddes' ideas as the Le Play/Geddes method, a fully-realised development from Le Play's original theories. Branford and Geddes believed that the 'thinking machines' provided the essential starting point and the underpinning of the whole enterprise.

Geddes' 'thinking machines' and notations

Geddes' 'thinking machines' were, however, to prove his Achilles' heel. He placed great store by his new methodology, even though he was aware of some of its shortcomings. But he was unable to persuade others to follow what he hoped was the path towards solving the problems that arose. At an early stage in his efforts to create totally new ways of studying social phenomena, he set his face resolutely against mathematical quantification or algebraic formulation. His belief in instinct and observation as the true sources of scientific discovery, led him to try and develop a method which was visually manipulative and encompassed all aspects of social life. His 'machines' were like charts with which he could plot the evolutionary spiral of the life-forces. It was an arresting idea. But his problems were always the selection of factors, their quantification and definition. He tended, when in difficulty, to rely on his own imagination.

Geddes' 'thinking machines' are really only of interest because of the insight they provide into the ways in which Geddes brought together the formative influences of his youth and tried to make of them a coherent social philosophy. For him they remained the means of support in his idiosyncratic and often lonely path through the social sciences. He was able to develop his own ideas only with their help,

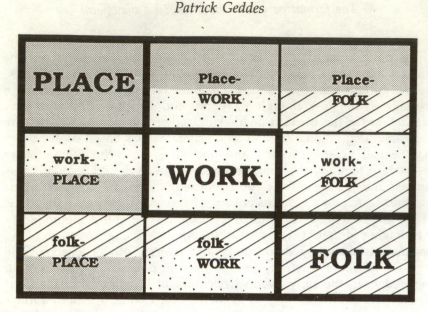

Figure 2.3   The Le Playist diagram: Place, Work, Folk

his 'idea middens' as he used to call them. No one else, however, could be similarly inspired because the structure and development of each 'chart' or 'graph' was entirely personal to Geddes and his particular experiences and knowledge. The fact that he was seeking a scientific method for studying unquantifiable phenomena or, even more importantly, the relationships between phenomena, and also the understanding of them, marks him as a would-be precursor of modern methods of 'systems' analysis and cybernetics. But his ambitions were always greater and the methodology much weaker.

Geddes' most famous 'thinking machine' was his earliest one which he invented in 1879: the translation of Lieu, Travail, Famille into Place, Work, Folk, which he placed in a sequence of squares made by folding a piece of paper three times. Figure 2.3 illustrates the various ways in which he tried to interconnect Place, Work and Folk in this simple way. A later diagram, Figure 2.4, using the same basic form as the earlier one, was built up substituting the social science disciplines relevant to a study of the interconnections between Place, Work and Folk.

Geddes was particularly anxious as a generaliser and synthesiser of knowledge, to emphasise the interrelatedness of all knowledge. Research for future understanding of social evolution, he believed, would have to take place at the interstices between specialisms. Here the interdependence of one with the other was most obvious and that was the central perception which would alert students to evolutionary trends. Comte's hierarchy of the sciences was the starting point for

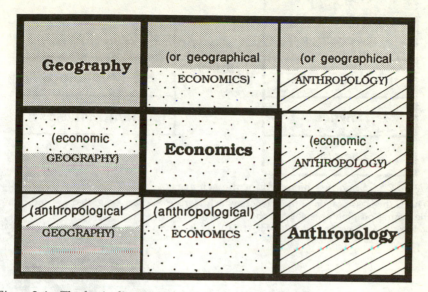

Figure 2.4   The basic diagram related to the social sciences

another basic 'thinking machine', though the order of the sciences Geddes used was that put forward by Herbert Spencer. Spencer recognised five fundamental sciences: sociology, psychology, biology, physics and chemistry. Of these biology, as the science of nature and the continuance of life, is always an intermediate between the other great sciences (see Figure 2.5).

Geddes then drew up another of his graphs in which Spencer's five sciences (made into four as he suggests physics and chemistry run into each other) are made to chart the relationship of the major disciplines with the social fabric, environment and society (see Figure 2.6). Geddes' use of his methodology was, rather incredibly, most disastrous and most successful when he tried to apply it to the study of cities. The successes came from the value of the diagrammatic form to suggest the complexities of city life which Geddes was able to use in his exhibition work and as an insight to guide him in his own practical planning activities. Figures 2.7 is an example of two of his basic 'city' diagrams.

Disasters abounded when he tried to use his idiosyncratic ideas to fill the charts about city life before academic and more critical audiences. What he was trying to reveal was the 'life-force' of the city, the relationships between physical and social phenomena which created the cultural context for change. With the French Le Playist school, he was moving towards a belief that culture determined the form of change and thus the evolution or regression of city life. But his diagrams of city life, when presented most fully in London for the first

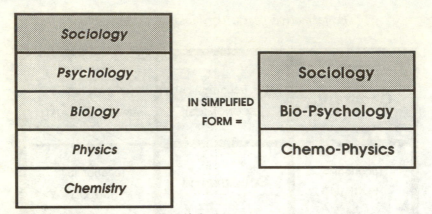

Sociology		
Psychology	IN SIMPLIFIED FORM =	Sociology
Biology		Bio-Psychology
Physics		Chemo-Physics
Chemistry		

Figure 2.5   Spencer's hierarchy of the sciences

'This is the classification of the sciences which Herbert Spencer suggested . . . indicating the central position of biology in the hierarchy.'

Source: J. A. Thomson and P. Geddes (1931) *Life: Outlines of General Biology*, Vol. 2, London: Williams & Norgate, p. 1241.

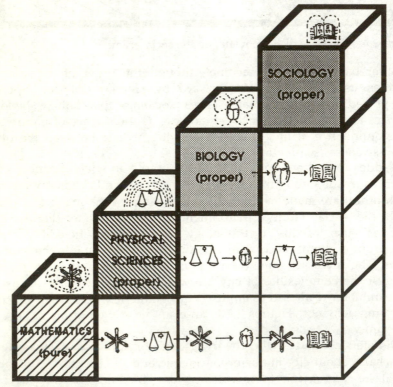

Figure 2.6   Geddes' stairway of knowledge

'It is easy to see on this diagrammatic stairway of knowledge the succession of these four distinctive sciences, mathematical, physical, biological and social, and to realise their respective dependence on their predecessors.'

Source: J. A. Thomson and P. Geddes (1931) *Life: Outlines of General Biology*, Vol. 2, London: Williams & Norgate, p. 1303.

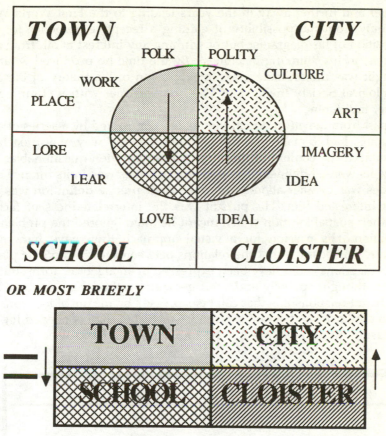

Figure 2.7   The city completed: town, school, cloister and city proper

'In the course of this four-fold analysis, it is plain that we have reached the very converse
. . . of that geographical determinism with which we started, and that we have returned
to a view corresponding to the popular one (of People, Affairs, Places) . . . It is time,
therefore, to bring these together towards the needed synthesis.'

Source: P. Geddes (1906) 'Civics: an applied sociology', Part II, *Sociological Papers*, (ed.) V. V.
           Branford, London: Macmillan, p. 90.

time at the Second Conference of the British Sociological Society in
1905, proved to be totally unappealing. Geddes' chance of a wider
audience for his ideas was effectively cut off by his insistence that his
'thinking machines' were the only way forward to a study of civics. He
believed they had great educational value in shaping people's
perceptions and thus refining their instincts about complex problems.
But it is not hard to see, even from a handful of Geddes' simplest, most
basic diagrams, just how unhelpful they were as educational tools.
When he began refining them and adding psychological factors and
symbolic references to the Greek Gods as manifestations of the life-
force, the lack of communication becomes obvious. Geddes moved

[49]

further and further away in the years leading to the First World War, not only from the possibility of gaining a receptive audience for the exposition of his diagrams, but of winning any interest at all, from any quarter, in his 'thinking machines'. By the time he produced what he thought was the ultimate graphic expression of his theory of life at a Sociological Society meeting in 1914, he was met with a stunning, if polite, indifference.[64]

Apart from the effort which the reader, confronted by a series of odd diagrams, had to make in order to follow Geddes' way of approaching his subject, the content of these machines was often questionable. His concepts were ill-defined. The sociology of cities was in its infancy and Geddes was cavalier about his terms. Their precise definition was not what interested him. His pursuit was the interrelatedness of factors and their spatial location which meant he often ignored the problem of definition. His passion for a visual approach then gave him more trouble. It is possible that the relations between the concepts he places in his diagrams could have been expressed in an algebraic form. But he always thought spatially and used geometry rather than algebra. The number of relationships that can be expressed by means of locations on a two-dimensional diagram is far more restricted than is needed for the increasing number of concepts he found it essential to use.

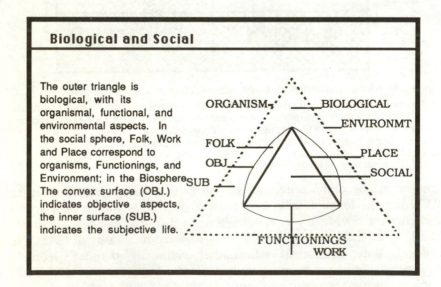

Figure 2.8   The Biological and the Social

'This is a spherical triangle, of which the outer convex side may represent the objective and physiological; the inner concave side the subjective and psychological.'
Source: J. A. Thomson and P. Geddes (1931) *Life: Outlines of General Biology*, Vol. 2, London: Williams & Norgate, p. 1253.

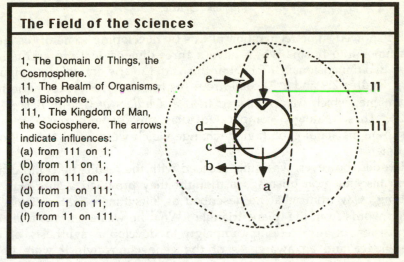

The Field of the Sciences

1, The Domain of Things, the Cosmosphere.
11, The Realm of Organisms, the Biosphere.
111, The Kingdom of Man, the Sociosphere. The arrows indicate influences:
(a) from 111 on 1;
(b) from 11 on 1;
(c) from 111 on 1;
(d) from 1 on 111;
(e) from 1 on 11;
(f) from 11 on 111.

Figure 2.9   The field of the sciences

'Biology has a large number of facts and ideas which the sociologist, if he is wise, will take account of, and the social reformer, if he is alert, will utilise . . . . To begin with, let us give attention for a moment to a simple graph, which suggests 1) the cosmosphere, the world of things, from solar systems to dewdrops; 2) the biosphere, clothing the earth with a thin living envelope, the realm of organisms; and 3) within that again, the kingdom of man, the sociosphere, the world of human affairs and institutions.'

Source: J. A. Thomson and P. Geddes (1931) *Life: Outlines of General Biology*, Vol. 2, London: Williams & Norgate, p. 1240.

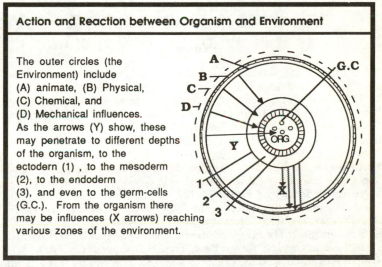

Action and Reaction between Organism and Environment

The outer circles (the Environment) include
(A) animate, (B) Physical, (C) Chemical, and (D) Mechanical influences.
As the arrows (Y) show, these may penetrate to different depths of the organism, to the ectodern (1) , to the mesoderm (2), to the endoderm (3), and even to the germ-cells (G.C). From the organism there may be influences (X arrows) reaching various zones of the environment.

Figure 2.10   Action and reaction between organism and environment

'It is a mistake to think that this relation of the organism to the environment is at all an easy subject. Suppose we draw a circle, place the organism in the middle, and try to see in what different ways the organism may influence the environment, and how in its turn the environment may play upon the organism. See the organism throughout its life running the gauntlet of never-ending environmental influences – mechanical, chemical, physical, animate. These influences take many forms.'

Source: J. A. Thomson and P. Geddes (1931) *Life: Outlines of General Biology*, Vol. 2, London: Williams & Norgate, p. 1255.

He showed some recognition of this by developing some of his later machines into prisms which gave a three-dimensional representation on a two-dimensional surface, but this is not nearly enough (see Figure 2.8). Figures 2.9 and 2.10 come from his last major effort to refine his technique, which was published in *Life, Outlines in General Biology* in 1931. These strange, complicated diagrams depend on a set of assumptions about evolutionary change which were, by that time, out of date.

Geddes, however, was quite pleased with the results he could draw from his biological prisms. He thought they provided a sure way of getting away from what he described as 'illegitimate materialism', in other words from the 'plane triangle'. What he was after was a visual expression of his life-long campaign to develop a synthesis of all knowledge and an awareness of the connections which were both possible and vital between various branches of knowledge. This was something his diagrams and thinking machines helped him to work at, and ultimately the value of his methodology lies in the guidance it gave him in the development of his ideas. But it is clear that Geddes' hopes of generating a new intellectual discipline were abortive. The promise of the biological sciences in the 1860s and 1870s to provide a conceptual basis for understanding the world was always elusive.

Notes

1. J.W. Burrow (1966) *Evolution and Society: a study in Victorian social theory*, Cambridge: Cambridge University Press, pp.101–36.
2. W. Durant (1962) *Outlines of Philosophy: Plato to Russell*, London: Ernest Benn, pp.386–7; C. Sherrington (1963) *Man in his Nature*, Cambridge: Cambridge University Press, pp.120–41.
3. Quoted from C. Bibby (1959) *T.H. Huxley: scientist, humanist and educator 1825–95*, London: Watts, p.90.
4. Much later in life he was to describe the mystical doctrines of 'vitalism' as 'too simply metaphysical vagueness' – P. Geddes (1930) 'What is Mysticism?', *Sociological Review* 22: 157.
5. For debate on defining a new cosmology see R.G. Collingwood (1945) *The Idea of Nature*, Oxford: Clarendon Press, pp.134–40.
6. K. Stephen (1922) *The Misuse of Mind: a study of Bergson's attack on intellectualism*, London: Kegan Paul, Trench Trubner & Co. For a sympathetic discussion of Bergson's work see H.J. Muller (1964) *Science and Criticism: the humanistic tradition in contemporary thought*, New Haven: Yale University Press, pp.246–50.
7. C. Lloyd Morgan was appointed in 1883 as lecturer in geology and zoology at Bristol University College, becoming the Principal of the College in 1887. His main works were (1894) *An Introduction to Comparative Psychology*, London, Walter Scott; (1896) *Habit and Instinct*, London:

E.Arnold; (1912) *Instinct and Experience*, London: Methuen; and (1923) *Emergent Evolution*, London: Williams & Norgate.

8. Geddes wrote of his mother's desire that he enter the church in an article 'Huxley as teacher', Supplement to *Nature*, 9 May, 1925, p.741.

9. Small has already been compared to another British pioneer of sociology, L.T. Hobhouse, who was, of course, extremely antithetical to Geddes − J.E. Owen (1974) *L.T. Hobhouse, Sociologist* London: Nelson, pp. 208−9. The point is the common cultural context they all shared. L.F. Ward's work (1883) *Dynamic Sociology*, New York: D. Appleton, contained many ideas about energy, evolution and the need for conscious control for future development with which Geddes was in full agreement. See also L.F. Ward (1898) *Outlines of Sociology* New York, London: Macmillan.

10. D. Southgate (1982) *University Education in Dundee − a Centenary History*, University of Dundee.

11. P. Geddes (1915) *Cities in Evolution: an introduction to the town planning movement and to the study of civics*, London: Williams & Norgate, p.99.

12. He made a last gesture of support when reviewing a book by I.O. Evans (1930) *Woodcraft and World Service Studies in Unorthodox Education*, in an article 'Scouting and Woodcraft − Present and Possible' *Sociological Review* 22: 274−7.

13. A final statement of a view first held in his youth: J.A. Thomson and P. Geddes (1931) *Life: Outlines of General Biology*, London: Williams & Norgate, vol.2, p.1388−9.

14. D. Herbertson (1946) *The Life of Frédéric Le Play*, Ledbury: LePlay House. Written in the 1890s, chapters 1 to 4 were published in the *Sociological Review* 12, (1920) and 13, (1921). The whole work was published in 38, (1946). Reprinted as a volume, edited by V.V.Branford and A.Farquarson, in 1946, p.107.

15. P.Geddes and J.A.Thomson (1911) *Evolution*, London: Williams & Norgate, p.112.

16. P. Boardman (1978) *The Worlds of Patrick Geddes: biologist, town-planner, re-educator, peace-warrier*, London: Lund Humphries, pp.8−19.

17. O.J.R. Howarth (1931) *The British Association: A retrospect 1831−1931*, London: British Association, p.95. See also H.E. Meller (1976) *Leisure and the Changing City*, London: Routledge & Kegan Paul, p.33.

18. P. Geddes (1903) 'A Naturalist's Society and its Work', *Scottish Geographical Magazine* 19: 89.

19. Four years later he was elected Rector of Aberdeen University.

20. Sir John G. Kerr (1952−3) 'Edinburgh in the History of Zoological Science', *Advancement of Science* 9: 22.

21. Ibid., p.21.

22. P. Boardman, op.cit., p.29.

23. H.G. Wells (1984 edn) *Experiment in Autobiography*, London: Faber, vol.I, p.199. Geddes wrote of Huxley's impact on him in 'Huxley as Teacher', Supplement to *Nature* 9 May, 1925.

24. C. Bibby, op.cit., pp.18−23.

25. J. Arthur Thomson (1925) 'Huxley as Evolutionist', supplement to *Nature* 9 May, p.718.

26. H. Spencer (1873) *The Study of Sociology* London: Williams & Norgate, p.71 quoted from J.D.Y. Peel (1971) *Herbert Spencer: the evolution of a sociologist*, London: Heinemann, p.161.

27. W.M. Simon (1960) 'Herbert Spencer and the "Social Organism"' *Journal of*

the History of Ideas 21: 294–9.
28. S. Eisen (1968) 'Frederic Harrison and Herbert Spencer: embattled unbelievers', *Victorian Studies* XII: 33–56.
29. For a discussion of Geddes' debt to positivism, see J.P. Reilly (1972) *Early Social Thought of Patrick Geddes*, unpublished PhD. thesis, New York: University of Columbia, chapter 1.
30. For a descriptive account of the organisations and individuals influenced by Ruskin in the 1870s and 1880s, see W.H.G. Armytage (1961) *Heaven's Below: Utopian Experiments in England*, London: Routledge & Kegan Paul, pp.289–304.
31. P. Geddes (1884) *John Ruskin, Economist*, (pamphlet) Edinburgh, p.36. See also chapter 3.
32. Ibid., p.39.
33. Ibid., p.37.
34. This was to be Geddes' only original contribution to the natural sciences. He also became an expert on marine stations, being sent by the British Association for the Advancement of Science to Naples to the zoological station there. He returned to Scotland to Aberdeen University to help set up a marine station there. In 1884 he helped Professor Murray with the Edinburgh marine station at Granton.
35. C. Bibby, op.cit., p.142.
36. T.N. Clark (1973) *Prophets and Patrons: the French university and the emergence of the social sciences*, Cambridge, Mass.: Harvard University Press, p.17.
37. Ibid., p.21.
38. P. Geddes, 'Paris University', unpublished manuscript, Rutgers University Library, New Brunswick, NJ. – quoted by Abbie Ziffren (1972) 'Biography of Patrick Geddes', in Marshall Stalley (ed.) *Patrick Geddes: spokesman for man and the environment*, New Brunswick: Rutgers University Press, p.9.
39. P. Geddes and V.V. Branford (1919) *Our Social Inheritance*, London: Williams & Norgate, (the 'Making of the Future' series) pp.344–5.
40. T.N. Clark, op.cit., pp.104–11.
41. H. Stuart Hughes (1959) *Consciousness and Society: the reorientation of European social thought 1890–1930*, London: MacGibbon & Kee, especially chapters 2 and 8.
42. P. Abrams (1968) *The Origins of British Sociology 1834–1914*, Chicago: Chicago University Press, pp.101–53.
43. M.Z. Brooke (1970) *Le Play: engineer and social scientist: the life and work of Frederic Le Play*, London: Longman, pp.116–7.
44. Ibid., p.78.
45. His system of classification was republished in diagramatic form in 1914 with the title *La Nomenclature des faits sociaux d'après Henri de Tourville*, M.Z. Brooke, op.cit., p.178.
46. Ibid., p.19.
47. L. Febvre (1925) *Geographical Introduction to History*, London: Kegan Paul, Trench Trubner & Co., p.15.
48. Ibid., p.16
49. T.N. Clark, op.cit., pp.162–95.
50. For a discussion of Kropotkin's and Geddes' views on this, their similarities, differences, and personal contact, see J.P. Reilly op.cit., pp.217–23.

51. P. Geddes (1905) 'A Great Geographer: Elisée Reclus 1830−1905, an obituary', *Scottish Geographical Magazine* XXI: 552.
52. One thesis was written − Aloo Dastur (1954) *Man and Environment: a study of Patrick Geddes' Valley Section*, Bombay: Popular Book Depot.
53. These included A. Demangeon (1905) *La Picardie et les régions voisines*; R. Blanchard (1906) *La Flandre*; C. Vallaux (1907) *La Basse-Bretagne: étude de géographie humaine*; J. Sion (1909) *Les Paysans de la Normandie Orientale*; M. Sorre (1913) *Les Pyrénées Méditerranéennes*.
54. T.N. Clark, op.cit., pp.107−8.
55. This school served as one of John Dewey's models in founding the University of Chicago Laboratory School.
56. B.M. Ward (1934) *Reddie of Abbotsholme*, London: George Allen & Unwin, pp.48−77; P.Searby (1989) 'The New School and the New Life: Cecil Reddie (1858−1932) and the early years of Abbotsholme School', *History of Education* 18 (1): pp.1−21.
57. Reddie had been a student at Edinburgh between 1878 and 1882 when he first met Geddes. He came back to Edinburgh in 1888−9, renewing contacts with Geddes immediately prior to founding Abbotsholme in 1889. He asked Geddes to inspect and report on the school in 1904 when the school was beset with problems. A copy of Geddes' report, 'On his Inspection and Examination of Abbotsholme School July 11th−18th 1904', (23 pages), is in the Abbotsholme School Archives, Ref. Eb(2)
58. E. Demolins (1898 edn. trans. L.B.Lavigne) *Anglo-Saxon superiority: to what is it due?*, London: Leadenhall Press.
59. Geddes' letters to his children provide an insight into their relationships. Geddes Papers MS10501, 10502, 10503, NLS.
60. T.N. Clark, op.cit., p.103.
61. An analysis of Comte's contribution to sociology is made in Kenneth Thompson (1976) *Auguste Comte: the foundation of sociology*, London: Nelson.
62. Geddes was instrumental in helping to organise the University Extension Movement in Scotland in the 1880s, and he wrote an article in *The Scottish Review* in which he claimed he was working towards the realisation of his new ideas on university education − P. Mairet (1957) *Pioneer of Sociology: the life and letters of Patrick Geddes*, London: Lund Humphries, p.58. One of Geddes' fullest statements of his views on universities is in vol.II of the *Indore Report* (1918).
63. P. Geddes (1920) 'Essentials of Sociology in Relation to Economics', *Indian Journal of Economics* (Allahabad) vol.III, pt I, p.12.
64. The reporters from newspapers sent to cover Geddes' lecture refused to report it.

CHAPTER 3

Practical experiments in social evolution

GEDDES' ENTHUSIASM WAS TO ENSURE THAT FROM THE earliest moment of his return to Edinburgh at the end of 1879 he was to pursue his interests in the social sciences on a voluntary basis. His problem was that his biological viewpoint, his use of his 'thinking machines', and his belief that sociology was a discipline which must grow through an interaction of thought and action, made him an outsider in the debate about sociological studies. He was to spend the decade of the 1880s developing his ideas and testing them in a specific context, the city of Edinburgh. In 1880 he was 26 years old and his energies were by no means absorbed by the work he was able to get as a demonstrator in zoology and lecturer in natural history at the School of Medicine. He was thus able to undertake voluntary social work in his spare time. He was to enter this field at a time when social questions were being debated by students everywhere. It was in the 1880s that Geddes' hero, Ruskin, was Slade Professor of Art at Oxford, teaching his students, not the finer points of aesthetics, but the nobility of manual labour and the ideal of citizenship.

Practical activity for the benefit of the community as a whole seemed the only moral response to the distressing spectacle of great poverty to be found in British cities one century after the beginning of industrialisation and the massive increase in the growth of wealth.[1] Right from the earliest days in the 1880s, however, in all his practical activities Geddes was always looking beyond immediate objectives, towards medium- and long-term goals. This sense of vision, and the search for the evolutionary potential of present activities for the future was what he wanted to explore and reveal to others. But first he had to establish his theoretical base for a study of social change, and for this he drew on his experiences of the debates in the social sciences as much as his biological viewpoint. He cast himself in the role of a student of social evolution and his activities in these years, both theory and practice, were experiments in pursuing this goal.

The British Association and developments in the social sciences

Geddes believed, probably correctly, that the best fora for learning about the theoretical developments in the social sciences in Britain were the annual congresses of the British Association for the Advancement of Science and Art.[2] He was a regular attender at these congresses in the late 1870s and 1880s. This enabled him to witness at first hand the confidence with which the leading natural scientists of the day used their knowledge to assess the progress that was being made in the social sciences. In Section E of the congress devoted to geographical studies, the discussions of these years were leading to the birth of what was to become known as 'the New Geography'.[3] In 1875 Richard Strachey, who was to become president of the Royal Geographical Society in the late 1880s, had already outlined his views on the subject. He wanted geographical studies to be more scientific, and he wanted the study of geography to present an interpretation of present conditions through their physical evolution. This would require classification and comparison of the detailed facts which had been so carefully collected in the course of the nineteenth century and had been the major purpose of geographical studies to date. At the 1881 Congress the president of Section E, Sir J.D. Hooker, devoted his address to emphasising the interrelated nature of the geographical and scientific work of the great pioneers of the nineteenth century. He claimed natural scientists such as Humboldt and Darwin, de Candolle and A.R. Wallace, as natural scientists with an interest in geography. By 1884 another young man of Geddes' generation, an Oxford graduate, H.J. Mackinder, had decided to try and give form to the 'New Geography' by lecturing on the subject to extramural classes.[4]

Quite overshadowing the developments in geography, however, was the struggle that was taking place in Section F, between the economists and the sociologists. At the 1876 Congress a number of natural scientists, including Francis Galton, had mounted an attack on the economists and statisticians of Section F on the grounds that they failed to approach their subject in a truly scientific manner, and that their explanations of current phenomena were totally unsatisfactory. The section managed to survive, but a couple of years later the president of the section, Mr J.K. Ingram, made a speech about the nature of economics and sociology which was to become the touchstone for a whole generation of young intellectuals seeking a way forward from the impasse created by this debate.[5] Ingram, a Comtist from Dublin, accepted that the old deductive moral philosophy had failed. He spoke of the 'notorious discord and sterility of modern economics', and the need for a new sociology based on a synthesis of knowledge, in which specific problems would be seen as parts of a larger whole. In his call for a new sociology he went even further than the critics of Section F in condemning statistics as unequal to sharing a position alongside sociology. He said:

> It is impossible to vindicate for statistics the character of a science: they constitute only one of the aids or adminicula of science. The ascertainment and systematic arrangement of numerical facts is useful in many branches of research, but, till law emerges, there is no science; and the law when it does emerge, takes its place in a science whose function it is to deal with the particular class of phenomena to which the facts belong.[6]

At the 1881 Congress Geddes was to present his first paper to Section F. The paper was entitled 'Economics and Statistics, viewed from the standpoint of the preliminary sciences'. He tried to take on the challenge outlined by Ingram and to point to a way of making statistics more scientific. Although it was his first essay in the social sciences, it was a typical example of both Geddes' originality and shrewdness, and his irresistible urge to impress his audience. He not only reviewed the existing state of statistics, he also offered a comprehensive method of classifying statistical knowledge in categories taken from the natural sciences.[7] Then, to prove just how effective his method was, he proceeded to fit the subject matter of all the other papers at that particular congress into his classification. He made his mark, yet his development of statistics was completely different to the mathematical route being pursued by Francis Galton. The difference between Galton and Geddes on this matter is at the heart of the difference between the Eugenic movement and Civics movement that they each launched respectively in 1904. Galton was seeking proof of the principle of genetic inheritance.[8] Geddes, on the other hand, was seeking to prove that human society could be classified and thus understood as a living unity, however complicated or diverse the manifestations of human life.

Having thus dealt with the problems of statistics, in an entirely satisfactory manner in his own view, Geddes turned his attention next to the thornier problem of economics itself. In 1884 he gave a series of lectures on the subject before the Royal Society of Edinburgh. He took as his title 'An Analysis of the Principles of Economics'. The ideas he put here give an indication of his ability to think radically. However, having established his position, he was never again to depart from it or to develop his ideas further. This was due partly to the method he chose for working out his ideas, using his own 'thinking machines' which were not designed for the detailed testing of general hypotheses, and partly because he used these ideas to justify his political stance which was a dominant factor influencing the way he chose to work.[9] He tried to place himself above politics. He wanted to put across the message that politics of whatever kind were irrelevant when it came to training people to adapt to modern conditions, which as an evolutionist was the way he saw modern social problems. Society was a great social machine which was becoming ever more complex, differentiated, and advanced.

> But the social machine, which is nobody knows how old, nobody knows how complex in its vast and innumerable ramifications — does any one think of repairing it? Wholesale, without understanding it — yes; that's politics: but in detail, city by city, no; that would only be practical economics; and people aren't interested in that.[10]

Debate over such issues as the ownership of the means of production could seem irrelevant if the focus of attention is the quality of life. After all, after the revolution, what then? People still need air, light, food, shelter, education, culture and social organisations in every locality.

To justify this analysis, Geddes wanted to approach economics from the point of view of the preliminary sciences. He thus tried to look at the subject matter first from the position of the physicist and chemist; then the biologist and psychologist; and finally, from the overall perspective of the Comtist sociologist. His lectures were an attempt to do this systematically. They give the impression of boldness and sparkle which helped to captivate his audience and, subsequently, readers of the printed pamphlet.[11] But for all his ingenuity, there were still many problems. Each stage in the reorganisation of the study of economics required new ways of measuring to gain some kind of precision for the discipline. At one point, in the section on the physical principles of economics, he tries to suggest that this was a study of certain forms of matter in motion. To compound the difficulties of this extraordinary suggestion, he airily ignores the fact that measuring this had hardly been solved in the natural sciences. Geddes, searching the literature, had uncovered an article by an Edinburgh professor, Professor Tait on 'The Sources of Energy in Nature', which had been published in the 1860s, and this he suggested was the starting point for such an approach to economics. His confidence was based on the inspiration he got from the ideas of W.S. Jevons. Jevons's work was a particular source of inspiration to Geddes, as he greatly admired Jevons's originality and open-mindedness in the pursuit of complex problems.[12] But Geddes by no means had Jevons's grasp of the principles of economics. Geddes' pursuit of irrelevant and counter-relevant concepts which he then illustrated with individual case studies was one of the major obstacles to a successful outcome of his own attempts at theorising.

What he sought to establish was a new perspective. The main point he wished to make about the physical principles of economics was that they revealed the processes of production and consumption as one vast mechanical process, absorbing and dissipating energy. However, no natural scientist would be prepared to accept an analysis of society as functioning like an automaton. Geddes, therefore, turned to the biological principles of economics, to put life back into the picture. He quoted the use made by economists of terms in the natural sciences to show that this was not altogether a new departure. Terms such as 'parasitism', 'competition', 'laws of population', 'social organisation' are all terms which originated in the natural sciences. The biological approach to economics could reveal the impact of specialisation of function upon the organism. This involved the modification of the organism by the environment. Most obviously, deprivation of food, light, air − the conditions of poverty − produced ill effects on the organism. Less obviously, conditions of luxury could also lead to degeneration. Degeneration in the natural environment was found amongst organisms which led a life of repose, with an abundant food supply and no external challenge. These were the kinds of conditions which reduced an independent organism, capable of development, to the inferior position of a parasite. In biological evolution, the key factor for further adaptation was the nervous system, which must be constantly stimulated in order to evolve to ever higher levels. How much more so must this be in human society when the organism was already so highly complex. Geddes was ready to state, therefore, that the key objective of the biological principles of economics was not food and shelter but culture and education. From this he was ready to extrapolate one of his pedagogic syllogisms: social evolution depended on art.[13]

Even Geddes, however, cannot ignore the fact that one of the most formative of life experiences stems from work and different occupations. While accepting this, he uses an ingenious argument to give an unconventional view of the nature of the labour market. He suggests that the more specialised the different functions of the organisms in a highly complex society, the less they compete against each other and, therefore, the most sophisticated society has a labour market of non-competing groups. This was, of course, diametrically opposite to the Marxist position. Future change Geddes saw in terms, not of conflict between the ownership of the means of production and labour, but in terms of a shift away from productive methods which exploited the masses, to methods which demanded ever more skills from individuals. In this way, the evil trend, apparent since the Industrial Revolution, which had subjected human beings to the machine, would be reversed, and machines would be developed to serve mankind. In such an environment there would be no rationale for competition between factors of production. It was a typically Geddesian kind of reasoning, part perceptive, part confused, since although the demand for more skills from individuals was happening, there was no necessary reason why this would lead to a reduction in competition. Sometimes it might; equally sometimes it might not.

But by the time he gets to the psychological principles of economics, Geddes is thoroughly enjoying himself. He attacks as complete nonsense the old concept in moral philosophy of pleasure and pain as the arbiters for the economic action of man. Instead of the pursuit of pleasure being the strong motivating force, he suggests that, in fact, pleasure is the product of luxury and degeneration. Not improbably, the physical process of degenerating was one of the most pleasurable of organic processes.[14] To replace the pleasure/pain principle, Geddes substitutes the idea of wants and desires as the psychology of action. This approach was one which, in fact, came quite close to some similar ideas that Alfred Marshall was having at the time on motivation in economic life.[15] But his concern was never the implications of these particular insights and how they could be worked out in detail. Instead he leapt straight to what he believed was the core objective of economics as a whole. This he saw very much in terms of supporting a new kind of sociology: a sociology which would lead to informed action to improve social life and thus the chances of successful social evolution in the future. He pours scorn on the idea that the social sciences should be dedicated to the study of time-honoured theories about 'human nature' or 'economic man'. He ends with a definition of what he believes should be the functional nature of economic studies. It is not just a question of the production of wealth:

> The problem, in fact, inverts itself, becoming not merely how to fill bellies, but how to place brains in the conditions most favourable to their development and activity, and so the problem of practical psychological economics passes into that of education.[16]

At the time when Geddes was giving these lectures, the need to explore new approaches to economics was widely recognised. In the late 1870s and 1880s, what has been described as an 'economics movement' was taking place which contained many different strands and contributions.[17] On the surface, the subject appeared to lose its coherence under the impact of the attack on its unscientific nature. Geddes was like many other would-be economists in seeking to make the subject into a practical one. Some of the new work of this period was not directed towards a restatement of economic theory, but towards practical and empirical studies on such subjects as land reform, Ireland, trade depressions, or technical education. Even while Alfred Marshall was devoting himself to the task of developing a new theoretical framework for the discipline, some of his published work was dedicated to practical matters. The book he wrote with his wife, mainly with the needs of extramural students in mind, was entitled *Elements of Economics of Industry*. A survey of all these developments was made by H.S. Foxwell, one of Marshall's students, in an article published in 1887. In this survey Foxwell suggests that the classical approach to economics, with, at its base, a mechanical and amoral concept of economic relations, was banished for ever. All those who used to attack the old moral economy: the philanthropists, the artists, and the church, were now united in helping to create the new. The emphasis was far less on abstract theory, far more on how technological and economic advances had come about.[18] The study of economics had an educational purpose. Attempts were made to extend opportunities to a broader cross-section of society to study the subject, especially the two largest groups hitherto excluded from higher education, the working classes and women. Theories about the importance of economics as a subject and the importance of adult education became entwined. At the centre were the universities, and the new university colleges developing in provincial centres.[19]

This was not an entirely new development of the 1880s. Since the 1840s the Christian Socialists, especially F.D. Maurice and Charles Kingsley, had tried to offer educational opportunities to working men as aids to helping them understand themselves and the society in which they lived.[20] A handful of outstanding and influential churchmen had begun to establish a liberal tradition of British education. Maurice's initiatives were improved upon by Cardinal Newman in the 1850s in the course of his work of re-establishing the Catholic Church in modern Britain. Newman founded a new university college in Ireland, University College Dublin, which pioneered the first programme of extramural university extension lectures in 1854.[21] Almost thirty years later Geddes was to begin his career as an extramural lecturer outside Edinburgh with frequent visits to Dublin. The idea of university extension work had been taken up by Oxford and Cambridge in the late 1860s, and in the course of the next two decades many provincial cities, visited by the peripatetic university lecturers, established university colleges of their own. The emphasis in this latter period was very much on the social sciences. Alfred Marshall, for example, was Principal of the newly-founded University College at Bristol before he went back to Cambridge. H.J. Mackinder developed geographical studies in extramural lectures, and one of his many subsequent posts was Principal of Reading University College.[22]

Education and social change: the impact of nationalism and culture

Geddes began his extramural work in Ireland after he had attracted attention with his paper to Section F in 1881. Mr Ingram himself invited him to Dublin and Geddes established a long-term friendship with him. His contacts with Ireland were an important factor in his quest to understand the dynamic factors in social evolution. He became particularly fascinated by the concept of nationalism and its potential as a means of generating the emotional commitment necessary for change. Ingram introduced him to the leaders of the literary and cultural scene in Dublin. The Irish Home Rule Party under Parnell seemed to have a real chance to secure Irish independence. There was an excitement which spilled over into every aspect of Irish life, from educational reform to artistic achievement. The crucial connection between nationalism, cultural identity and social endeavour impressed itself indelibly on Geddes' receptive (and Scottish) mind. He became fascinated not by the politics of nationalism but the culture. He found himself in sympathy with the idea of using Celtic history and culture as a means of building up a sense of separateness and individuality amongst the subject nations: the Irish, the Welsh and the Scots in the United Kingdom. There was a growing revulsion against the cultural domination of the English which seemed to grow in inverse proportion to the tightening of the imperial power over colonial countries brought by better means of communication and transport. The imperial pretensions of the English seemed to have no limit. In 1876 Disraeli had secured for Queen Victoria the title of Empress of India, and in 1878 he had returned from the Congress of Berlin, having triumphantly asserted British world leadership.[23] In these years London had become not only the dominant capital of the British Isles, but also a world city. Culture, power and influence radiated out from the metropolitan centre of England as never before.

In the early 1880s, Geddes was mainly observing rather than trying to interpret this phenomenon. He became interested in the relationship between nationalism and the demand for higher education in the Celtic countries. In Wales, for instance, the conscious effort to perceive and define the elements of a modern Welsh civilisation was in the hands of the small groups of men and women who developed the University Colleges of Wales. The first college, Aberystwyth, had been built up on a shoe-string budget from very humble beginnings in 1872 to become, in the 1880s, a centre of nationalistic fervour. It enjoyed, in the historian Kenneth Morgan's phrase, 'its own organic relationship with the Welsh people and their social culture'.[24] This self-conscious development of a Welsh culture was the result of the fact that the Welsh nation had undergone a major migratory movement from the north to the south, from the hills and the rural economy of the north, to the valleys and the industrial economy of the south. Welsh culture and traditional values had to be recreated afresh, and the Welsh set about this task quite deliberately. The National Eisteddfod was reconstituted in 1881 and new studies were made of Welsh language, Welsh history, and Welsh culture. Geddes, in his travelling, had been made aware of these developments. Later he was to find a close friend and collaborator in H.J. Fleure, who went to Aberystwyth in 1897 as a student in the natural sciences, and stayed to play an important part in building up the University College and, at the same time, to make a major contribution to the development of human geography.

Back in Edinburgh Geddes did not find much evidence of nationalistic stirrings. This difference has been attributed by a recent historian to the fact that Scotland's union with England had been a voluntary agreement and the major institutions of state, particularly the church and the legal system, had been left intact.[25] Since the Act of Union the Edinburgh bourgeoisie had continued to administer national affairs, insulated from any pressure demanding change or fresh assertions of national identity. In fact, in the nineteenth century the problem of national identity had taken second place to the great divide opening up between Edinburgh, the preserver of Scottish national institutions, and Glasgow, the centre of industrialisation. The latter had become a frightening giant of a city more than twice the size of Edinburgh in 1901, with a culture and tradition transformed by its industrial nature, and the many immigrants from Ireland and elsewhere, who helped to swell its numbers. Scottish national identity was thus highly fragmented, not only between the English-speaking Lowlands and Gaelic-speaking Highlands, but also between Edinburgh and Glasgow, the two most dominant Lowland cities. Edinburgh remained the most important cultural centre of the country in terms of patronage, yet it was a culture based on the taste and aspirations of an administrative bourgeoisie. Scottish patronage of the arts thus tended to foster an arch conservatism, and artists, architects, and writers wishing to succeed had to turn their backs on Glasgow and on coming to terms with an industrialised, urbanised society. They drew their inspiration instead, either from a romanticised view of Scotland's past, or from fashions imported from England, such as the Arts and Crafts Movement. Edinburgh's society thus had a rarified and inbred atmosphere. Geddes' attempts to break through this barrier were never wholly successful, partly because he was receptive to these same sources of inspiration.[26] But in the 1890s he, more than any other individual, managed to create a sense of Scottish nationalism, which he wished to use in the cause of promoting higher levels of social evolution.

Education and social change: the impact of art and late Victorian neo-Romanticism

In the 1880s, whilst he was conscious of the power of nationalism, his energies were directed more specifically to discovering the source of the emotional forces which he believed lay behind what Bergson was to call 'creative evolution'. Taking Ruskin's words to heart, Geddes was amongst those who were caught up in a strong current of neo-Romanticism, in which art and science were united in the service of man. Emotionally, Geddes found romanticism a great personal release, and its anti-intellectual nature could be subsumed by the natural scientist's awareness that not all of the natural world was yet understood. Romantics could be united by their sense of wonder. The biographer of one of Geddes' most romantic collaborators, William Sharp, has written that perhaps

> the single most characteristic motive behind the new romanticism of the later nineteenth century is the sense of wonder opened to the artist through modern science. The power of delicate, sensitive observation, in itself so much underlying Ruskinian aesthetics and what Ruskin admired in the Pre-Raphaelites, joins the scientific and artistic temperaments at this time unequivocally to one another. The measure of 'modernity' and relevance is the extent to which the artist admits and rejoices in the discoveries made by scientific enquiry.[27]

Geddes' evolutionary perspectives made him particularly responsive to these ideas, and he wanted to unite in himself the consciousness of artist and scientist. Since his student days in London he had spent many hours visiting art galleries and exhibitions, and in 1887 he published a special guide to the Jubilee Exhibition held in Manchester that year. His pamphlet was entitled *Every Man his own Art Critic*, and in it Geddes revealed his romanticism, his belief in the unity of science and art and a further feature, which was a token of the neo-Romantic style, the use of symbols and visible imagery. He repeated the exercise at the Glasgow exhibition a year later making his position even more explicit. He wrote:

> despite all his faults and failings, therefore, the painter is not far behind his fellows in science and literature; for him as for them the whole present world is well-nigh become a possession; the past also not only is coming back with unexpected completeness, but is becoming peculiarly his invisible imagery; and soon, even before leaving the gallery, we shall see him not only mirroring for us much that is best of past and present, but casting aside the curtains of the future, and imaging for us upon his magic window the unending drama of the ascent of man.[28]

Education and social change: the social reform movement

In these views Geddes was contributing his support to a widespread movement of Ruskinian followers, who banded themselves together to pursue the cause of the 'ascent of man'.[29] Such an 'ascent' appeared to require a two-pronged attack: on bourgeois social conventions on the one hand, and the appalling social conditions to be found in large cities on the other. Sometimes the two were connected, as in the University Settlement Movement as it was developed by Canon Barnett at Toynbee Hall. There was a rash of new and often small societies in many provincial towns and centres, those with university colleges or colleges of higher education tending to take a prominent role. These societies took many forms, but their responses can be more broadly generalised into three large, and rather amorphous, categories: there were those who opted out of the 'capitalist system'; those who wanted to reform the system; and those who wanted to overthrow it. Representative examples of those who fell into these categories were: amongst the 'opters out', the 'fringe' groups, anarchists, and believers in Utopian communities based on the ideas of Ruskin and Tolstoy; amongst the second group, the political and social reformers, such as the Fabians and Canon Barnett's graduates at Toynbee Hall; and amongst the third, the revolutionaries who tended to be more politically committed to socialism and Marxism, members of the Social Democratic Federation, William Morris's Socialist League, and the propagandists for trade unionism amongst the unskilled. However, there were no hard or fast lines between these categories. Many of those involved in developing one or other of these 'solutions' shared ideas in common with others. Most of them were drawn from the educated middle classes, and many claimed a common inspiration for their search in the writings of Carlyle and Ruskin.[30]

Perhaps the best example of this kind of activity was the Fellowship of the New Life, founded in 1882 by the peripatetic Scottish philosopher, Dr Thomas Davidson. Davidson, described as possessing the 'perfervid emotional Scottish temperament carried almost or quite to the point of genius', hoped that his society would provide a new social context for its members. The objective of Utopia was to be sought in many ways. There were three elements that seemed common to most members: a desire to create a new educational system designed to promote 'the whole man'; the desire to explore a new response to feminism and sexual relations (which involved a direct challenge to many social conventions); and a desire to pursue 'culture' in the Arnoldian sense as the highest manifestation of human life. While Geddes did not join the Fellowship, largely because it was based in London, he had much direct and indirect contact with members. Havelock Ellis, an early member, was to be the general editor of the *Contemporary Science Series* in which Geddes' first monograph, written with J. Arthur Thomson, on *The Evolution of Sex* was published. A fellow student of Thomson's at Edinburgh, and another young disciple of Geddes, was Cecil Reddie, who went on to make his contribution to education by founding the Fellowship school, Abbotsholme, the Progressive Public School in Derbyshire with the help of Edward Carpenter.

In 1884, the Fellowship of the New Life had to face the moral challenge of socialism and it split, the new group calling itself the Fabian Society. The two societies continued, in George Bernard Shaw's immortal phrase: 'one to sit among the dandelions, the other to organise the docks'.[31] Geddes' sympathies remained with the parent society on whose precepts he had modelled his own activities, and he retained for the rest of his life considerable scorn for the Fabian approach to social problems. He was hostile to a centralised state and welfare policies, believing always that the individual had to be the focus of policy, not the masses. No state machine, he believed, could control or develop the interaction of individual with environment, which was the only path for future progress. By making this stand Geddes was to take himself outside the political debate in which the future social progress of the nation was actually worked out.[32] This was not because he was not well informed about the debate. Edinburgh was on the circuit of the lecture tours of politicians and propagandists. The growing concern over social conditions, social progress, and the nature of capitalist society, ensured that the invitations to speakers continued to flow.

Geddes often offered hospitality to visiting speakers including Positivists such as Professor Beesly, artists such as Walter Crane, and on one memorable occasion, William Morris. Geddes' old friend, James

Mavor, a Professor of Political Economy at Glasgow and later Toronto, describes in his autobiography what happened on Morris's visit. After the lecture Morris spent the evening with Geddes and his friends:

> Geddes was vivacious and suggestive as always. His general standpoint on social questions was that of Comte, although he had many original views on such questions, as on all others he touched with his acute intelligence. I do not remember at this distance of time what it was he said that roused Morris's fury. I thought at the time Morris did not quite understand Geddes' point. He was certainly not familiar with Geddes' elusive style or with the philosophical and scientific background which Geddes presupposed. I ventured to try and explain Geddes' position and, in doing so, no doubt in some degree advocated it. Morris turned upon me with a roar, shaking his fist at me across the table, and blazing with magnificent leonine passion. 'You!' he said 'Geddes knows no better; but you! you know; and yet you say these things'. I roared with laughter, and after a while, Morris calmed down.[33]

Practical social work

Geddes took the fateful step of ignoring the political debate. He concentrated his energies instead on becoming an informed sociologist undertaking practical action. In his evolutionary perspective, present political tensions were of little importance, although current events of all kinds were vital subjects of study. This paradox was at the heart of much of the confusion Geddes created in the minds of those who tried to follow him.[34] While social questions were still a matter of political debate, Geddes was seeking practical solutions that could be implemented immediately. This was the message he tried to put across in another influential pamphlet, 'On the Conditions of Progress of the Capitalist and the Labourer' published in 1886. In his apolitical approach he believed that 'real' change was brought about by the fruitful interaction of social processes and spatial form. State intervention of whatever kind could only be clumsy and harmful because it upset delicate balances about which little was known. Significant details were different in each particular context. Every city in its region, he believed, should be an autonomous unit responsible for its own development, though sharing economic and cultural links with others. While developing his views in his series of pamphlets and lectures, he set out in the 1880s to give Edinburgh's individual response to modern social change a unique and Positivist flavour.

He did not have to look far for his first challenge. In the early 1880s, the burning contemporary issue was the housing of the working classes. The national debate on the matter was strong enough to stimulate the government to appoint a Royal Commission in 1884 to investigate the matter, and as a sign of the importance attached to its deliberations, the Prince of Wales was appointed to serve as a member of it.[35] Edinburgh had played an important part in the national debate. The city had pioneered public health reform by means of an Improvement Scheme set up by special Act of Parliament in 1867. The Act set up an Improvement Trust with powers to demolish insanitary areas in the city centre.[36] This drastic action had been taken because Edinburgh's problems were particularly severe. The rapid decline and deterioration of the Old Town, as the eighteenth century New Town absorbed the upper and professional classes, had accelerated in the course of the mid years of the nineteenth century. Former aristocratic and bourgeois homes in the Old Town became divided into one-room tenements. The building density was already extremely high in the Old Town for historical reasons and this, coupled with severe overcrowding and a complete lack of sanitary facilities, created conditions of much squalor and disease. The poor could not escape. There were few working-class suburbs. The students of Edinburgh's Medical School were not only given an excellent training, they also had direct experience of most of the diseases known to man which were to be found on their doorstep.

In the classic tradition of social initiatives, the worst conditions produced the pioneers of the public health movement. Edinburgh, under its Lord Provost, William Chambers, initiated the great Improvement Scheme. This pre-dated in its powers the national legislation which was to follow in 1875. The Improvement Trust set up to administer the scheme met regularly between 1867 and 1889. The aim was to bring light and air to the insanitary areas by large-scale demolition. Most of the work was completed by the mid 1870s and the Trust's main function after that was merely a tidying up operation. In 1883 William Chambers died and he was given a hero's funeral as the saviour of Edinburgh. But in the midst of all the civic pride in his achievements, a few ugly facts began to raise their heads. Improvement had been seen in physical terms: wholesale demolition, road improvement, new layout of building plots. What had been forgotten was the human factor. The poor evicted from their insanitary homes did not go away. They moved to adjacent areas which soon became as polluted as their former homes. Perhaps Improvement Schemes, far from solving public health problems, actually perpetuated them.[37] Land which had been improved rose in value. Rents of property built on it thus rose also. The poor could never be rehoused in the same location. Perhaps public health, after a certain minimum level in the provision of sewers and a water supply, depended on people changing their habits and their life-style. How was that to be achieved? It was the major issue of concern to the Royal Commissioners of 1884. Did the pigs make the sty or did the sty make the pigs, and if the latter could legislation effect any changes? The answers that the Royal Commission came up with were, in fact, inconclusive. Legislation and building were considered adequate. The problem was a social one. The real problem was poverty and the need for the casual labourer, particularly, to live close to the potential market for his or her labour.[38]

As the first report of the Commission was being drawn up, Charles Booth began his great survey of the East End of London. He was to find in the course of it that those living in poverty were those whose wages were too low; where the head of the household was sick or dead; where women were bringing up families on their own; where families were large.[39] The nature of these revelations demanded a fresh response. For the fight against poverty and bad housing, the Socialists put their faith in the organisation of labour and their effort into developing the trade union movement.[40] The Fabians wanted more and better state intervention and administration, hoping to deal with social problems through state bureaucracy.[41] The philanthropists redoubled their efforts to reach individuals, and the case-study work of the London Charity Organisation Society was taken up by ever more voluntary groups.[42] In most major cities, however, areas (often those closest to the centre) tended to remain in the grip of a cycle of deprivation and decay.

In all this welter of activity and concern, Geddes struck out on his own, to seek a solution to the problems of insanitary housing. His target was immediate and permanent civic betterment. He started from the premise that cities flourished or declined according to the people who lived in them. The period immediately prior to the long process of decline in Old Edinburgh had also been the heyday of the university, when it had been known as the 'Athens' of the North. 1883, the year of Chambers's death, was also the tercentenary of the university. Illustrious scholars of the past, especially Adam Smith and his School, were remembered at the centenary celebrations. Geddes actually wrote a small pamphlet giving biographical details of the famous alumni.[43] In his involvement with the glorious days of the eighteenth century, he brought to the slum areas of Old Edinburgh a vision of the past, when these old courts had housed the famous scholars. Edinburgh, the scholars, and the university had all flourished together. For Geddes this had to be the starting point. Yet in practice he found himself confronted with the physical problems of an insanitary area. For guidance on how to treat these, he looked as usual for inspiration from Ruskin, and he found it in Ruskin's support for the work of Miss Octavia Hill. Geddes' contact with the philanthropic world in Edinburgh had been relatively slight until he met Miss Anna Morton. As the daughter of a Liverpool merchant, she had undertaken philanthropic work in Liverpool, which was a progressive city in the philanthropic world.[44] Patrick and Anna were to fall in love and to get married in 1886. The years of their courtship, however, were spent working together on the problems of Old Edinburgh. Anna encouraged him to be responsive to Miss Hill's work. Octavia Hill, of all the housing reformers, accepted that it was as important to reform the people as it was to improve the quality of their housing. Geddes went to London especially to visit Miss Hill and to look at her work in Marylebone.

Since Geddes was interested in the interaction of the organism and the environment, this made him extremely sympathetic to her views. He believed that for the city to flourish every man, woman, and child should have a chance to reach his or her full potential. This meant, on a physical plane, improving the quality of nurturing. Since women were the prime nurturers of life, all improvements in this area should be modelled around their lives. They should be constantly encouraged, both by their environment, and by careful reconstitution of nurturing traditions, to grow and develop in this role.[45] Octavia Hill and her sister Miranda had, with Ruskin's financial aid, been undertaking practical work to achieve these ends over the previous twenty years. Miss Hill had been an important witness before the Royal Commission on Working Class Housing of 1884, since she was prepared to suggest how bad housing might be improved despite widespread poverty. From her earliest work in the slum court in Marylebone which Ruskin had purchased for her in 1864, Octavia Hill had set herself the task of reaching the poorest of the working classes to be found in city centres, those least able to withstand the economic forces which were depressing their living conditions and creating slums. She published a record of her approach and activities *Homes of the London Poor* in 1875, and a second edition of this was brought out in 1883. In it, she makes four points very clearly. First, as a landlord's agent she nevertheless approached her work very much in the spirit she felt appertained to the countryside where the local squire would have responsibility for and personally know his tenants. In other words, what she was attempting in her work was to reconstitute the best cultural and social traditions she believed had existed in the past in the changed circumstances of the present.[46]

Next, she put her faith in personal contact between landlord or agent, and tenant. That contact had to be on a business footing but informed by direct concern for the welfare of the tenants. Miss Hill invented the role of the philanthropic lady rent collector who had a business relationship with everyone, insisting on the regular payment of rent who yet, through this contact, could become the friend, helper and adviser of her tenants. In this latter capacity, Miss Hill makes her third suggestion: the way to improve the living conditions of the poor was to teach them how to clean and to repair their own homes. She allocated a proportion of the rent for a maintenance fund and she employed her 'deserving' tenants when they were unemployed, on lime-washing, roof or window repairs, or tending to the cleanliness and conservation of staircases and courtyards. Her contention was that: 'You cannot deal with the people and the houses separately. The principle on which the whole work rests is that the inhabitants and their surrounding must be improved together. It has never yet failed to succeed'.[47]

Finally, Miss Hill suggests that it is of equal importance to care for the emotional and recreational needs of her tenants. She believed that it was vital to the maintenance of higher standards of behaviour and an improved environment to introduce elements of joy into the lives of her tenants. She subscribed to current theories on leisure, introducing facilities for 'rational' recreation such as reading rooms and places to meet outside the pub. But more than that, she wanted to recast and incorporate festivities of the rural past as focal points on a regular basis for anticipation and enjoyment. She held May Day celebrations in her courtyard, with maypole and dancers, and Christmas festivities. Her sister Miranda founded a society especially to promote such activities, the Kyrle Society in 1875.[48] It was instantly popular and branches were set up in many towns and cities, including Glasgow. Apart from the maypole dancing and other activities, Miss Hill believed the year-round practical way of introducing joy into her courtyards was to plant them with trees and creepers, thus softening the harsh lines of wall and pavement. The effect was that visitors to Miss Hill's courts could see immediately the difference between those managed by her and adjacent ones. By the 1880s, Octavia Hill-type schemes were running in many places, and there was some interest in her methods in Scottish cities such as Glasgow and Edinburgh.[49]

Geddes was very impressed by the success of her techniques. Her businesslike approach, 'five per cent philanthropy', and her desire to deal with organism and environment together, were particularly appealing to him. He reported his observations on her work to the group of friends he was working with in Edinburgh. This small group, drawn mostly from amongst staff and students of Edinburgh University and the local intelligentsia, had formed themselves into a debating society for the purpose of inviting visiting speakers. Under the influence of James Oliphant and Geddes, the group began to organise itself more formally in order to carry out practical schemes in

Edinburgh. They formed themselves into an Environment Society which was, in the words of James Oliphant, 'a scheme for the organisation of all benevolent enterprise! But its special aims are to provide or rather improve existing material surroundings, by decorating halls and schools, planting open spaces, providing musical and other entertainments for the people, etcetera, etcetera'.[50] In some ways these aims were echoed in the work of another Ruskinian disciple, T.C. Horsfall of Manchester. He wrote a pamphlet in 1884 entitled 'The Means Needed for Improving the Condition of the Lowest Classes in Towns', advocating more religion, recreation, and better housing for the poor; and went on to found his Art Museum in Ancoats in 1886. In Edinburgh the little Environment Society, under Geddes' influence, turned its attention to the unhealthiest area of Edinburgh which was the ward of St Giles in the Old City, which was also the location of the university.

Philanthropic activities and the renovation of Old Edinburgh

There were already in Edinburgh, a number of voluntary organisations concerned with the health and sanitary conditions of the poorer areas of the city. A Health Association had been founded in 1880, when it was recognised that the Improvement Trust was not going to eliminate disease and squalor.[51] This organisation ceased to function in 1884, but its missionary activities were carried on by a group anxious to protect the middle classes from infection (since disease was no respecter of social status), who called themselves 'The Sanitary Protection Society'. For those who put their belief in better municipal administration, a Social and Sanitary Society was formed to liaise with the city's Public Health Committee. This voluntary group, heavily supported by clergymen, also hoped to bring permanent improvement to the poor. Those of the Edinburgh working classes who could afford to help themselves, formed associations to provide new housing, and their efforts were supplemented by the Edinburgh Association for Improving the Dwellings for the Poor in 1885.

Geddes and his friends did not lend their support to any of these initiatives. They were, however, able to capitalise on this current concern over the sanitary and housing conditions of the poor, to get support for their own activities. At first, these had hardly gone beyond finding young artists to carve statues and provide drinking fountains in St Giles. Geddes had involved himself and his friends in trying to overcome the dilapidation and neglect of the old housing by cultivating any waste ground, making small gardens and planting trees and creepers. But as Geddes' ideas grew more ambitious, it became obvious in the months following his London visit that the Environment Society

[73]

needed to be recast on a more formal basis to mobilise support for his schemes. Here the expertise in the philanthropic world of the Morton sisters, Mrs Oliphant and the future Mrs Geddes, was of some use. Since the 1860s Liverpool, their home town, had pioneered new ways of organising the relief of poverty, placing emphasis on two major developments: the centralised organisation of charitable activity, and second, under the leadership of the Rathbones, a new kind of sensitivity to the actual problems of the poor.[52] These developments were echoed in the setting up of London's Charity Organisation Society in 1869.[53] Octavia Hill and the young Rev. Samuel Barnett were founder members of this latter society. Other cities, where the leaders of philanthropic activity were responsive to these changes, began to adopt similar methods. For example, under the guidance of Henrietta Carey, in the context of a lively tradition of philanthropy by Nonconformist families, Nottingham totally reorganised its charitable activity and founded an umbrella organisation – the Nottingham Town and County Social Guild, to co-ordinate effort.

This organisation set up in 1875 refined the new responses to philanthropy quite specifically. On the one hand, established charitable activity was placed under the supervision of the Nottingham Society for Organising Charity in collaboration with the Poor Law Authorities; on the other, new solutions to combat poverty, tentatively looking beyond the shortcomings of the individual to the influence of the environment, were to be co-ordinated by the Nottingham Town and County Social Guild. In practice, the same people were involved in both, but the Social Guild, with its emphasis on environmental factors, followed the lead of the Hill sisters and became involved in housing management and environmental improvements of the kind promoted by the Kyrle Society. This latter society, the brainchild of Miranda Hill, had been named after John Kyrle, the Man of Ross, who had been noted for his efforts to provide amenities in his native town. The society and its many branches outside London aimed to bring pleasure into dull lives by such means as beautifying elementary schools with decorative panelling, arranging for flowers to be grown in window boxes and back yards, and organising wholesome entertainment in the form of Happy Evenings.

In Glasgow the Kyrle Society quickly became a dominant philanthropic body. The emphasis on cultural activities soon lessened as it became involved in the problems of housing, slums, and the basic needs of relief, as well-meaning philanthropists came into touch with the realities of life for Glasgow's poor during the period of severe depression in the city's economic fortunes, which had begun in the mid 1870s and continued until the 1890s. But in Edinburgh, in the aftermath of Chambers's Improvement Scheme, the context was different. The prospect of making some impact on the Old Town using the methods of the Hill sisters seemed good. It was feasible that the little Environment Society could be made the nucleus for such an attempt, and its members consciously chose the Nottingham Town and County Social Guild as their model.[54] To the existing activities of the Environment Society of beautifying the Old Town with gardens and stone carvings, the prospect of acquiring housing for philanthropic management was immediately added.

Initially the divergence between Geddes' intentions of promoting social evolution and the philanthropic mould within which he worked was not clear. The Edinburgh Social Union formed in January 1885, while never a dominant philanthropic body, was to attract a number of civic philanthropists who served it loyally over the years. Having played a leading role in its foundation, however, Geddes was soon to leave it. He had no interest whatsoever in the tedious and time-consuming process of housing management along Octavia Hill lines. That was the role for the volunteer lady philanthropists who were able to devote a great deal of time and effort to this work. Instead, Geddes had a vision of the reconstruction of the Old Town, physically and socially, in such a way as to produce a new cultural environment. Then he believed the city and its citizens would flourish together, each he hoped, achieving ever higher stages of evolution. Most of his hopes were pinned on the university, still located in the Old Town, and over the next couple of years he began to concentrate on the possibility of bringing students back to live in close proximity to the very places where the great scholars of the eighteenth century had lived when Edinburgh had been a leading cultural centre in the world.

As he withdrew from the Edinburgh Social Union, Geddes made a dramatic personal move. He took his newly-married wife Anna to live in a tenement block in James Court in St Giles Ward. He rented all the one-roomed tenements on the top floor of the block and made them into a flat. Few of his contemporaries understood his action. To go and live amongst the poor had become a symbol of religious commitment in the 1880s, with the proliferation of the University Settlement movement and other experiments in London and elsewhere. But Geddes did not fit that pattern at all. His friend, James Mavor, suggested an alternative that he, as a student of Russia, knew much about. This was the rather unsuccessful 'V Narod' or 'To The People' movement in Russia, in which young intellectuals in the mid 1870s, had tried to identify themselves with the people by living amongst them and by politicising them. Mavor came to the conclusion about Geddes' move that the experiment was interesting and not destitute of a practical side: but it did not afford the vital touch of the Russian example, because 'the intellectuals' did not establish organic relations with their surroundings. They revived the eighteenth century traditions of the Lawn Market, but these fitted in rather awkwardly with those at the close of the nineteenth.[55]

Geddes' concern was not with the problems of poverty. Luxury not want and quality of life not the provision of bare necessities, was the challenge which Ruskin had so forcefully formulated and which seemed most important to the budding evolutionary sociologist. The objective of Geddes' work in the old tenements in Edinburgh was to involve his helpers in a dynamic relationship with their environment, to encourage their own personal growth, and to contribute to the transmission of cultural values of the highest order from one generation to the next. Geddes' wildest fancies were held in check by his wife and most important collaborator, Anna Morton. In her 31 years of marriage to Geddes, she gave him not only lifelong devotion and support in often difficult circumstances, but also practical administrative help with his ventures, which gave him public credibility. Her first child, her daughter Norah, was born in the slum tenement, James' Court, yet Anna continued to work for her husband's schemes. She managed the financial arrangements which not only enabled them to rent their own flat but to lease three others on The Mound which Geddes wanted to use to house university students.

Seven students moved in on 1 May 1887, and these students formed the nucleus of what was to become University Hall, Edinburgh University's first Hall of Residence for students. From the start Geddes stipulated that these residences should be self-governing and rules and regulations should be drawn up and administered by the students themselves. The pioneering spirit of the venture, Geddes hoped, would also inspire the students to work voluntarily, providing time and labour to continue the kind of work that Geddes had begun. He was not to be disappointed. In the atmosphere of high moral endeavour which influenced the student body in the mid 1880s, and sent Oxford graduates to the East End of London, Edinburgh students were eager to become involved. J. Arthur Thomson, A.J. Herbertson, John Ross, Victor Branford, T.R. Marr, Edward McGegan, and many others were drawn to Geddes and helped him in his work. More tenements were quickly acquired to supplement the three original student tenements on The Mound, until the entire tenement block was taken over. Geddes initially focused his attention on the area around the university and adjacent to the Castle, Castle Hill, and the Lawnmarket. He then began operations at the other end of the Old

Town near Holyrood Palace, hoping eventually that one day the improved areas would meet. He was working however, in areas where the dilapidation was extreme. The great age of many of the buildings, their misuse and multioccupation, the lack of sanitary facilities, and the presence of the sick and the poverty-stricken, made the task daunting. Riddle's Court, acquired in 1889, could not be used for seven years as a students' residence because of the deplorable state of the surrounding property and 'the perpetual occurrence of nuisances of every kind'.[56]

What Geddes achieved in the decade 1886–96 was quite remarkable. He worked alongside the Edinburgh Social Union which continued to acquire properties and manage them along Octavia Hill lines. Geddes found ways, however, of financing his own activities independently. He was able to use a number of methods. He raised capital from private investors on the basis of Octavia Hill's 'five per cent philanthropy', and bought up properties. From 1892, Edinburgh Town Council began to apply the regulations of the 1890 Housing Act which made public monies available for improvement schemes in run-down areas.[57] This funding helped to pay for demolitions. The Town Council had divided up the unhealthy areas of the city into ten districts, and assigned individuals to manage the work in particular areas. Geddes was made responsible for the Lawnmarket, and also unofficially took action in the area at Riddle's Court. There was little formal regulation of his activities which were carried out on an *ad hoc* basis. There was an extraordinary amateurishness about the whole proceedings, with Geddes often taking personal liability for the financial transactions which he was undertaking on behalf of the Town Council. The boundaries between public action and private philanthropy were blurred, and the fact that Geddes used many of his larger acquisitions as self-regulating Halls of Residence for students confused the issue still further. He took over a block in St Giles Street, formerly occupied by the *Scottish Leader* newspaper in 1895, and opened it as St Giles House for twenty residents. Blackie House was opened soon afterwards, an amalgamation of old houses on The Mound, and hostels for ladies were provided in the Lawnmarket. As well as restoration work, he also undertook new building. In 1890 he had acquired Ramsay Lodge and Ramsay Gardens; and a couple of years later began building a block of flats on the Castle Esplanade which were to be known as Ramsay Gardens, financing it from loans from the future purchasers of the flats. This did not work out altogether satisfactorily as Geddes had an altercation with one of his would-be purchasers. This left him with a burden of debt which became progressively more onerous over the next few years. However, despite these financial problems, Geddes moved with his family into one of the flats as Anna had had her second baby in 1891. His activities altered the famous skyline of Old

Edinburgh as viewed from Princes Street in the New Town, but the romantic style of his new building blended well with the old. During this time the number of students housed in this area increased about fifteenfold, and the total population of the largest university hall rose from 10 to above 200.

The optimism that this engendered amongst Geddes and his friends thus had some grounds. They worked along in the wake of the activities of the city's improvement schemes, as narrow closes were opened up and dilapidated property removed. The old Chambers Improvement Trust had been responsible for some improvements in the central areas around Cockburn Street and Cowgate, and around North Bridge. Geddes' voluntary workers provided the individual touch to bring to life again these newly-treated areas. But despite all this activity much remained to be done, and what had been done began to be an unbearable financial burden on Geddes personally. Eventually in 1896 a financial crisis occurred and Geddes' friends rallied to help him by setting up a Town and Gown Association to take over the responsibility for financial liability. The record of the properties to be put in its charge gives some idea of the extent of Geddes' activities. He had acquired residential houses, student halls, workmen's dwellings, shops and building sites to the value of £41,900; he had developed new properties to the value of £6,155, and he had expended a further £4,500 in furnishing student halls and undertaking the many other small improvement schemes related to his civic betterment programme. Geddes hoped that the Town and Gown Association would carry on the work, not only in Edinburgh, but also in other Scottish university cities, and an Advisory Committee was actually set up in Glasgow. Like many other initiatives put forward by Geddes in the course of his career, however, this one also petered out, and no such extension of the Town and Gown activity flourished beyond the boundaries of Edinburgh.

But his objective had never been just the renovation of the old parts of the city and the housing of students. In 1892 Geddes acquired the old Observatory at the end of the Castle Esplanade, with its camera obscura on the roof as an attraction for visitors. He was not sure what to do with it at first, but once again turned to his experience of Ruskin's initiatives for guidance. When a student at London in the 1870s, Geddes had visited Sheffield to view Ruskin's Museum for working men which Ruskin had encouraged as a means of preserving the cultural traditions and standards of local craftsmanship.[58] It had been part of a scheme to set up a small-scale self-supporting community settlement. Geddes was briefly involved in the problems which developed between the communitarians and their new leader Mr W.H. Riley, international socialist and teetotaller, as he wrote letters on behalf of the former to Ruskin. But what he remembered of this was the little museum on the hill dedicated to nurturing the cultural traditions of the region. From this seed Geddes was to produce his regional museum, the Outlook Tower, which will be discussed more fully in the next chapter.

Education and Social Change: the 'living organism'

The fact remained that, with his Outlook Tower, and in his work in the Old City, only Geddes and one or two of his closest friends and disciples appeared to understand fully what he was trying to do. He himself never lost sight of the fact that his activities were part of a scientific experiment in social evolution. The Halls of Residence and the Outlook Tower were not ends in themselves. They were the means of educating the young about social and cultural change and making them more self-aware. Geddes was always seeking new ways of furthering these ends to promote what he called 'higher and higher individuation'[59] and thus social evolution. In 1889, after a visit to Paris for the exhibition of that year, he conceived the idea of resuscitating the Franco-Scots college which had flourished at the University of Paris in the eighteenth century. Aware that most Edinburgh students were drawn from Scotland, Geddes believed that travel to foreign centres of higher education was an essential element in their personal development. Pursuing his idea to engineer social evolution by using not only the best traditions of the past, but the spatial form in which they were nurtured, he tried, but failed, to secure the very same building in Paris for his project. His friend Mavor wrote to him that such a scheme was a good idea, but he pointed out that Geddes must be dreaming if he thought he could revive the conditions of the eighteenth century and follow in the footsteps of the great scholars: 'Has the air of James' Court so mesmerised you that you must follow their steps in every particular?'[60] Links between Edinburgh and Paris were set up and a number of young students, including A. J. Herbertson, were introduced to the developments of the social sciences in France.[61]

In his search for finding ways of promoting social evolution there was one vital area which dominated Geddes' thinking throughout the 1880s. This was the relationship between the sexes. The feminist movement of this decade created a strong undercurrent of support for Geddes and his schemes. The administration of student halls was undertaken by women; the Edinburgh Social Union was largely run by women; women worked in the gardens and civic betterment schemes. The progressive young ladies who lived in the ladies' hostels in the Lawnmarket provided a fertile recruiting ground for voluntary social workers, and women swelled the ranks of those attending the Summer Meetings. It was in these small, 'progressive' circles that the great debate on 'Women and their role in society' was carried on.[62] It was a debate in which biological factors provided the arguments, and which no evolutionist could ignore. The central question was to determine the significance of sex differences in the evolution of the higher forms of life. The problem for all concerned, Geddes included, was that not enough was known about the biological determinants of sex characteristics. As Jill Conway has pointed out: 'It was not until 1901 that sex-linked characteristics were understood to be tied to the sex chromosomes, and not until 1903 that the working of hormones in human physiology was understood'.[63] The challenge for the biologists was to interpret their knowledge of evolution in the natural sciences to contribute authoritative arguments to the debate. Geddes, as an evolutionist and sociologist, believed he was uniquely qualified to undertake this. In the course of the 1880s, as he had ceased his work in microbiology, he had developed his expertise in the study of physiology, morphology, and cell-theory.

During the decade Geddes and his former student friend and collaborator, J. Arthur Thomson, kept themselves abreast of developments in the theory of organic evolution. Together they worked on bringing current knowledge to the layman by writing a series of articles for the Edinburgh-based *Chambers Encyclopaedia*, and for the *Encyclopaedia Britannica*. (In the former, the topics covered were Biology, Botany, Environment and Evolution; in the latter, Reproduction, Sex, Variation and Selection and Darwinian Evolution.) Darwin had published his views on the relationship between the concept of evolution in nature and human society in 1873 in his work, *The Expression of the Emotions in Man and Animals*. He argued that anatomy, behaviour, structure and function were intimately linked and that man, like other living creatures, evolved through natural selection. J. Arthur Thomson found this view convincing, but Geddes had serious reservations. While in form humans may be related to the primates, in one dimension there was no comparability. In evolutionary terms, mankind has leapt forward in terms of intelligence and mental evolution beyond comparison with any other species. The socio-biologists had therefore to construct a biogram of human evolution using data from human societies rather than from the animal world. Man is a product of culture as much as of nature.

Geddes' forays into socio-biology were part of a wider movement, in the late nineteenth century, to establish the scientific study of man and his behaviour patterns which was to become defined eventually as social anthropology.[64] In the 1870s and 1880s, when the theory of biological evolution was paramount, attempts were made to construct

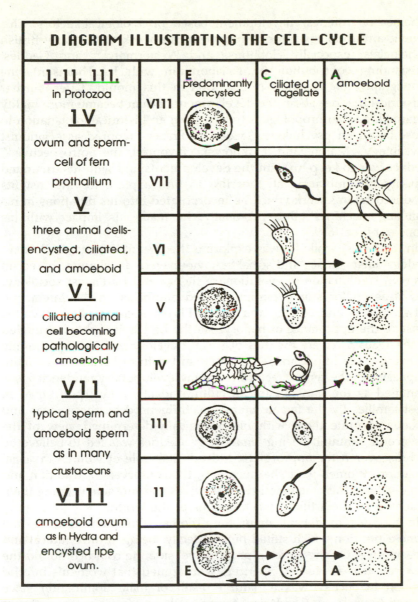

Figure 3.1

Cell theory: all organisms are made up of cells, and start from cells. Geddes was not interested in trying to go further with this into the realm of embryology. His purpose was 'the restatement of the theory of organic evolution . . . (that not of indefinite but definite variation, with progress and survival essentially through the subordination of individual struggle and development to species-maintaining ends.) It becomes the first practical application of the biological sciences to map out the two paths of organic progress . . . utmost degeneracy and the highest progress and blossoming of life.'

Source: P. Geddes and J. A. Thomson (1889) *The Evolution of Sex*, London: Walter Scott, p. 86 and preface.

[81]

schemes of unilinear development based on biological factors. This genetic anthropology informed much of the work of the early British sociologists, especially Hobhouse and Westermarck[65] and Geddes' outstanding contribution (in collaboration with J.A.Thomson), the monograph, *The Evolution of Sex*, was in this mould. Later, further ahistorical theories about the development of man became more highly regarded, and anthropologists turned from an historical explanation to those based on psychology. Tylor and Frazer favoured associationist psychology. Marett and Malinowski favoured the introspective.[66] Geddes was to keep up with the developments and with his unlimited capacity for embracing all theories, to find a place for the insights produced by this work which he incorporated into his notations in his 'thinking machines'. But, personally, he made his impact with his monograph of 1889.

In the 1880s, while he was exploring theories of 'organic evolution', Geddes' ideas on sex and social behaviour were developing, based on his own combination of evolutionary theory and Le Playist sociology. His ideas on 'organic evolution' placed emphasis on the interaction between male and female, and his Le Playist sociology kept alive his sense of the importance of history. As the Le Playist School struggled in the 1880s to create theories of social development based on family structures, the importance of history and culture became ever more emphasised. The central factor in shaping evolutionary tendencies was identified as the transmission of cultural ideals, a task performed in most families by the female members. Influenced by these ideas and anxious to unite them with his biologically-determined view of the essence of femininity being 'anabolic',[67] Geddes was led to believe he had discovered the appropriate, indeed vital, role of women in social evolution. Women were the nurturers, the conservers of tradition and moral values, their biological roles as wives and mothers keeping them untainted by the artificial machinations of the world.

In the current debate about the 'Woman Question' Geddes had found a position which suited him perfectly, being at once radical and conservative. On the radical, 'progressive' side, he could confound the views of such as Herbert Spencer, who argued that women's inferior position in society was a natural result of male domination since earliest times.[68] In Geddesian terms such an argument completely underestimated the crucial role of women in shaping evolutionary trends. Women's role as wives and mothers meant that they needed male protection but, in return, their nurturing tendencies shaped the economic and social environment, creating ever higher levels of civilisation. On the conservative side, Geddes' views fitted in most comfortably with conventional nineteenth century ideas on women's position in society.[69] His dislike for Fabianism extended to radical feminism and the campaign to gain votes for women. Women's ability to transform the world would not come from direct involvement in the contaminating world of politics but through their innate biological role in the private sphere. Change would be produced by thousands of individual women operating independently, not by collective effort. One of Geddes' most quoted phrases on this was his comment that 'What was decided among prehistoric protozoa cannot be annulled by acts of parliament'.[70]

When the *Evolution of Sex* was published in 1889, it brought the authors, particularly Geddes himself, considerable notoriety as an explicit statement on sexual matters. Geddes had not confined himself to outlining the significance of women's sexual role, he had also proferred advice on how they should conduct their lives. In *Evolution of Sex* he was still relatively restrained, and confined his discussion mostly to the 'population question' and the views of the Rev. Thomas Malthus. Again he was able to appear progressive with his arguments in favour of the control of fertilisation. What he wanted to advocate on the grounds of achieving the best trends for the social evolution of the future was 'prudence *after* marriage'. He argued forcefully, 'The vague feeling that control of fertilisation is "interfering with nature" in some utterly unwarrantable fashion, cannot be consistently stated by those who live in the midst of our highly artificial civilisation'.[71] He argued for a time span of at least two years between children (he and Anna managed to achieve a four-year gap between their children), so that each child would have the undivided attention of the mother in the crucial formative years. To achieve this, he was prepared to outline briefly the main methods of birth control explicitly. It was a brave act to publish this only twelve years after the Bradlaugh/Besant trial for obscenity because of their responsibility for the publication of a pamphlet containing similar information. But by the end of the 1880s, the work of others such as Havelock Ellis and Edward Carpenter, and the dominance of evolutionary theory on intellectual life, had created a new level of toleration for serious studies of sexuality.[72]

Encouraged by his success with *The Evolution of Sex*, Geddes was to continue over the next few years to develop his ideas on the 'Woman Question' even further. He met head-on the current debate about the 'Two Spheres' and whether or not women should struggle for a wider role outside their homes in what was conventionally considered to be the 'male sphere'. He was quite adamant that they should not. Biologically women were nurturers, and any attempt to move away from that role would be counter-productive in their quest for personal fulfilment, though 'nurturing' could be interpreted liberally to include activities outside the home which fell into this category. Geddes used the term 'civicism' to cover all the activities women could usefully, even essentially, perform for the sake of the urban community as a whole. He was, however, quite clear about the ultimate fulfilment for any highly civilised and educated woman who understood her own nature. It had to be an enhanced ability to enjoy romantic love on the highest level.

He produced an argument of potent emotional strength for the 'New Woman' of the 1890s.[73] He suggested that, in evolutionary terms, the chances for a modern woman of achieving a romantic love relationship were very much higher than they had ever been in history. Whereas in medieval times the sublime relationship between Heloise and Abelard had been so exceptional that it had become the subject of legend and fable, in modern times levels of culture had so progressed that such attachments were now widely possible. Personal happiness could be built on this and it had to be the proper basis for marriage. Geddes' Summer Schools in Edinburgh in the 1890s became noted for the number of romantic attachments which developed between the participants, and a number of marriages, including that of one of Geddes' close collaborators, A.J. Herbertson, ensued. The 'high thinking and plain living' of the little circle of Geddes' Edinburgh friends was sustained at many points by the free rein given to a highly moral, self-disciplined but intense pursuit of romantic love.[74]

The *Evolution of Sex* was Geddes' most important publication during his lifetime, though it is hardly read now, and pales into insignificance alongside the influence of *Cities in Evolution*. He had written it after nearly a decade of intense activity exploring the nature of evolution and social evolution in the natural and social sciences. It was the decade when he had worked closest with his friend and disciple J. Arthur Thomson, who always managed to increase Geddes' productivity in terms of publications. His reward was to be the Chair in Dundee, endowed for him by J. Martin White under special conditions. He had gained for himself an academic position, yet at the same time, total freedom to do his socio-biological experiments. Apart from creating a biological garden at Dundee, in which he would demonstrate his evolutionary ideas, he did not put down any roots in the city. He antagonised his colleagues and he did not care to work on Dundee's economic and social problems. Instead he retained his base in Edinburgh and continued, amongst his circle of friends, to develop all the educational and social projects he had already begun. His days of exploration in the natural sciences as a research worker were now over. His attention turned entirely to the dissemination of his ideas.

Notes

1. See H.E.Meller (1976) *Leisure and the Changing City*, London: Routledge & Kegan Paul, chapter 6, 'The "civilising mission" to the poor'.
2. O.J.R. Howarth (1931) *The British Association: a retrospect 1831–1931*, London: British Association.
3. L.M. Cantor 1960–61 'Halford Mackinder: pioneer of adult education', *Rowley House Papers* 3: 24–9.
4. Brian W. Blouet (1975) 'Sir Halford Mackinder 1861–1947: some new perspectives', *Research Paper*, 13, pamphlet, School of Geography, University of Oxford, pp.8–11, and J.F.Unstead (1947) 'H.J.Mackinder and the "new" geography', *Geographical Journal* CX.
5. Reprinted in full in P. Abrams (1968) *The Origins of British Sociology: 1834–1914*, Chicago: University of Chicago, pp.177–98.
6. Ibid., p.193.
7. P. Geddes (1881) 'The Classification of Statistics and its Results', *Proceedings of the Royal Society of Edinburgh*, XI: 295–322. Abstract: 'Economics and Statistics, viewed from the standpoint of the preliminary sciences', *Nature* XXIV, (1881).
8. D.A. MacKenzie (1981) *Statistics in Britain, 1865–1930: the social construction of scientific knowledge*, Edinburgh: Edinburgh University Press, pp.51–72.
9. P. Geddes (1884) 'An Analysis of the Principles of Economics', *Proceedings of the Royal Society of Edinburgh* XII: 943–80.
10. Quoted from a later pamphlet, P. Geddes (1886) 'On the Conditions of Progress of the Capitalist and the Labourer', *'Claims of Labour' Lectures*, no.3, Edinburgh Co-operative Printing Co., p.24.
11. V.V. Branford claimed that these pamphlets accounted 'for a good deal of my mental equipment in economics', letter to Geddes, 28 July 1900, Geddes Papers MS10556, NLS.
12. P. Geddes, 'An Analysis', op.cit., p.17.
13. He repeated this message again and again in the early pamphlets, for example, 'On the conditions.', op.cit., pp.33–4.
14. A favourite theme: the extent to which ease in life and material comforts lead to parasitism and degeneration:
    And thus have done with the current obsessions of the money world, of most ease with least labour, of getting something for nothing; perhaps above all, of that seeking after the assured life of petty, sedentary functionarism, which is becoming a main curse of civilisation. P. Geddes and J.A. Thomson (1911) *Evolution*, London: Williams & Norgate, p.111.
15. R.N. Soffer (1978) *Ethics and Society in England: the revolution in the social sciences, 1870–1914*, Berkeley: University of California Press, pp.69–89.
16. P. Geddes, 'An Analysis', op.cit., p.38.
17. H.S. Foxwell (1887) 'The Economic Movement in England', *Quarterly Journal of Economics* XII: 84–103.
18. Foxwell mentions Geddes and his biological approach to economics, commenting that he 'has done good service by criticising economic methods and results in the light of the latest biological analogies', ibid., p.94.
19. W.H.G. Armytage (1955) *Civic Universities: aspects of a British tradition* London: Benn.
20. J.F.C. Harrison (1961) *Learning and Living*, London: Routledge & Kegan Paul, pp.92–94.

21. J.H. Newman (1852) *The Idea of a University*, quoted in J.E. Thomas (1985) 'Wisdom and the City', Inaugural Lecture, Department of Adult Education, University of Nottingham, p.7.
22. Blouet, op.cit., p.11.
23. R. Blake (1969) *Disraeli*, London: University Paperback edn, pp.653–5; John Bowle (1977) *The Imperial Achievement: the rise and transformation of the British Empire*, Harmondsworth: Pelican Books, pp.357–75.
24. K.O. Morgan (1981) *Rebirth of a Nation: Wales, 1880–1980*, New York and Oxford: Oxford University Press, p.106.
25. C. Harvie (1977) *Scotland and Nationalism: Scottish Society and Politics, 1707–1977*, London: Allen & Unwin, pp.135–45.
26. H.E. Meller (1983) 'Patrick Geddes and "City Development in Scotland"', in Christopher J. Carter (ed.) *Proceedings of Symposium: Art, Design and the Quality of Life in Turn of the Century Scotland, 1890–1910*, Dundee: Duncan of Jordanstone College of Art, 13, Perth Road, Dundee DD1 4HT, pp.25–48.
27. Flavia Alaya (1970) *William Sharp – 'Fiona Macleod', 1855–1905*, Cambridge, Mass.: Harvard University Press, p.86.
28. P. Geddes (1886) *Everyman His Own Art Critic: an introduction to the study of pictures*, Glasgow Exhibition, (pamphlet), Edinburgh: William Brown, p.22.
29. W.H.G. Armytage (1961) *Heavens Below: Utopian Experiments in England, 1560–1960*, London: Routledge & Kegan Paul, pp.290–304.
30. Ibid., pp.305–50.
31. Ibid., p.330.
32. B.B. Gilbert (1973) *The Evolution of National Insurance in Great Britain: the origins of the Welfare State*, London: Michael Joseph, pp.21–58.
33. J. Mavor (1923) *My Windows on the Street of the World*, vol.1, London and Toronto: J.M.Dent, p.199.
34. This was most clearly the case with T.R. Marr, one of Geddes' most ardent supporters all his life, yet who always wanted to engage Geddes in working for social reform.
35. A. Wohl (1977) *The Eternal Slum: housing and social policy in Victorian London*, London: Edward Arnold, chapter 8, 'The Bitter Cry'.
36. P.J. Smith (1980) 'Planning as Environmental Improvement: slum clearance in Victorian Edinburgh', in A.Sutcliffe (ed.) *The Rise of Modern Urban Planning, 1800–1914*, London: Mansell, pp.99–134.
37. Ibid., p.125.
38. *Report of the Royal Commission on the Housing of the Working Classes* (1884–5), cd.4547.
39. C. Booth (1889) *Life and Labour of the People of London*, vol.1. London: Macmillan, pp.131–55.
40. Henry Pelling (1968) *A Short History of the Labour Party*, London: Macmillan, 3rd edn, pp.1–18.
41. Beatrice Webb (1948) *Our Partnership*, London: Longmans, Green, pp.103–18.
42. C.L. Mowat (1961) *The Charity Organisation Society 1869–1913: its ideas and work*, London: Methuen, pp.105–110.
43. P. Geddes (ed.) (1886) *Viri Illustres* (pamphlet), Edinburgh: Pentland.
44. Margaret B. Simey (1951) *Charitable Effort in Liverpool in the Nineteenth Century*, Liverpool: University Press, pp.98–112.
45. Theory enunciated in P. Geddes and J.A. Thompson (1889) *The Evolution*

of Sex, London: Walter Scott, pp.270–81.
46. Octavia Hill (1970) *Homes of the London Poor*, London: Frank Cass reprint, pp.38–9.
47. Ibid., p.51.
48. *Extracts from Octavia Hill's 'Letters to Fellow Workers' 1864–1911*, compiled by her niece, Elinor Southwood Ouvry, London: The Adelphi Book Shop 1933, p.17.
49. O. Checkland (1980) *Philanthropy in Victorian Scotland: social welfare and the voluntary principle*, Edinburgh: John Donald, pp.298–314.
50. Letter from James Oliphant to Anna Morton (Geddes' future wife) 16 November 1884, Geddes Papers MS10503, NLS.
51. Hector Macdonald (1971) 'Public Health in Edinburgh in the late Nineteenth Century', unpublished PhD thesis, University of Edinburgh, chapter 10.
52. M. Simey, op.cit., p.86.
53. C.L. Mowat, op.cit., chapter 1.
54. Minutes of the inaugural meeting of the Edinburgh Social Union, 6 January 1885, Local History Collection, Central Reference Library, Edinburgh. For Nottingham Town and County Social Guild see H.E. Meller (ed.) (1971) *Nottingham in the 1880s: a study in social change*, Nottingham: Adult Education pamphlet, pp.22–7.
55. James Mavor, op.cit., p.160.
56. Prospectus of the Town and Gown Association, Edinburgh, 1896, p.7.
57. M. Cuthbert (1987) 'The Concept of the Outlook Tower in the Work of Patrick Geddes', unpublished MA thesis, University of St Andrews, p.31.
58. W.H.G. Armytage, *Heaven's Below*, op.cit., pp.293–9.
59. P. Geddes (1886) 'On the Conditions of Progress of the Capitalist and the Labourer', *'Claims of Labour' Lectures*, no.3, Edinburgh Co-operative Printing Co., pp.28–9.
60. J. Mavor to P.Geddes, 1 December 1889, Geddes Papers MS10569, NLS.
61. See p.127.
62. David Rubinstein (1986) *Before the Suffragettes: Women's Emancipation in the 1890's*, Brighton: Harvester Press, pp.3–11.
63. Jill Conway (1970) 'Stereotypes of Femininity in a Theory of Sexual Evolution', in M. Vicinus (ed.) *Suffer and Be Still: women in the Victorian Age*, London: Methuen paperback edn 1980, pp.140–2.
64. E.E. Evans-Pritchard (1972) *Social Anthropology*, London: Routledge & Kegan Paul, reprinted edn, pp.21–42.
65. John Owen (1974) *L.T.Hobhouse: Sociologist*, London: Nelson, p.182.
66. E.E. Evans-Pritchard, op.cit., pp.33–63.
67. In Geddes' cell-theory: The early growth of the cell, the increasing bulk of contained protoplasm, the accumulation of nutritive material, correspond to a predominance of protoplasmic processes, which are constructive or *anabolic*. The growing disproportion between the mass and surface must however imply a relative decrease of anabolism. Yet the life, or general metabolism, continues, and this entails a gradually increasing preponderance of destructive processes, of *katabolism*. (Anabolism corresponds to female characteristics, katabolism to male. *Evolution of Sex*, London: Walter Scott, 1889, p.223).
68. Conway, op.cit., p.141.
69. Brian Harrison (1978) *Separate Spheres: the opposition to women's suffrage in Britain*, London: Croom Helm.

70. Geddes and Thomson, op.cit., p.267.
71. Ibid., pp.292—3.
72. The Evolution of Sex was the first of 'The Contemporary Science Series' edited by Havelock Ellis. This highly successful series was to include Elie Reclus's *Primitive Folk*, C. Lloyd Morgan's *An Introduction to Comparative Psychology* and Havelock Ellis's own *The Criminal* and *Man and Woman*. See P. Grosskurth (1981) *Havelock Ellis: a biography*, London: Quartet Books, p.114.
73. Rubinstein, op.cit., pp.24—37.
74. Geddes and Thomson, op.cit., p.267.

CHAPTER 4

Museums,
actual and possible

IN THE FIFTEEN YEARS BETWEEN 1889 AND 1904, GEDDES discovered that his many 'evolutionary' activities had begun to take him in one specific direction: towards the foundation of a new kind of museum movement. It was to be unlike any other museum movement in that the visitors to the museum became participators in its life, its aim being the evolutionary one of helping people and place, organism and environment, to be brought into a closer and more fruitful relationship. Geddes had an anarchic vision of the individual development of people and place. The community, taking responsibility for its own future, he believed, would want its own culture-institute, its Outlook Tower, its powerhouse, to co-ordinate all the activities in developing the interrelations of Place, Work, Folk. The precise nature of this museum movement, however, had to remain ill-defined. The opportunities of the moment had to be seized, in true evolutionary fashion, to help find new ways of encouraging people to interact with place. The problem with evolutionary activities was that they could not be predetermined. For most of the fifteen years Geddes was experimenting with his prototype museum in Edinburgh, he was not sure what his next move would be.

He was perhaps fortunate that he found, current in his own society and readily invoked, a concept of idealism and dedication as vague and as ill-defined as his own mission to mankind. It was the ideal of 'citizenship', an ideal which had particular connotations to those concerned with poverty and the social problems of cities.[1] Beyond the philanthropists and social workers, the value of 'citizenship' was debated by philosophers and social scientists as a major challenge of modern society.[2] The pioneering British sociologist, L.T. Hobhouse, gave a definition of citizenship in his monograph, *Morals in Evolution* (1906). Citizenship was the final product of the latest stage of civilisation. He suggested that, since earliest times, the evolution of moral order had rested on the nature of the society. In an era before civilisations were established, moral obligations between individual and individual were determined by the group, the clan, or tribe. In contemporary society, where there was mass urbanisation, then the principle of citizenship served the same purpose. In evolutionary terms, citizenship 'allows freedom to the individual and a flexibility to the whole structure. It involves the concept of "common good" and in its later stages demonstrates the possibility of a world state'.[3] Geddes was totally eclectic in his use of the concept. On the one hand, he was to work alongside philanthropists undertaking practical social work in the name of 'citizenship'; on the other, he was to claim a moral justification for his ideas as vital to evolutionary progress. In this respect he paralleled (on a much smaller scale), the position adopted by Canon Barnett in his work in his University Settlement in the East End.[4]

Both men saw social problems in cities in terms of social relations and contemporary culture as a whole. They were against the strong trend in philanthropic work, personified by the professional casework of the Charity Organisation Society of London,[5] of finding specific solutions for particular problems of individuals. Both wanted to change the whole cultural context of the city by promoting new social relationships through practical activities and both had recourse to the ideal of citizenship.

Geddes and Barnett were nurturing their ideas on cities and citizenship at a time when the demand for effective responses to urban problems had never been higher.[6] This demand was intensified by a wider aspect of mass urbanisation which was penetrating the thinking of many individuals in the second half of the century. This was the special cultural significance of a nation living mainly in cities. The shift of the basis of a whole nation from being a mainly rural- to a mainly urban-dwelling people was enough to excite the imagination of those who saw social progress in terms of the future of modern civilisation itself.[7] The relationship between a national civilisation and its component parts, the great individual cities, made such abstract concepts as civilisation seem more tangible. The essence of city life was that it was artificial and man-created, and as such it was not impossible to believe that it could be man-directed in the future. The power of this idea was to penetrate the social and political life of the nation at a myriad of points. Practical issues such as the demolition of slums, the provision of better housing, the relief of poverty, the improvement of education, thus became caught up in a wider debate on the nature of society and its direction in the future.

In political terms, it appeared to be a debate about the growth of state administration and the raising of the resources to pay for new state services.[8] Geddes and Barnett to differing degrees had reservations. Barnett was an old-fashioned Liberal with a healthy distrust of state intervention, while Geddes was particularly conscious of socialist thinking and the justification it gave to the Fabian prescription for a bureaucratic state.[9] Geddes, however, was not interested in prescriptions, he wanted a cure. State intervention was too clumsy, it could upset the ecological balance, and in the end produce more harm than good, whatever the original intentions. Not enough was known about how social networks in cities operated, how people understood their environment, what kind of improvements would transform their lives. Geddes and Barnett were united in the view that the prime necessity was to investigate how people lived and to experiment on ways of improving the cultural environment as a precursor of a better future.[10]

The practical experiments that they developed to pursue their aims were, in Barnett's case, Toynbee Hall, the first University Settlement; in Geddes' case, his Outlook Tower, a civic and regional museum. Both these institutions were described by contemporaries as sociological laboratories.[11] The activities that they promoted, however, were not strictly comparable either in scale or in content. As far as the former was concerned, Barnett had the support of Oxford and Cambridge Universities; he attracted undergraduates who were to make their mark in the political life of the nation, and his experiments were carried out in the full glare of publicity in London. The Outlook Tower, on the other hand, had no support from the University of Edinburgh; the undergraduates who helped to keep it going were personally loyal to Geddes, and publicity about its actions was strictly limited. As far as practical activities were concerned, the administration of these institutions was completely different. The activities and experiments that were put in hand stemmed from the personal views of Geddes and Barnett on social evolution.

Barnett saw society in class terms and based his work on promoting good class relationships between rich and poor which he believed had been destroyed by large-scale urbanisation. Like Miss Hill, he utilised an idealised view of the past, and he felt that the Settlement at Toynbee Hall, by bringing university men to the East End, a society deserted by the middle classes, would be able to achieve what 'revolutions, missions and money had failed to do'.[12] It would foster good personal relationships which would create trust between the Two Nations and provide the necessary knowledge for the social and political leaders to act in the interests of all. This would inspire the kind of 'citizenship' which might produce a leader who would discover new means of solving social problems as yet beyond the understanding of contemporary society. William Beveridge, one of the key architects of the Welfare State, was to serve his apprenticeship as a deputy warden of Toynbee Hall. In terms of learning how society works, Barnett's message was 'know your local community then you can understand society'.[13]

Geddes, on the other hand, started from a completely different premiss. His socio-biological perspective directed him to concentrate on improving both the organism and the environment by controlling the interaction of one with the other. What he wanted to do was to train young people to understand their environment so that they could interpret the direction of evolutionary trends and reinforce the most promising ones. To do this, of course, was no easy matter. It required an ability to look at a specific practical context from every conceivable viewpoint. To meet this demand it was essential that students were trained as scientific observers to gather the visual information about every aspect of a particular place. Such trained observation enabled a multi-faceted response to the complexities of the environment and its interaction with human society. But even more was required. The 'bud' hunters of the future needed the emotional inspiration of the artist to enhance their chances of success. They therefore needed on the one hand the most complete education both in the arts and the sciences, on the other, a moral commitment to serve the community. The message Geddes wanted to put across was: 'know your region and you can understand the world'.

Regional education and the Edinburgh Summer Meetings

The question was, how? Geddes' answer was to study its geography and its history and this was the initial function of the work undertaken at the Outlook Tower. What he was pioneering was the study of place and people which hitherto had been largely ignored in the formal educational system. The relevant disciplines of geography and geology, economic history, the natural and social sciences were either non-existent or barely established in academic form in any institution in Britain, and Geddes was one of a small number of academics trying to remedy this. In the 1880s, H.J. Mackinder in geographical studies and Toynbee in history, had been making their mark.[14] As far as Geddes was concerned however, there was a difference. He believed that the reform of academic studies was not enough. There needed to be a synthesis of all new knowledge and such knowledge needed to be based on experience as much as theory. The Outlook Tower as a regional study centre was to give form to educational activities of a totally new kind, outside the confines of conventional academic study. In the late 1880s and 1890s, as he became involved with the spread of the University Extension movement to Scotland,[15] he began to formulate his ideas on higher education and the theory of knowledge he had been slowly developing over the previous decade.

In the course of the five years from 1890–95, the elements of his approach to education were refined both in theory and in practice, the latter mainly in the Edinburgh Summer Meetings. As usual, it was a piecemeal and evolutionary process. Geddes began to build upon the courses in natural history that he gave at the marine station at Granton from 1885 onwards for the benefit of elementary school teachers. In 1889 and 1890 he added a lecture course dealing with the application of the idea of evolution to social as well as biological studies. In 1891 the school was moved to Edinburgh, and utilised the new student Halls of Residence. The next year the subjects covered were extended. Geddes reached a full understanding of what he thought the new developments in the content and method of education should be in the years 1893–5 when the Summer School was held in the Normal School of the Training College. The school was particularly successful at this time. The Outlook Tower had begun to function as a regional museum and Geddes was able to gain a grant from the town council, and was able to attract the support of foreign scholars including his old friend from Paris days, M. Demolins, the social scientist, and the eminent geographer M. Elisée Reclus. The number of courses offered at the school was increased to include philosophy and social science, history and geography, as well as the natural sciences.

Geddes was convinced that conventional methods of study produced apathy amongst students. Their creative faculties were blunted by the arid academic diet they were offered, and the last shreds of interest in any subject were killed by the threat of examinations. He wanted to concentrate on stimulating interest, inspiring enthusiasm, and thus releasing the potential creativity of every student. This interest could be sustained only if the student was actively involved in his or her studies beyond book study. This meant that there had to be practical activities such as laboratory work and field studies; that the student should not be allowed to specialise too narrowly in any one field without being aware of what was going on in other disciplines; and that all students should be trained to learn independently through observation. The starting point for interdisciplinary studies and the training of the eye was Geddes' own 'thinking machines'. He had developed them constantly over the past decade and was ready to use them to present information in graphic form. He hoped that by this method, he would be able not only to transmit information but also to highlight the connections between disciplines, between ideas, between movements, all at the same time. The walls of the Outlook Tower were covered with material to which Geddes applied a 'thinking machine' as the key to its understanding. He believed that students trained in this manner would naturally develop a practical response to the economic and social problems that society faced, and that they would work to achieve both immediate and long-term solutions for these. This became the major objective of the Outlook Tower and its related activities.

Geddes felt that he had developed nothing less than a new philosophy of education, and he was at pains to explain this to the overseas students he tried to recruit for the Edinburgh Summer Meetings to bring a more cosmopolitan air to the proceedings. It is worth quoting Geddes' prospectus for his Summer Meetings at some length as this proved one of the clearer statements of his educational ideas:

Starting from the familiar idea of working from the concrete to the abstract, from the senses toward the intellect, it is attempted in each subject of study (1) to freshen the student's mind by a wealth of impressions; (2) to introduce him to the advancing literature of the subject; (3) to supply him with the means of summarizing, arranging and more clearly thinking out these accumulations of observation and reading. Hence (1) the insistence on demonstrations, experiment and field excursions; (2) the introduction in several subjects of the seminar, which, with its guidance to the world of books and activity in using them, is so marked a strength of the German university; (3) the extended use of graphic methods.

The student, though first of all freshened as an observer, is regarded not as a receptacle for information, but as a possible producer of independent thought. Hence the examination method, everywhere falling into such merited disrepute, is here definitely abandoned; a keener stimulus, even a more satisfactory test of progress being found in accustoming a student to take part in his own education by attention first to the increase and systemisation of his materials, next to the occasional contribution of his best results to the common stock of class notes and summaries, and thence to fuller collaboration with his teacher.

[94]

Passing from the manner to the matter of education, it is attempted (1) not merely to offer a series of special courses, each of adequate thoroughness, but (2) to keep up as far as possible a parallelism of treatment and (3) to coordinate these parallel courses into a larger whole. Hence the general courses addressed to all students, dealing specially with the history of civilisation, the historical development of the sciences, their general principles and mutual relations. The present theme is, in fact, an attempt to work in theory towards the organisation of knowledge, and in practice towards the more rational arrangement of curricula studies.

The legitimate claim of the man of science is affirmed by the very existence and method of these courses; yet the corresponding claim of the scholar and humanist that, whatever be the progress of natural science, the study of man must remain supreme, is also recognised; witness that subordination of biology to the social sciences which is a characteristic feature of the present scheme. . .

Education is not merely by and for the sake of thought, but in a still higher degree by and for the sake of action; hence each course of scientific study is not merely related to those dealing with the other sciences, but to an even more immediate degree to the corresponding arts of life.

Each study must thus seek its highest result, not in a mere destructive analysis as of flower or verse, but in a constructive synthesis, it may be a work of art; hence these beginnings of library and museum, of garden or of gallery. The prominence given to the school of art is thus explained; the study of landscape and animal life being definitely associated with the school of natural science and that of figure with anthropology and history, the student thus working for the artist and the artist for the student. Hence also the association of a course of literature.

At this point the highest principle comes into view. Everyone recognises in theory that the efficiency of a scheme of education is tested by its reality for the preparation of life; and on this alone the present scheme might base its claim for trial, since it seeks to fit the student for some of the higher activities of life by actually sharing them. He is invited to become not a mere passive auditor but an active collaborator.

How in this way the individual and competitive spur to study becomes more than replaced by the cooperative and social one, or how this is strengthened by the selection of appropriate practical work among the possibilities above enumerated, needs no detailed explanation. This choice of course, depends on the previous training and actual preferences of the student, whether artistic, literary or scientific; while the former advantage can only be properly realised as he sees the decorative work in progress, or feels the collective store of educational material enriched by some contribution of his own. . .

Thought then, does not exist by and for itself, as is too much the view underlying the old order of education, nor has it merely application to life, as is enforced by the dominant school of technical educationalists. It arises from life and widens in proportion to its range, not only of observations but of action, and even of social intercourse. Hence the advantage of associated residence. Vivendo discimus.[16]

'Vivendo discimus' was Geddes' motto which he had carved over the door of his museum, the Outlook Tower. The visitors to the Summer Meetings felt themselves to be an elite in Britain, studying subjects not yet incorporated in British universities, in ways which were patently unconventional. Yet the attendance at the Summer Meetings was never very large. It peaked in 1893 when about 120 people attended the School, but most years it was considerably less. Geddes had been the pioneer of Summer Schools in Britain, but only by a very short margin.[17] In the late 1880s the development of the university extension movement had led to a widespread demand for vacation courses for further study. In 1888 a number of educationalists in the universities and public schools, including Michael Sadler, the Rev. John Percival (a founding member of the University College of Bristol), and the Rev. Dr Paton of Nottingham, had met to discuss the feasibility of establishing a Summer School along the lines of the American Chautauqua gathering.[18] In 1890 the first was held in Oxford, 900 people attending. In 1891, as the numbers grew ever larger, a cycle was started, each meeting over the next few years being devoted to a particular historic period in chronological sequence. A course for university extension students was begun in Cambridge devoted to chemistry, offering laboratory work at an elementary and advanced level. The Cambridge course was influenced by events at Oxford, and the Summer Schools at both places became popular events in the course of the 1890s. The most popular subjects tended to be history, philosophy, literature, economics, and art. Since many of those attending were teachers, there were courses in the history of education and methods of teaching.

Geddes' schools in Edinburgh never reached those numbers. It was the special features of the Edinburgh Summer Meeting, however, which created a small band of regular followers, both students and lecturers, who provided the core of support each year. These features included the emphasis Geddes began to give to geographical studies; the cosmopolitan nature of the meetings; and the factor which was less tangible but possibly the most important, the radical, the unconventional, highly moral, yet exciting and 'liberated', atmosphere that Geddes managed to create at his meetings.[19] The publication of the *The Evolution of Sex* had earned him some notoriety. Numbers of young women eager to be amongst the 'advanced' of their generation flocked to the school each year, and their presence contributed not a little to its success. They made essential contributions, especially to the artistic, cultural and social activities. A leading role was played by Mrs Geddes who hosted many parties, organised the events such as the dramatic re-enactments of episodes in history in costume[20] and she also offered direct entertainment herself as she was a gifted pianist. Music of the Highlands and Islands was performed by Mrs Kennedy-Fraser who, with her father, was a pioneering collector of Celtic folk songs. Stimulating favourable trends in personal evolution could and did extend to interacting more with other students, and a number of romantic attachments blossomed at the Edinburgh Summer Meetings.

Geddes did his best to sustain this exciting atmosphere. Each day was carefully structured. He deliberately tried to engage the emotion of his visitors as much as their intellect. He developed a technique which he continued to use many years after these heady days in Edinburgh, when he was to hold summer meetings in London, Dublin, Madras, Darjeeling, and elsewhere. The meetings were organised around lectures in the mornings, outings and rambles in the afternoon, and social and cultural activities in the evenings. Geddes gave an introductory lecture, usually on Edinburgh, and on social evolution, firing his audience with his portrayal of its civic past and present. He would also walk the participants around the city, sometimes at night after supper, sometimes in the early morning, climbing up Arthur's Seat to view the city in the early morning mist. He would then proceed to give lectures every day, working through the great world civilisations of the past and their social evolution, concluding on the last day with a discussion of the present. Then the focus was redirected on to Edinburgh, past, present, and future. The sociological substance of his analysis of world civilisations he took from the French Le Playist school, with which he was in ever closer contact in the 1890s. He was able to persuade M. Demolins to come to Edinburgh for the summer school of 1893 and 1895.[21]

Regional education and national identity

The summer meetings of 1893–5, however, were made especially exciting because Geddes threw himself into developing, in a self-conscious way, a new sense of Scottish nationalism. For more than a decade he had witnessed the creative inspiration of Irish nationalism in Dublin.[22]

He conceived the idea of trying to forge links between Edinburgh and Dublin to give form to a pan-Celtic consciousness. An academic justification for such an ambition was at hand. The currently fashionable search for the historical origins of nations, as pursued by the Professor of History at Edinburgh, Professor Stuart-Glennie (a friend of Geddes), had alerted Geddes to the close evolutionary ties between Irish and Scottish Celtic culture.[23] Perhaps fortunately for Geddes' purposes, historical scholarship though had not yet advanced to the point of exploding the myths of Ossian and the supposedly glittering civilisation that had flowered in Scotland back in the mists of time.[24] For Geddes and his circle, neo-Romantic views of such a past were more important than mere facts. What he needed was to attract the support of a man of letters to sponsor and co-ordinate a new Celtic literary and publishing venture. His choice lighted upon William Sharp, a Scotsman of the same age as Geddes, who had been earning a living on the fringes of the literary world in London.

The recruitment of William Sharp provides an insight into the nature of Geddes' activities in the early 1890s. Bent on seeking a higher evolutionary level, Geddes had become caught up in an emotional critique of contemporary society. He may have been pursuing practical ways of making his museum movement a means for achieving higher levels of social evolution, but the issues this raised forced him to confront larger issues about the nature of modern culture and society. He was not afraid to address himself to the greatest cultural dilemma of the current age: the bifurcation between the arts and the sciences, the great fault-line in contemporary knowledge. Herbert Spencer had tried to build a bridge by being an academic polymath.[25] Many others had tried different routes. The pioneers of 'scientific' geography such as H.J. Mackinder believed that geography was an interdisciplinary subject which would provide the link between the two.[26] Geddes, while taking on board all that others had to offer, was more convinced that the coming together of the sciences and the arts could take place only in a specific context, with practitioners of both co-operating with each other for common ends.

When he recruited Sharp to come to Edinburgh, he believed he was getting the services of a creative artist who would share his perspective and complement his scientific work in an artistic way. The initial

[98]

common ground with William Sharp was their shared national
identity. Beyond that Sharp fully shared Geddes' neo-Romanticism, his
hostility to the cultural domination of England, and an emotional
commitment to an almost mystical romanticised Celtic past.[27] Geddes
thought he would be an ideal collaborator. Sharp had already tried,
using his own contacts, to launch a Scottish literary magazine, the
Scottish Arts Review which had only lasted during 1888 and 1889. It was
a handsomely printed publication with a Celtic/Scots emphasis, and it
had contained a miscellany of art, poetry, short stories and criticism. In
Italy, in 1890, however, Sharp had met Mrs Edith Wingate Rinder, a
member of the Geddes' Edinburgh circle, and he had formed a deeply
romantic attachment to her. She introduced him to one of her great
interests: the work of Maeterlinck and the Belgian School of
Impressionistic Dramatists. Maeterlinck was a controversial figure in
the strongly nationalistic Franco-Flemish movement as he wrote in
French and not in Flemish. Sharp began to write articles in his defence,
suggesting that the use of a 'universal tongue' instead of a 'provincial
dialect' such as Flemish did not invalidate his contribution to the
Flemish nationalist school.[28]

This strand was to have a personal significance for Sharp himself as
his interest in themes of Scotland's Celtic past deepened. He wanted to
develop his own work as an expression of Scottish cultural identity and
yet, with the example of the Belgians before him, he was convinced
that English and not Gaelic should be the medium which he should
use. In the early 1890s he had severed his links with London, and he
was delighted to respond to Geddes' invitation to come to Edinburgh
and collaborate with him on a publishing venture. In 1893 Geddes set
up a small publishing firm, Patrick Geddes, Colleagues and Company,
and Sharp was made a partner and managing director. Briefly, in this
congenial atmosphere, Sharp's literary talent blossomed, and he
brought to himself and to Geddes fame and a reputation which gave a
lustre to Geddes' Edinburgh activities. Yet the form of this achievement
was symbolic of the whole enterprise. Sharp gained his reputation
using the pseudonym of a woman, Fiona Macleod, and it was 'her'
work which captivated the critics and led to international acclaim.
Fiona Macleod's most famous works, *The Sin Eater and Other Tales*, *The
Washer of the Good*, and *From Hills of Dream* were all published by
Patrick Geddes, Colleagues and Company in 1895 and 1896. They are
all forgotten now since they belong to a genre which is almost
inaccessible to the modern reader. As Sharp's biographer, Dr Alaya,
comments, 'in dealing with Sharp and his era . . . the present day
reader is confronting revisionary romanticism from the vantage point
of even further revision'.[29]

The coming together of Sharp and Geddes in the early 1890s was thus providential for both men. For a while the scientist and writer were united by their sympathy for certain neo-Romantic principles. They not only shared an interest in Scottish culture, they were both deeply interested in, and had written about, the subject of sex and gender. It was only to the Geddes' that William Sharp revealed the fact that he was the author of the works of 'Fiona Macleod', which remained a secret until the last day of his life. The imaginative core of Sharp's work was built on a series of tensions created by a conflict of identity. He added to his imaginative exploration of the man/woman dichotomy a further conflict between his nationalism and his sense of cosmopolitanism. Again this was something he shared with Geddes. They had a special sense of nationalism deliberately non-political, deliberately rejecting a narrowly nationalist perspective, and adopting as the key to all further development, a paradoxical commitment to cosmopolitanism. The paradox was resolved in that their sense of national identity was built on a perception of place, and it was a romantic sensitivity to place which was the key to cosmopolitanism.[30] Such a sense of place was for both Geddes and Sharp heightened by their reading of the work of social anthropologists and human geographers, especially the work of Elie Reclus whose *Primitive Folk* of 1891 deeply affected Sharp.

Their unity of outlook fed Geddes' optimism that a 'Celtic revival' could be engineered in Edinburgh, his Summer Meetings collecting sympathetic support, the Outlook Tower sustaining initiatives, and the publishing firm giving permanence to their work. A Celtic Library series was projected and a northern seasonal issued in 1895, entitled *The Evergreen*. The four volumes devoted to the seasons illustrate all the neo-Romantic principles of nationalism and cosmopolitanism that Geddes and Sharp were anxious to promote. Each volume was divided up into sections on the appropriate season in Nature, in Life, in the North, in the world. Elisée Reclus contributed to the book of Autumn. His piece was an optimistic, non-factual fantasy entitled 'La Cité du Bon Accord'. It was the kind of sentiment which inspired the whole project. Sharp, for all his enthusiasm for the Celtic past, was a little cautious about *The Evergreen*. He wrote in a letter to Geddes: 'Much of it I like but some of it seems to me to lack distinctiveness as well as distinction'. The enterprise was slated by critics, especially in London, who were particularly savage about the lack of artistic talent displayed in the illustrations.[31]

The 'Celtic revival' might have sunk ignominiously if it had not been for Fiona Macleod. However, Sharp's association with Geddes lasted for less than two years. The projected quarterly, *Celtic World*, never emerged, and Sharp's lack of application to matters of mundane administration, coupled with Geddes' overcommitment in other projects, meant that the whole project foundered. Sharp was paid off by being given shares in the Town and Gown Association set up in 1896 which he was then annoyed to find he could not sell. Acrimony and difficulties ended the attempt to 'centralise all Celtic works of Scottish, Irish and Welsh in Edinburgh'.[32] Geddes' contacts with Sharp and discussion about the 'Celtic Revival' and Scottish nationalism, however, played an important part in the development of his theory of civics. For Geddes it was a reaffirmation of the importance of place, but given a special meaning. The evolution of the Summer Meetings and the Outlook Tower over this period had helped him to interpret place, work, folk, according to the developing social sciences, geography, economics, and social anthropology, whilst the romantic nationalism of his Celtic revival fed his imagination and gave him the emotional means for uniting studies of geography and history.

As he was to write of Scottish history, there were three different planes on which it could be studied: in terms of the 'simple patriotism' of Scottish chauvinists; in terms of the English viewpoint; or, as he recommended, a third way:

> in which Scotland is viewed . . . at length truly and fully seen upon the general map as the region which in geography, race and history of the whole most fully represents and epitomises North-Western Europe and which is hence one of the richest and most fascinating fields, for the student of social geography, of general and comparative history, whatever his country or race may be.[33]

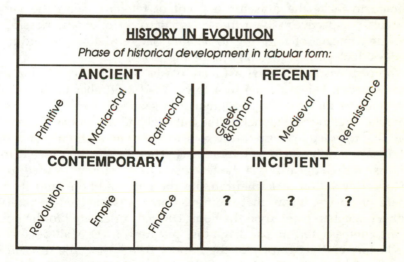

Figure 4.1   History in Evolution

'The inception of the opening future may be increasingly defined since all these apparently predicted phases are already incipient among us, and are thus really matters of observed fact, of social embryology, let us say, in short, of city development.'
Source: P. Geddes (1906) 'Civics: an applied sociology', Part II, *Sociological Papers*, (ed.) V. V. Branford, London: Macmillan, p. 108.

This mixture of the geographical and the historical, the regional, national, and cosmopolitan, became the unique combination which Geddes developed in his work at the Outlook Tower. It was obviously not a mixture which was easy to communicate to those who did not share Geddes' evolutionary ideas. Furthermore, his use of evolutionary theory in support of his idiosyncratic views became ever harder to sustain in the 1890s. Evolutionary theory had been in a constant state of flux and new knowledge was undermining even the certainties of Geddes' biological viewpoint.

From the regional to the cosmopolitan perspective: the evolution of activities at the Outlook Tower

In terms of social theory, Geddes found himself in something of an impasse by the mid 1890s. It was in this context that the development of activities at the Outlook Tower became even more crucial to him as a means of escape. He had always held that progress in understanding came through a combination of thought and action. Since theory was getting more and more difficult, the way forward had to be the further development of his museum. So far he had established a 'social laboratory' and educational activities. He had tried to deepen the experience of people of their local region. Now perhaps the time was ripe to go beyond the region to a new understanding of the world. He wanted to make the museum a point of reference with the widest possible perspective. Like a first-class library, it might be possible to provide a means of referring to all knowledge (defined according to Geddes' biological understanding of the word).

The organisation of such a function was already there in embryo. The museum, being located in a tower, had a number of floors. The exhibition of material was designed in a sequence to lead the visitor from the local to the regional, to the national and the global. The visitor was able to go straight up to the roof to the camera obscura and view Edinburgh and its region by its means. Then began the descent from this vision back to earth (via the tea-room) to the storey devoted to the historical evolution of Edinburgh, its present conditions and its best future prospects. Below that the next storey was devoted to Scotland with a huge map painted on the floor, correctly orientated to the points of the compass and maps, diagrams, pictures on the wall devoted to the history and geography of the nation. Below that, the next storey was devoted to the Empire and English-speaking countries, with a special alcove allotted to the United States. Below that there was Europe and on the ground floor the World. In Geddes' mind, the Outlook Tower became the prototype of museums of the future, actual and possible[34]

Geography provided the unifying theme to connect the disparate elements of the Tower. Equally important, however, was Geddes' use of symbolism. He took the neo-Romantic propensity to use symbols to suggest deeper meanings to its ultimate extreme. All visitors to the Outlook Tower were instructed to view the great stained-glass window that Geddes had designed to symbolise his view of life. It was the Arbor Saeculorum, The Tree of Life. The different branches of knowledge were shown to have a Lamarckian common root, and the small figures on the branches, the pioneers of new knowledge, were giving the fruits of their work to the world. Between the branches was a great mist, which acted like a barrier between the different specialisms. Geddes wanted to convey the need for synthesis and to show that for lack of this effort, knowledge could never be applied. The whole window was a typical instance of Geddes' use of symbolism to convey what he believed was his 'practical' approach to Life.[35] The more outrageous his symbolic flights of fancy became, the more he insisted he had a practical purpose in mind. In later life he was to utilise one of his earliest loves, Greek mythology, as the appropriate source to illustrate symbolically his theory of Life.[36]

From the earliest days of the Outlook Tower and before, Geddes had been using artistic symbolism as a way of conveying to others the implications of his work. The demand he created for the artistic realisation of his symbols had created a minor outburst of artistic endeavour in the city. Geddes had recruited the services of an Edinburgh student, John Duncan, to run what was to be called the Old Edinburgh School of Art, initially from University Hall and then from the Outlook Tower. Geddes' ambition was that the artists would provide the ideals for regeneration, mostly through the allegorical meaning of their work.[37] His aim was to release art from the imperative of being commercially viable by supporting artists with the proceeds of his building operations. Then their activities would provide a vital link in the chain of evolutionary activities at the Tower.[38] His whole range of activities were thus synthesised into one operation. The exhibitions and maps of Edinburgh provided a survey or description of the area; the building operations in the Old Town were the field of action; the Summer Meetings provided a new kind of evolutionary education; the publishing ventures, a new cultural identity; and the Art School provided the ideals.

In this welter of activity, Geddes optimistically believed that the next stage in the proceedings would naturally emerge. But practical activity required direction, and meanwhile he was not unwilling to make a few decisions himself. For example, he decided that the Old Edinburgh School of Art had to divide its time between constructing vast friezes depicting evolutionary stages of past civilisations on the one hand, and

[103]

on the other, designing ornaments which could stand on the kitchen mantelpiece in the refurbished tenement blocks. John Duncan managed to keep the school going between 1892 and 1900, and he was responsible for the best illustrations published in *The Evergreen*. Geddes' patronage, however, was obviously not sufficient to secure for him a satisfactory career, and Geddes managed to get him a more permanent and lucrative post on his visit to the USA in 1899 when he recommended Duncan as art teacher at the Parker School in Chicago. But in 1895, flushed with the success of the publishing activities of Patrick Geddes Colleagues and Company, Geddes felt that he was really producing regional activity of an evolutionary kind. What he felt he had not achieved was the promotion of world-vision and world citizenship. This proved a much more difficult activity to initiate.

Help was at hand, however, in the figure of M. Elisée Reclus who came to the Summer School in 1893 and 1895.[39] In the mid 1890s, Reclus was at the peak of his fame and influence as a geographer, educationalist, and anarchist. The outstanding quality of his work had helped to offset the notoriety of his past as revolutionary and Communard. Exiled for his political views from France in the 1870s, he had begun his life's work which was to produce a twenty-volume geographical study of the world, entitled *Géographie Universelle: La Terre et les Hommes*. On its completion he was awarded the Gold Medal by the London Royal Geographical Society in 1894. Yet the past, and his political stance as an anarchist, still dogged his footsteps. In 1894 he was made Professor of Comparative Geography at the University of Brussels, only to be expelled by the rector shortly afterwards for his political views. Reclus subsequently became a member of the Université Nouvelle set up in Brussels. Here, with his colleagues, he tried to pioneer new ways of developing higher education with the emphasis on synthesising knowledge instead of pursuing specialisation.[40]

His own path to the social sciences had been idiosyncratic and cosmopolitan. He had worked at many different occupations, he had travelled the world, he had studied in centres of excellence in European universities. These experiences, shared to a lesser extent by Geddes, made both men totally dissatisfied with current attempts to create higher academic standards in particular disciplines. The independent scholar needed a breadth of vision, and without it the specialist became a victim of his specialisation, unable to see the implications for society of even his own work. Reclus, again like Geddes, was a propagandist beyond higher education, to the public at large. In 1895 he began a long campaign to realise the building of a huge globe for the projected World Exhibition of Paris in 1900. He believed such a globe had scientific value because it could be designed in such a way that it would be possible to keep it up to date, incorporating all new knowledge from geographical explorations and surveys.[41].

The ultimate objective of the enterprise was, however, as much emotional as scientific. Reclus wanted to inspire the visitors to his globe with a new understanding of the world and the economic and social forces which shaped their lives. He wanted to give meaning to the concept of world citizenship in a way which would cut across the political divisions of nation states and reach out to the people, to the unchanging human heart that beat beneath the differences of race and creed. It was a deeply emotional message. Geddes understood it perfectly, as it coincided with his own. He wrote of the globe:

> Instead of a book, were it the best, the latest, here was now the most monumental of museums, the most simple of observatories, the microcosm of the macrocosm itself . . . but this was no mere scientific model in its institute, but the image, the shrine, and temple of the Earth-mother, and its expositor no longer a modern professor in his chair, but an arch-Druid at sacrifice within his circle of mighty stone, an Eastern Mage, initiator to cosmic mysteries. . . . The future of its accomplishments . . . no longer solely cosmic, but henceforth primarily human − the unity of the world now the basis and symbol of the brotherhood of man upon it: science is an art, geography and labour uniting into a reign of peace and goodwill.[42]

Geddes was fascinated by the globe project. There were reasonable grounds to hope that it might be realised, although, of course, the construction of the globe would be very costly. However, at the 1889 exhibition in Paris there had been a globe which measured 120 feet in circumference, which was roughly on the scale of 1:1,000,000. But it was not an accurate model. What Reclus wanted was to make a globe of at least twice that size and use its construction as a means of developing the science of sphaerography to keep pace with developments in cartography. A Swiss geographer, M. Perron, was working on an accurate relief model of Switzerland for the world exposition on the scale of 1:1,000,000. The problem was the cost. Geddes had never been deterred in all his activities by financial considerations, and he was to pursue energetically all means of realising the globe project. His commitment to it, though, stemmed as much from his own dilemma about what to do next at the Outlook Tower as from his admiration for Reclus. Trying to work out evolutionary theory in the social sciences was proving very confusing. The vast interconnected evolutionary process seemed to defy analysis. Geddes wrote 'come back to our spiral of evolution which is not concerned with place but time and we have its first indication (for it is to a "thinking machine" to which we may come back in all our difficulties). Its spiral always pointing the way'.[43] Following the way demanded practical activity in the hope that the experience gained would elucidate new theory. Hence Geddes' readiness to dissipate his energies in as many different directions as he could invent, to the extent that he not only confused others, he also confused himself. The globe project came at an opportune moment for him, and Geddes wanted to promote a campaign for the Outlook Tower in conjunction with the globe. Yet even one of Geddes' most fervent admirers, T. R. Marr,[44] who took on the lease personally of the Outlook Tower between 1896 and 1901, and worked dedicatedly to make a success of it, wrote to him 'it is very difficult to lay hold of people who will readily be taken by the Tower and Globe notion – especially by any general account of it'.[45] By 1896 Geddes had found himself in the midst of many serious problems.

The Outlook Tower as a prototype for a new kind of municipal museum

In the previous decade his property speculations and work in Old Edinburgh had kept him in a whirl of activity which had its own momentum. All this was brought to an end when his resources became overstrained and he was baled out by the formation of the Town and Gown Association. When it was set up, Geddes' property had to be systematically valued, and at the end there was a shortfall of more than £2,000 between the value of his loans and the valuation of the property for which Geddes was personally responsible. It was a salutary experience and he was forced temporarily to stop his almost compulsive habits of property speculation. In this relative lull in building activities in Old Edinburgh, the cultural activities at the Outlook Tower took on a new significance. Geddes began to see that, as the creator and curator of a new style of museum, he might find a more congenial future career for himself than as a part-time Professor of Botany. In 1892 a national Museums Association had been established in recognition of the fact that many towns and cities, mostly in the last quarter of the century, were acquiring museums.[46]

In the 1890s, at the early conferences of the Museum Association, the debate focused largely on what kind of museum was most appropriate in the new context of the modern large city. In 1895 the director of the United States National Museum, Professor Browngoode, was invited to give a paper on the nature and scope of museums, and Geddes found his survey of current practice the best summary to date of the museum movement. Geddes wrote that in Browngoode's paper:

> the growth of the museum ideal, of the public appreciation of the material value of collections and still more of their higher functions, along with libraries, reading-rooms and parks, (as 'passionless reformers') is vigorously outlined; and the paper ends with the

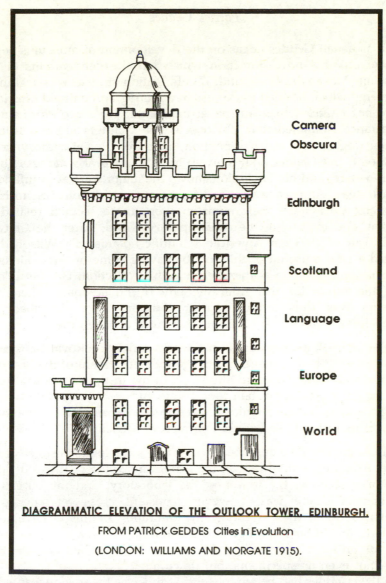

Camera
Obscura

Edinburgh

Scotland

Language

Europe

World

DIAGRAMMATIC ELEVATION OF THE OUTLOOK TOWER, EDINBURGH.

FROM PATRICK GEDDES Cities in Evolution

(LONDON: WILLIAMS AND NORGATE 1915).

Figure 4.2   The Outlook Tower, Edinburgh

'The Outlook Tower in diagrammatic elevation, with indications of uses of its storeys – as observatory, summer school, etc., of regional and civic surveys, with their widening relations, and with corresponding practical initiatives.'
Source: P. Geddes (1915) *Cities in Evolution*, London: Williams & Norgate, p. 324.

proposition that 'the degree of civilisation to which any nation, city or province has attained is best shown by the character of its public museums, and the liberality with which they are maintained'.

Professor Browngoode also made the point to which Geddes was particularly receptive: 'a finished museum is a dead museum, and a dead museum is a useless museum'.[47]

But as usual, Geddes' ideas on the development of museums began to take off in a novel direction which left the deliberations of the Museums Movement far behind. A culmination of ideas was beginning to emerge which he tried to express in a monograph entitled *Museums, Actual and Possible* which was never published. The reason for its non-appearance is immediately obvious as the chapters are constructed without logic, with much repetition, and from a very idiosyncratic viewpoint. But Geddes shows an up-to-date grasp of what was going on elsewhere, and his discussion shows the major sources of influence on his ideas. America was a particular source of inspiration to him, especially since much new museum work there seemed related to charting and explaining the contemporary world. After the Chicago World Fair of 1893, an American scientific botanist, Dr Wilson, had created a new museum of this sort by saving some of the exhibits of the Chicago Exhibition as the nucleus of his collection. He was able to raise the funds for this from the city of Philadelphia where the museum was thus located. Victor Branford later described the Philadelphia Commercial Museum as a cross between the

> Imperial Institute (organised on a world scale), the Royal Geograph-ical Society, the Commercial Intelligence Department of the Board of Trade and of the Foreign Office, . . . throw in the commercial page of the daily press, and then concentrate and weld these into a single organisation, but we should still be short of exhausting the manifold functions of the Philadelphia Commercial Museum.

The museum could be described in short as 'a catalogue raisonée of the world, from the point of view of the American exporter'. It was divided into three sections, the museum, the laboratory, and the bureau of information. There was, however, constant cross referencing of material between these sections. In admiration of this achievement, Branford quoted Sir W.H. Fowler's dictum that 'what a museum really depends on for its success and usefulness is not its buildings, not its cases, not even its specimens, but its curator'.[48]

By 1895 Geddes had begun to feel that this was the role he wanted to play. The time seemed right for an experiment in museum-making dedicated to social evolution, bringing together art and science (in this case especially the work of Ruskin and Le Play), and instigating what might become a new museum movement, or at least a centre of sociological studies in the UK. In Europe a number of initiatives had been set in motion to encourage the study of the social sciences, of which two particularly interested Geddes because of their efforts to collect and classify social knowledge. The first was the Musée Sociale of Paris; the second, the International Bibliographical Institute in Brussels. The Musée Sociale was a prestigious institution which was indirectly connected to government circles, though it was financed by a wealthy private benefactor. It played a leading part in disseminating ideas on social reform in French political circles under the leadership of Jules Siegfried, and it gave a permanent venue to the kind of exhibition work, meetings, and discussions that had formerly been pioneered by the Le Playist school at the Paris Exhibitions of 1867, 1878, and 1889. Its large building incorporated a library, display places, and meeting rooms. Geddes could not hope to emulate this metropolitan institution, and in some respects it was not his intention to do so. He suggests that it was 'rather an excellent reference library, post graduate school of economics and statistics than a sociological museum proper'.[49] Geddes wanted to synthesise modern knowledge rather than just provide a reference library of the social sciences.

The activities of Paul Otlet in Brussels were to fire his imagination in this respect.[50] The nationalist revival in Belgium had had a considerable effect on higher education, and the cultural life of the nation, which had seemed to encourage the work of individuals in many different endeavours. Paul Otlet and Geddes did not meet until the World Fair 1900. But Otlet was rather a Geddesian figure in his determination to devote his life in a practical way to an immense undertaking in pursuit of an altruistic cosmopolitan idea.[51] He conceived a need to create an international bibliography of all knowledge, and to facilitate the exchange of knowledge on an international basis. He established in 1889 an International Office of Bibliography devoted initially to the classification of works on sociology, and in 1895 he had, with the help of his collaborator Henri La Fontaine, beaten the Royal Society of London in a race for world recognition of his organisation at Brussels, as the international organisation for co-ordinating efforts for producing a world bibliography. They had discovered the American Dewey system of classification and, with amendments, wanted to make it the basis of their international work. In 1895, having gained the support of the Belgian government, the International Office of Bibliography was born.[52]

At the centre of the debate about setting up such an enterprise was the question of classification. Geddes was vitally interested in this issue as he saw how it might affect his own work in the Outlook Tower. At the back of his mind was a vision of a museum system which would complement the libraries. Modern knowledge was to be found in the libraries, but Geddes knew that the application of that knowledge in practical circumstances would require training and experience. If he could invent a museum which could be an active study centre and not just a repository of artefacts, this would be the perfect complement to the role of the first-class library. What made this project (which was nothing less than the visual presentation of all knowledge) feasible was the concept of evolution. The classification of material could theoretically be done according to evolutionary theory. To determine exactly how, would be the main task of the curator of such a museum.

From Outlook Tower to Index Museum: the prospect of international concerns

Gradually over the next five years or so, Geddes began to dream of what he was to call an Index Museum.[53] Giving full rein to his imagination, he began to piece together his ideas by drawing on the appropriate cultural traditions of Edinburgh, and by giving himself further practical experience by extending the activities of the Outlook Tower to achieve an adequate prototype of his vision. Geddes had already been involved for a number of years with what he considered was Edinburgh's best literary traditions in the dissemination of knowledge. In the course of the nineteenth century Edinburgh's publishing firms had made a large contribution to extending the market for knowledge by providing standard scientific works and reference books. Since the great days of Blackwood's *Edinburgh Encyclopaedia* published between 1808 and 1830, several firms had published encyclopaedias, the firm of Chambers reaching out to the ever greater numbers of the literate public with their volume *Information for the People* in 1857 and later encyclopaedias. Another Edinburgh firm produced the *Globe Encyclopaedia of Universal Information* in six volumes between 1870 and 1881, with an illustrated version being produced between 1890 and 1893.[54]

Geddes had contributed scientific articles for *Chambers Encyclopaedia* and the *Encyclopaedia Britannica*.[55] He was familiar with the problems of organising the material in such works. But his Romantic attachment to the eighteenth century and the Encyclopaedists further reinforced his idea to start from the format of an encyclopaedia when working out his ideas on an Index Museum. Such a museum, he wrote is

> first of all, more than an ordinary museum, it is not only an encyclopaedia but an *Encyclopaedia Graphica*. That is, we may think of it as an Encyclopaedia of which the articles may be imagined printed separately, and with their illustrations and maps condensed and displayed as an orderly series of labels; labels to which specimens are then as far as possible supplied, so that over and above the description, the image, the interpretation of the thing, you can see the thing itself in reality if possible, or in reproduction or model as the case may be.

Just as the planning of a conventional encyclopaedia was a complicated task, so the planning of an *Encyclopaedia Graphica* created many problems. 'Most difficult of all, the plan, the order must be no longer alphabetical or numerical, but rational; that is, in conformity at once with reason and observation, with philosophy and the order of nature'.[56]

It was with this idea in mind that he saw the possibilities of uniting the globe project with his Outlook Tower. The globe would provide the geographical starting point, whilst the exhibitions in the Tower would illustrate the evolutionary patterns of specific environments. Geddes was fortunate in the mid-1890s that his activities in Edinburgh, especially the Halls of Residence and the Summer Meetings, had gathered together a number of enthusiastic young men who pledged themselves with high resolve to work for his objectives. While J. Arthur Thomson continued to support Geddes, writing articles for *The Evergreen*, and lecturing at the Summer Meetings, there were a number of others who helped to keep the Outlook Tower going in Geddes' many absences, and to implement his evolutionary ideas as they occurred.[57] For a couple of years Geddes had sustained support at the Tower from Thomson, from A.J. Herbertson, and from John Duncan, with many others offering help for shorter intervals. From 1896 to 1901 the management of the Tower was taken on by another impecunious student, T.R. Marr, who struggled to make all the activity financially viable. Marr kept the Outlook Tower going single-handed until the end of 1898 when the Town and Gown Association provided funds for an assistantship. This was given to Edward McGegan.[58]

The correspondence between Marr, McGegan, and Geddes paints a picture of the Outlook Tower which showed that its activities were very far from Geddes' ideal. It was woefully short of funds. The upkeep of the Tower cost about £400 a year, and its annual income was about £100, more than half of which came from the visitors to the camera obscura. Marr wrote 'five years experience of it has saddled me with a load of debt which will take a long while to work off, even granting that creditors are complacent'.[59] He estimated that to run the Outlook Tower on a satisfactory basis would require £750 a year, £250 to £300 as a salary for the director, £100 for the secretary, £50 or so for a typist, £15—20 each for some boy apprentices, special workers for models as required, a minimum of £200 over two years for materials, and an organised body of workers recruited from friends, working men, teachers, etc. to develop the Tower and be 'real' workers. This group could only be formed and held by the work of a man who gave himself entirely to the task of running the museum and caring for its objectives both in spirit and in fact. This, Marr suggested, was plainly the role that Geddes should be undertaking.[60]

But, ever circuitous in his way forward, presumably keeping to the vision of an evolutionary spiral, Geddes was not prepared to stay in Edinburgh. Ideas of further developing both his civic museum movement and his own abilities constantly drew him away from Edinburgh. He was fortunate again that he had the help of two young chartered accountants, ex-Edinburgh students, Victor Branford and

John Ross. Branford and Ross both worked for Geddes through the Town and Gown Association, and Branford the loyal disciple, was ready to try to use his business expertise on behalf of Geddes and to find ways of raising money for projects. In 1896 he had helped float a company designed to provide money for an economic enterprise in Cyprus. This so-called Eastern and Colonial Association, had, as its object, to raise the capital to finance Geddes in his attempts to bring relief to the Armenian refugees in Cyprus who were in flight from Turkish brutality.[61] Always concerned with contemporary affairs as the raw material for the exhibitions of the future, Geddes saw in this international crisis a chance to gain some personal experience. He needed to test his ideas in an alien environment, and to prove himself a practical operator. The news of the plight of the Armenians had broken during the Edinburgh Summer Meeting of 1896. It became the main topic of discussion, especially amongst members of the Current Events Club, one of Geddes' earliest organisations at the Outlook Tower.[62] The outcome of the concern over the Armenians and Branford's expertise meant that enough capital was raised to fund a pilot project in Cyprus to be set up under Geddes' direction. In the autumn of 1896 Geddes and his wife set off for Cyprus to find practical ways of helping the refugees.

Geddes, however, did not leave without help. Using his French contacts he had uncovered an Armenian who was a graduate of the School of Agriculture at the University of Montpellier. With this man, Mr Salmaslian, as his co-worker and interpreter, Geddes hoped to set up projects which were modelled on the kind of voluntary activities being undertaken by Sir Horace Plunkett Green and his Land Organisation Society in Ireland.[63] With the capital Geddes acquired land and then set about organising its more efficient utilisation. He used the technical expertise of Mr Salmaslian to find out ways of improving yields with better seeds, better irrigation, rotation of crops, and so on. He also put in hand a number of projects for rural industries, particularly his favourite one, the silk industry. The reliance of this activity on natural processes, and its demands for orderliness and cleanliness, made it a socio-biological ideal from Geddes' point of view. He tended on the whole to disregard the problem of demand or market forces. After three months of intense activity Geddes had to return home in time to go to Dundee for his period of teaching there. On his way back he stopped off in London where he attended an International Conference on Armenian Aid and was able to give the conference the benefit of his recent experience.[64] The publicity that he gained from this meant more money flowed into the Eastern and Colonial Association, and it was able to keep going for two or three years. However, the projects began to founder without the stimulus of his personal presence, and Mr Salmaslian, for all his training, proved to be a broken reed as an administrator. The directors of the Eastern and Colonial Association, in fact, sacked him on the grounds of negligence of his duties even though this was expressly against Geddes' wishes. With Mr Salmaslian's departure the viability of the Eastern and Colonial Association came to an end.[65] Geddes, however, had gained some practical experience from this venture, and his confidence in his approach was confirmed. He brought back much material from Cyprus to add to his collection in the Outlook Tower. It was to dominate the material devoted to Europe.

Edinburgh to Paris in 1900: from regional museum to hopes for an international Index museum

By this time the ever closer prospect of the World Exposition in Paris gave Geddes the hope that it might be possible to raise the money for Elisée Reclus's globe and to improve the Outlook Tower's exhibition in its 'world' room. He began to devote himself to contacting potential supporters. Once again he went to the Congresses of the British Association which he had last attended regularly in the late 1870s and 1880s, and he undertook his first lecture tours of the United States, in 1898 and 1899, hoping to find useful contacts there. His impact on the States though, was rather muted. His reputation as an eccentric Professor of Botany who had published a book on sex helped him to draw full audiences for his lectures. But his almost inaudible discussion of the differences between anabolic and katabolic organisms, and his recourse to his 'thinking machines', soon lost his audiences. Geddes found those most receptive to his ideas amongst the promoters of Settlements in the United States, especially Jane Addams of Hull House, and his contacts with the sociologists in Chicago University. A member of the Faculty at Chicago, Charles Zueblin, had been a regular visitor to the Edinburgh Summer Meetings, and he wrote a eulogy on the Outlook Tower as the world's first sociological laboratory which was published in the *American Journal of Sociology* in March 1899.

There was however, no American millionaire willing to fund the globe, and as Geddes began to see the possibility of realising this project at Paris fading, he decided to try and get funding to set up his own presence there. He conceived the idea that he might take his annual Summer Meeting to Paris for the duration of the Exhibition and run it as an international summer school. He began to see that the World Exposition might provide him with what he was looking for, the raw material for a great evolutionary study of the world, an Index Museum on the largest scale. He eventually got financial backing from

[113]

a Scottish industrialist from his own home town of Perth, Sir Robert Pullar. With this money in hand he was then able to approach the British Association and its French equivalent at their respective annual congresses at Dover and Boulogne in September 1899. He had already managed to set up an Anglo-American group interested in his project. With these international contacts he established an 'International Association for the Advancement of the Sciences, Art and Education'. This was to be the administrative umbrella for his Summer School which he was to hold in Paris for the duration of the World Fair.

The faithful few from the Edinburgh Outlook Tower backed him in this venture, Marr and McGegan playing a leading role in setting up an administrative structure. Within the financial resources available and the special circumstances of the fair, what they were able to achieve was remarkable. In the course of four months the school organised 134 courses. Visitors from all around the world with a serious interest in absorbing the information on display at the fair, found a course or lecture suited to their needs. Eight hundred classes were held in just 120 days, and the average attendance at these was between 40 and 50. Some were not well attended, mostly those put on by the Anglo-American group for the English-speaking tourist, whom Geddes accused of being more likely to seek entertainment than instruction. Some attracted audiences of above 300, usually because of a strong demand shown by the French. Geddes was to be found at all times, viewing the exhibition and the city of Paris from his temporary Outlook Tower in a gallery in the Trocadero, or taking small groups from pavilion to pavilion at the fair. In these excursions and meetings Geddes made many personal contacts with influential people from many different nations, as well as renewing contacts he had built up since his years as a wandering student. The exhibition was a watershed in gaining for him an international reputation.[66]

From France, he was to meet Henri Bergson, whose philosophical search for the moving force in evolution had been so closely paralleled by Huxley's students such as Geddes and C. Lloyd Morgan, now Principal of the University College of Bristol. From America Geddes renewed friendships with Jane Addams and Lester Ward, and others whom he had met on his lecture tour. Some American friends introduced him to Sister Nivedita, the Indian Swami Vivekananda's European disciple, and she was to become closely involved in the Paris Summer School, and a personal exploration of Geddes' ideas.[67] The Swami himself came to Paris, and he rekindled the old interest in Indian philosophy and religion that Geddes had shared in the early 1880s with members of the Fellowship of the New Life.[68] Geddes met Paul Otlet and Henri La Fontaine for the first time and a host of others. The threads of common interest which drew Geddes and many of these visitors together was twofold. On the one hand, the World Fair offered a view of social as well as economic progress, and Geddes' Le Playist approach to analysing this met with a wide response; on the other, there was a desire to form international organisations to promote the exchange of ideas on such matters. Geddes' international association itself was a testimony to what international effort could achieve.

Surrounded by evidence of widespread 'goodwill', Geddes conceived his most ambitious project yet. He wanted to save the national pavilions of the fair with their exhibits intact to form the nucleus of a vast Index Museum of the World. He wanted to do more than just follow the example set by the Chicago Fair of 1893 and the Philadelphia Commercial Museum. At Paris there was a chance to set up an international venture in a truly co-operative spirit. Each nation would donate its own exhibit. The Index Museum he foresaw would not be dedicated to the commercial enterprise of any one nation. It would be a powerhouse to generate ideas about how to achieve peaceful economic and social progress on a global scale. It was an international interpretation of Le Play's concept of 'social peace', though the French Le Playists were in no position themselves to support him. Their influence had been undermined on the theoretical side by the work of Durkheim and others seeking a different basis for the study of sociology. In practical endeavours they had been superseded by the Musée Sociale, at this time more concerned with industrial relations and specific problems of labour, rather than any vision of future social harmony.

Thus Geddes, the foreigner, with his international association, found himself in a unique position to argue his case. He did this with such vigour that he almost succeeded. Notwithstanding the blatantly competitive spirit between the economic exhibits of different nations, and the pavilion on war designed by Jean de Bloch of Poland, all nations shared some common ground in that, as they industrialised, they had to prepare their societies for a constant process of change, accelerated by the rapid growth of the multilateral trade of a world economy. The mood in Paris in 1900 was expansive, the World Fair was the largest ever held, it was a good year for business. Many nations agreed to Geddes' idea of offering their exhibits to form a permanent collection even if they were not altogether sure what an Index Museum was. The problem which scuppered the project was a legal one, concerning the land on which the fair had been held, and the future relationship of the proposed institution with the Paris authorities. The project would have required a vast capital investment to sustain and develop it even if much of the initial cost of the pavilions was borne by individual nations.[69]

The project, had it come off, would have provided Geddes with a full-time occupation for the rest of his life. When this prospect was snatched away he did his best over the next three years to retrieve or recreate at least part of it. The most outstanding exhibit, de Bloch's pavilion on 'war' Geddes managed to save, with a team of Edinburgh volunteers taking it personally to the War Museum at Lausanne. As for his own prospects, Geddes tried to resuscitate his International Association and to mount another international summer school at the next major exhibition which happened to be in Glasgow. This, of course, was a much smaller affair than the Paris Exhibition, but its virtue was that it would also be very cheap. Nevertheless, Geddes found that not only were the funds for his association virtually exhausted, but also the energies of his stalwart helpers, especially Marr and Ross, who had borne the brunt of the Paris venture since Thomson, Herbertson, and Duncan had all departed. Marr was induced to write several letters to Geddes in which he sought to bring to an end his direct involvement with Geddes' projects. He wrote to Geddes:

> We are in complete sympathy with your ideas. It is in the practical carrying-out and sustaining of them that we lose confidence in you. We believe, rightly or wrongly, that, left to ourselves, we could carry out one or more of your ideas; but the freedom from interference with the *method* used which is a sine-qua-non of such work, is just what you will never give. The blame is largely ours. You *do* promise and intend not to interfere with the administration but in the face of the actual working-out of your idea you find it impossible to resist.[70]

Geddes was contrite and tried to rekindle their interest by offering them a future ideal that together they might strive for:

> My view is, that the Association and its assembly have to be understood not as a summer school simply, as at present (no matter how successful that might be made but as an educational/scientific endeavour starting, it is true, from that democratic basis, it also from/to the highest University and Museum, etc. level/i.e. that of Paris) and therefore hence appealing to the university and scientific societies and museum movement everywhere.[71].

Marr, however, had had enough, and the Town and Gown Association released him from his seven-year lease in 1901 after five years. He remained a fervent admirer of Geddes, and he was in fact to carry out in Manchester a survey of housing that Geddes advocated but never actually carried out himself in Edinburgh.[72] Geddes was thus left with a hiatus in the administration of the Outlook Tower and no prospect of a curatorship in the immediate future. However, he had begun to see a new opportunity arising which might be the route to fresh recognition of his work with the Outlook Tower. Geddes made a bid to launch it as a Geographical Institute at a time when geographical studies were gaining wider recognition. His efforts were to fail, but his influence on geographical studies in Britain was to be extremely far reaching, making Geddes and his ideas one of the formative influences on the social sciences in Britain at the turn of the century. Geddes' path was beginning to cross with mainstream developments in the social sciences.

Notes

1. See T.H.S. Escott (1897) *Social Transformations of the Victorian Age: a survey of court and country*, London: Seeley, chapter entitled: 'Social Citizenship as a Moral Growth of Victorian England', pp.67–76.
2. M.Richter (1964) *The Politics of Conscience: T.H. Green and his Age*, London: Weidenfeld & Nicolson, pp.344–55.
3. John Owen (1974) *L.T.Hobhouse, Sociologist*, London: Nelson, p.93.
4. See H.E. Meller (ed.) (1979) *The Ideal City*, Leicester: Leicester University Press, Introduction.
5. Barnett had been a founder member of the C.O.S. but parted company from the society on the issue of state involvement and the need for environmental improvements – C.L. Mowat (1961) *The Charity Organisation Society, 1869–1913: its ideas and work*, London: Methuen, pp.117–130.
6. E.P. Hennock (1976) 'Poverty and Social Theory in England: the experience of the eighteen-eighties', *Social History* 1: pp.67–91.
7. M. Arnold (1969) *Culture and Anarchy*, Cambridge: Cambridge University Press, pb edn., reprint pp.35–6.
8. B.B. Gilbert (1966) *The Evolution of National Insurance in Great Britain: the origins of the welfare state*, London: Michael Joseph, pp.159–226.
9. B. Drake and M. Cole (eds) (1948) *Our Partnership by Beatrice Webb*, London: Longmans, Green, pp.316–421.
10. P. Hall (1984) 'Utopian Thought: a Framework for Social, Economic and Physical Planning', in P. Alexander and R. Gill (eds) *Utopias*, London: Duckworth, pp.189–90.
11. H.E. Meller, op.cit., p.22.
12. S.A. Barnett (1884) 'University Settlements', *Nineteenth Century*, February.
13. H.E. Meller, op.cit., p.12.
14. R.N. Soffer (1978) *Ethics and Society in England: the revolution in the social sciences 1870–1914*, Berkeley: University of California Press, pp.15–48.
15. See chapter 2, pp.62–4.
16. P. Geddes (1895) 'Edinburgh Summer Meeting', *Extension Bulletin*, no.9, University of the State, New York, July.
17. He began these courses in 1886 in Edinburgh. The next three years they were held at the Granton Marine station, and from 1891 most of the classes were again held in Edinburgh. Information given in the only extant copy of a periodical, *The Interpreter* 1, 3 August 1896.
18. Michael Sadler (1891) *University Extension: past, present and future*, London: Cassell.

19. For a firsthand account of the kind of enthusiasm Geddes could display see Israel Zangwill's description of walking through Edinburgh with him in P. Mairet, (1957) *A Pioneer of Sociology: life and letters of Patrick Geddes*, London: Lund Humphries, pp.65–9.

20. From these small beginnings Geddes was to develop a whole sequence of masques which he published as *The Masque of Learning and its many meanings: a pageant of education through the ages* (1912). These were to be performed in Edinburgh and London on a grand scale – A. Defries, (1927) *The Interpreter: Geddes, the Man and his Gospel*, London: Routledge & Kegan Paul, pp.41–51.

21. Other lecturers at the 1895 Summer School included: William Sharp (Life and Art); Dr Wenley (First Steps in Synthesis); Principal C. Lloyd Morgan (Philosophy); in the new department of Civics and Hygiene, Dr Dyer (Aspects of Citizenship) and Miss Jane Hey (Woman's Work); Professor J.A. Thomson and Mr Scott Elliott (Natural Sciences).

22. He had visited Dublin often, since his first invitation from Mr Ingram to lecture there. See chapter 2, pp.61–3.

23. J.S. Stuart-Glennie (1890) 'The Desirability of Treating History as a Science of Origins' *Transactions of the Royal Historical Society*, (second series) pp.229–40.

24. The Ossian myth was finally exploded in the late nineteenth century. The legend of Ossian had been propagated in 1762 by James Macpherson, Scottish poet. He 'discovered' Ossian's poems and published the epic 'Fingal'. There were, though, genuine Ossianic ballads which were Irish lyric and narrative ballads in the Scots/Irish Gaelic tradition.

25. Spencer used evolutionary theory to cover all the natural processes – J.D.Y. Peel (1971) *Herbert Spencer: the evolution of a sociologist*, London: Heinemann, pp.131–65.

26. Mackinder referred to the cultural divide as 'the abyss which is upsetting our culture' – quoted by K.C.Edwards (1961) 'The Mackinder Centenary in the East Midlands', *East Midland Geographer* 15, (June); 39–40.

27. Flavia Alaya (1970) *William Sharp – 'Fiona Macleod', 1855–1905*, Cambridge, Mass.: Harvard University Press, pp.146–72.

28. Edith Wingate Rinder was also the inspiration for 'Fiona Macleod', Sharp's most successful literary creation, ibid., pp.126–7.

29. Alaya, op.cit., p.11.

30. Ibid., p.60.

31. One of the most vociferous was H.G. Wells in a review published in *Nature*, 29 August 1895 – quoted in P.Kitchen (1975) *A Most Unsettling Person: an introduction to the ideas and life of Patrick Geddes* London: Victor Gollancz, pp.151–2.

32. William Sharp to Geddes, 21 January 1895, Geddes Papers MS10563, NLS.

33. P. Geddes (1902) 'Edinburgh and its Region, Geographic and Historical', *Scottish Geographical Magazine* 18 (June): 308–9.

34. He wrote the MS of an unpublished monograph, using the title *Museums, Actual and Possible*, Papers of Sir Patrick Geddes, University of Strathclyde.

35. Geddes' use of symbolism of this kind was probably strongest in the 1890s when he was setting up the Outlook Tower. It was an aesthetic device much in use at that time and extensively used by William Sharp in his work – Alaya, op.cit., pp.177–81, *The Evergreen* offers an example of the use of symbolism in nature.

36. Decorations of the Collège des Ecossais at Montpellier and the garden were inspired by Greek mythology.
37. See, for example, his *Report on the School of Art 1893–4*, typescript, Papers of Sir Patrick Geddes, University of Strathclyde.
38. Geddes (1888) published his views on the social significance of the artist in his pamphlet, 'Every Man His Own Art Critic: an introduction to the study of pictures' (Glasgow Exhibition), Edinburgh: William Brown, 1888.
39. Geddes had a more personal link with the Reclus family. Part of the reason for Reclus's trip to Edinburgh was to visit his nephew, Paul Reclus, who had been involved in a violent anarchist plot in 1893 and had had to flee from France. Geddes gave him refuge and he worked in the Outlook Tower using the pseudonym, George Guyou – J.P. Reilly (1972) *Early Social Thoughts of Patrick Geddes*, unpublished PhD, University of Columbia, New York, pp.215–17.
40. Marie Fleming (1977) *The Anarchist Way to Socialism: Elisée Reclus and nineteenth century anarchism*, London: Croom Helm, pp.234–6.
41. Elisée Reclus (1898) 'A Great Globe' *Geographical Journal* XII (4): 401–6.
42. P. Geddes (1905) 'A Great Geographer: Elisée Reclus, 1830–1905', *Scottish Geographical Magazine* XXI: 561.
43. He wrote this in his *Report on the School of Art, 1893–4*, op.cit., p.25.
44. T. Marr was a student at Edinburgh in the early 1890s. He became Geddes' assistant at Dundee in October 1894. The following year he undertook the management of the Outlook Tower until 1901. In 1901 he moved into the University Settlement movement, becoming co-warden with Alice Compton of the newly-amalgamated Manchester University Settlement, and T.C. Horsfall's Art Museum. He was warden until a breakdown in his health in 1909. He conducted a survey of housing in Manchester which was published as the first of two volumes, the second being T.C. Horsfall's much more celebrated work *The Improvement of the Dwellings and Surroundings of the People: the example of Germany* (1904). Subsequently, he lived and worked in France in the building construction industry. He met up again with Geddes in Montpellier in the late 1920s, and helped him with the business details of the project. He was a staff member of the Collège des Ecossais from 1930–39.
45. T. Marr to P. Geddes, 14 June 1898, Geddes Papers MS10566, NLS.
46. H.E. Meller, op.cit., (1976), pp.65–71.
47. P. Geddes, *Museums, Actual and Possible* (unpublished MS) op.cit., chapter IV, p.2.
48. V.V. Branford (1902) 'The Philadelphia Commercial Museum' *Scottish Geographical Magazine* XVIII: 243–7.
49. P.Geddes, *Museums, Actual and Possible* op.cit. Chapter II, p.11.
50. G.Lorphevre (1954) 'Henri La Fontaine, 1854–1943; Paul Otlet, 1868–1944' *Revue de la Documentation*, XXI p.89–96.
51. P.Uyttenhove (1985) 'Les efforts internationaux pour une Belgique moderne', extrait de *Resurgam. La Reconstruction en Belgique après 1914* (catalogue), Bruxelles: Crédit Communal de Belgique, p.47.
52. W.Boyd Rayward (1975) *The Universe of Information: the work of Paul Otlet for documentation and international organisation*, published for the International Federation for Documentation by the All-Union Institute for Scientific and Technical Information, Moscow, p.37–57.
53. He describes what he means by this term in *Museums, Actual and Possible*, op.cit. chapter VIII, 'The Index Museum on the largest scale: its

application in an International Exhibition'.

54. R.Allison (1964) *Encyclopaedias: their history through the ages*, London: Hafner, p.188.

55. He contributed entries on the following subjects: evolution, Darwinian evolution, sex, reproduction, variation and selection, parasitism and a number of botanical topics in the 1888−92 edition of *Chambers Encyclopaedia* and the 1875−89, 9th ed., of the *Encyclopaedia Britannica*.

56. P.Geddes, *Museums* etc. op.cit., chapter VI, p.1.

57. The Edinburgh based group was dominated by women: by far the most important were Mrs Geddes and her daughter, Norah. Others included Mrs Craigie Cunningham, Miss Homes, Miss Forbes, Miss Jardine, Miss McGegan, Miss Hay, and Miss Barker. Sir Thomas Whitson, and one or two other Edinburgh philanthropists, supported Geddes' initiatives; there were a number of university people including Professor J.Geikie (brother of Sir Archibald), Professor J.Stuart-Glennie, and Professor G. Baldwin Brown. A.P. Laurie, Principal of Heriot Watt College, and a number of clergymen, the Rev. Dr John Glasse, the Rev. John Kelman and the Rev. Professor Paterson and others. Most active support came from students, artists, and musicians such as R.N.Rudnose Brown, Phoebe Traquair, James Cadenhead, and Marjorie Kennedy-Fraser.

58. Edward McGegan, another Edinburgh student without a clear idea of a future career, worked at the Outlook Tower between 1898 and 1903, when he joined the Bournville Village Trust.

59. T.Marr to Geddes, 17 December 1900, Geddes Papers MS 10566, NLS.

60. T.Marr to Geddes, 4 March 1901, Geddes Papers MS 10566, NLS.

61. Papers relating to the Eastern and Colonial Association and its affairs are in the Papers of Sir Patrick Geddes, University of Strathclyde, 19/19/196.

62. Several of Geddes' students became enthusiastic about the Current Events Clubs, and the idea was exported by John Ross to the Passmore Edwards Settlement in London, and by Marr to Manchester when he became Warden of the University Settlement there.

63. T.West (1986) *Horace Plunkett, co-operation and politics: an Irish biography*, Washington DC.: The Catholic University of America Press.

64. Mr and Mrs Patrick Geddes, *Cyprus and its Power to Help the East*, Reprint of Report to the Conference, May 1897. Papers of Sir Patrick Geddes, University of Strathclyde, 19/23/197. See also article 'Cyprus, Actual and Possible', *Contemporary Review* June 1897.

65. Papers of the Eastern and Colonial Association, op.cit., 19/27/197.

66. Mairet, op.cit., chapter X; Boardman, op.cit., p.178−190.

67. See chapter 7 note 61.

68. There were a number of new translations of Indian religious texts in the 1880s; see, for example, 'Bhagavadgita' trans. by K.T. Telang (1882) *Sacred Books of the East* viii. 34.

69. Mairet, op.cit., Chapter X.

70. J.Ross and Marr to Geddes, 5 February 1901, Geddes Papers MS 10566, NLS.

71. Geddes to Marr and Ross, 9 March 1901, Geddes Papers MS 10566, NLS.

72. See note 44.

CHAPTER 5

Maverick
in the social sciences

Between 1880 and the first world war, in Europe and the USA, modern economics, modern history, sociology, geography, anthropology, and social psychology were all institutionalised and assuming their modern form.[1] Although they were to pursue independent lines of development, they were all, to some extent, influenced by developments in the natural sciences and the concept of evolution. In Britain, however, it is remarkable that there were fewer social scientists working within the mainstream of academic life in comparison with France, Germany and the USA. One of Britain's most influential social scientists at this time was Herbert Spencer who worked totally independently.[2] To some extent Spencer's mantle was to fall to Geddes who, with his manifold activities, gained for himself a national and international reputation in the social sciences before the First World War.[3] He too, remained totally independent, moving between different groups of people working in separate disciplines in the social sciences, both voluntarily and within institutions. He built on his reputation as a populariser of the natural sciences, and his fame as the author of *The Evolution of Sex*, to create for himself a similar role in the social sciences. He was able to do this because of his evolutionary perspective which helped him to try to relate all the social sciences together. But it is arguable that Geddes' attempt to popularise and integrate new disciplines was premature.

Geddes' contributions to the academic social sciences were thus, at best marginally illuminating, and at worst, counter-productive.[4] He did harm to his own academic reputation to the extent that he alienated those amongst the academic community who might have taken up and developed his ideas. In fact, by his many idiosyncracies, he cast himself and his ideas into the wilderness, where he remains in terms of modern scholarship, except when he is partially rescued from time to time by one of his disciples, either from amongst his contemporaries or from the small number of dedicated people who have tried to keep his ideas alive.[5] Yet he was to be very influential in at least two of the social sciences, geography and sociology, in Britain, and provided the philosophical basis for the practice of modern town planning at its inception. Geddes' achievements in all these areas were as considerable as his shortcomings. The value of his contribution lies as much in the questions he raised as in his success or failure in answering them himself. Geddes created a unique perspective for himself which many others have found stimulating. He was a maverick amongst British social scientists, a position he enjoyed, but for which he, perhaps deservedly, paid dearly.

Geddes' contribution to the 'new' geography

From this isolated position, he was able to attract and help individuals who did belong to the mainstream and were to affect the future of their disciplines. The eclectic nature of Geddes' espousal of 'regionalism' was a broad enough umbrella to cover any new initiatives in exploring and controlling the environment or educating the public. At the end of the century some of his earliest efforts were becoming recognised. Since the first Edinburgh Summer School at the Granton Marine Station, Geddes had been involved in trying to introduce nature studies into schools by training the teachers. He had been helped in this work by J. Arthur Thomson, and in 1899 nature study became an accredited subject under the new Scottish School Code of that year. This lead was followed in England in 1900. In 1902 a National Nature Study Exhibition was held to which Geddes had contributed some material from the Outlook Tower. The exhibition was displayed at the Cambridge Summer School of that year, and also at Geddes' own Summer Meeting in Edinburgh.

This success led Geddes to hope for more. Since the 1870s, he had been following keenly the developments in 'scientific' geography[6] and contributing to the development of the subject at the Outlook Tower.

When the London Royal Geographical Society finally set up a Geographical Institute at Oxford University in 1899, Geddes felt that this was an initiative that he could follow. Why not make the Outlook Tower into a geographical institute in recognition of the work that had been done there? After the failure of the globe and Tower campaign, this would bring new life to the Outlook Tower now beset with problems of debt and declining support. Geddes thought of recasting the Outlook Tower as a regional centre for geographical studies and as an educational centre for geography teachers. Then with his usual optimism he began to see not just teachers of geography but social scientists of the future coming to the Outlook Tower where they would be trained in geography and history, economics and sociology. From these students would come the social evolutionists who would be both willing and able to act as consultants on social and environmental planning. Starting with geographical studies, Geddes could see a whole new future growing up for his work in Edinburgh. His hopes were sustained by the great emotional appeal of the 'new' geography which had swept through institutions of higher education in France and in Germany during the 1890s and 1900s.[7] He believed that the Outlook Tower with its wide-ranging basis of activity was the appropriate kind of institution to meet the complexity of the challenge of environmental studies.

As his visions grew, so did his ambitions. Soon the Outlook Tower on its own was not enough. He embarked on a major new propaganda campaign for a National Geographical Institute of which the Outlook Tower was only a small part. The Outlook Tower would be the section devoted to studies of Edinburgh and its region and would act as an assembly point for a number of local detailed investigations and surveys. But there was to be a new purpose built geographical institute which would be the national centre of studies in human geography. Geddes had been closely in touch with the developments in human geography that had been taking place in France and in Germany and had done much to introduce the work of some of the leading practitioners of it into Britain.[8] He thus believed himself to be experienced enough to manage such an enterprise. He also believed that Edinburgh was the right place for it to be located.

Some of his optimism had firm foundations. Edinburgh was uniquely well endowed by both its natural environment and the kinds of activities which had flourished there to offer support for such an institute. The Royal Scottish Geographical Society had been playing a part in cultivating geographical activities in Scotland for a number of years, and the *Scottish Geographical Magazine* had been published annually since 1884. Scottish explorers had contributed an honourable share to the work of geographical exploration through the century and were still doing so in the 1880s and 1890s.[9] The two areas of the world which were still relatively unknown were the beds of the great oceans and on land, the Antarctic. Exploration of the latter had been caught up in the late nineteenth century nationalistic jingoism which permeated the activities of the conquerors of the unknown. Reaching the South Pole was treated as a nationalist contest to display the racial superiority of those who got there first. Scott's tragic and disastrous expedition in 1911 trying to beat the Norwegians marked the peak of this particular phase.[10] In the late 1890s and 1900s Scottish nationalism had played a part in sustaining the efforts of young Scotsmen to prove their worth. An Edinburgh student in Geddes' circle, W.S. Bruce, was able to raise the funds to refit an old whaling ship as a vessel of scientific research and discovery. He named it the *Scotia* and sailed to the Antarctic, but his achievements were curtailed because of lack of funds.[11] He was able to discover and name one headland in Antarctica after his main patron for the 1910 expedition who was, in fact, the Prince of Monaco. But he was not able to keep the *Scotia* as an educational training ship. She ended her days funded by the British Government as an observational weather ship in the North Atlantic keeping track of the movement of ice-flows which could be a danger to shipping.

A much more important factor making Edinburgh a pioneering centre for the natural sciences and 'scientific' geography was the establishment of the Challenger Office, in the zoological department of the University. It was in the mid-1870s that Professor Wyville Thompson had gone on HMS *Challenger* as the director to undertake oceanographic surveys in the Antarctic.[12] Wyville Thompson's successor to the Chair of Natural History, John Murray, had been a young naturalist on board the *Challenger*. Between 1884 and 1895 he issued reports from the Challenger Office set up at the University to collate the information produced by the explorations. Geddes had paid a high price for his determination to go to London to study with Huxley. Had he stayed at Edinburgh it is very likely he might have become involved in the activities surrounding the *Challenger*, and he would then have been at the centre of a number of activities which reached out beyond the natural sciences to 'scientific' geography. When the actual work of exploration was completed in 1895 and John Murray issued his last report, which was in fact an outstanding contribution towards establishing the discipline of oceanography, the Challenger Office did not close. Instead it began to gain a function in promoting related disciplines dedicated to the study of the natural environment. Work was carried out there on climatology, geology, zoology, and botany. Murray was to build on these initiatives by setting up an observatory on Ben Nevis as a further outpost to collect data for the study of climatology.[13] Another observatory was set up at Fort William and hourly observations of meteorological elements were made there and at Ben Nevis by unpaid volunteers for many years.

Much of the work undertaken in these fields was done with very little funding and much personal dedication. At the centre, giving support to the many initiatives, was the Royal Society of Edinburgh, which drew much of its active support from the University. Professor P.G. Tait, who held the Chair of Natural Philosophy at Edinburgh,[14] was secretary of the Royal Society in the early 1880s. In 1883 Pofessor Tait applied to the Treasury for a grant to undertake a bathymetrical survey of Scotland's lochs and inland waters.[15] The money was not forthcoming but the work was carried out on a voluntary basis. It was considered important work, not only because of the scientific data collected, but as a means of training young investigators in the techniques of scientific survey. Between 1896 and 1900 Professor Murray, now Sir John Murray, helped to bring the work to a successful conclusion.

Scientific survey work in Scotland in fact had been well established since the early 1880s, with the rise to prominence of the Geikie brothers. Sir Archibald was to relinquish his Chair of Natural Science to his brother James, on taking up the Directorship of the Geological Survey of Great Britain in 1882.[16] Sir Archibald seems to have used his influence to prevent Geddes getting a Chair at Edinburgh in the 1880s, though brother James was a close personal friend of Geddes. At the Summer Meetings, however, and courses at the Outlook Tower, members of the Geological Survey, especially Mr James Goodchild, offered their services to Geddes over a number of years. The Geological Survey pioneered the use of new techniques in its work, and since 1889 a photographic department had been set up to supply series of photographs to illustrate the geological features of different parts of Scotland. The Geological Survey also contributed to the improvements in cartography that were developed in Edinburgh. Mapping the complicated ground in the North-West Highlands involved the development of new techniques, and the series of maps which were produced between 1883 and 1896, together with photographs, managed to illustrate, in an unprecedented way, the great movements which had resulted in the cracking of the earth's crust and the formation of mountains. This work made the region a classic example for study and a number of educational institutions were to use this material.

Geddes' own activities in the 1880s and 1890s had left him little time to participate in any of these developments. Yet he kept himself informed about what was going on through the meetings of the Royal Scottish Geographical Society and its publication, the *Scottish Geographical Magazine*. He was to make his special contribution by encouraging the young students he came in contact with to pursue voluntary work on these projects. In this way, he came across one young student, A.J. Herbertson, who was willing to work with him in the Outlook Tower. The two men developed a symbiotic relationship which was to mould Herbertson's career and clarify Geddes' ideas on the geographical potential of the Outlook Tower. They were a powerful combination. Herbertson, practical, conscientious and hard-working, and Geddes, mercurial, bubbling with ideas, but lacking the patience for practical application, ideally complemented each other. Herbertson was one of the group of Edinburgh students whose life, both intellectually and personally, was totally transformed by contact with Geddes. In return, he worked hard on Geddes' schemes, particularly the geographical aspects of the Outlook Tower, giving Geddes' ideas credibility in the 1890s when Geddes was at a formative stage in his thinking about the Tower. Herbertson was to become an influential pioneer of geography teaching in Britain.[17]

He had come up to Edinburgh University at the relatively advanced age of 21 in 1886. He then proceeded to have a most unconventional career. The first signs of Geddes' influence on him was that he remained a student for six years without ever graduating. He began by studying natural philosophy with Professor P.G. Tait and then won a scholarship in Experimental Physics which meant he had to act as a demonstrator in Professor Tait's laboratory. He went on to study geology with Professor James Geikie, and by 1891 was doing courses in advanced mathematics, natural philosophy, practical astronomy, agriculture and rural economy. To round out this eclectic fare, Geddes sent him to Paris to meet M. Demolins and the Le Playists, and to spend some time absorbing the developments in the social sciences in France. Herbertson was to prepare the English translation of de Rousier's monograph on *American Family Life* which had begun to throw doubt on the whole Le Playist thesis about the family as the fundamental social unit. Herbertson was thus a close spectator of the work of the Le Playist school just at the point when M. Demolins and others moved away from the family thesis and began to investigate the transmission of cultural values as the prime mover in the evolutionary change.[18]

Geddes had always had doubts about the family thesis and he had explored the possibility that botanical studies might prove a more satisfactory technique for studying the interaction of organisms with environment. Using his personal contacts again, he sent Herbertson to Montpellier in 1893 to his friend Professor Flahault, who had invented new techniques in surveying the incidence of vegetation in specific environments in ways which might reveal fresh insights into the relationship between the two. Herbertson was following very closely in Geddes' footsteps. He was particularly close to him in the three years immediately after the publication of the *Evolution of Sex* . In 1889 Herbertson had been in Paris with Geddes during the time of the Paris Exposition and the Fourth International Geographical Congress which was also held at that time. It is likely, since Herbertson was giving life to Geddes' dream of reviving the Franco-Scots College, that Herbertson attended the paper given to the Conference by Mr A. Silva White who was secretary of the Royal Scottish Geographical Society and editor of the *Scottish Geographical Magazine*. The subject of Mr Silva White's paper was 'The Achievements of Scotsmen during the Nineteenth Century in the Fields of Geographical Exploration and Research'.[19] Thus inspired, Herbertson went off for a semester to the University of Freiburg-im-Breisgau where he began work on his PhD thesis on the monthly rainfall over the land surface of the globe which he completed in 1898. However, in 1891–2, he went back to Scotland to be Geddes' demonstrator at Dundee, one of the first of the hand-picked people whom Geddes took to Dundee for a year or so and who subsequently became advocates of Geddesian natural philosophy, sometimes more successfully than Geddes himself. After embarking upon his meteorological research in 1892, Herbertson had had his name put forward to be a Fellow of the Royal Geographical Society in London, and in 1894 was offered the post of Lecturer in Political and Commercial Geography at Owens College Manchester, a post funded by the Royal Geographical Society and the Manchester Geographical Society. He was forced to teach a kind of economic and commercial geography very far from the perception of human and physical geography that he had been cultivating with Geddes. He attempted, using some of Geddes' slides, to introduce their approach to Manchester in a lecture on 'Edinburgh: a Study in Cause and Effect' but with little success.[20]

He was able to continue this work however, on his return to Edinburgh in 1896. Meanwhile, in the course of the 1890s the Royal Geographical Society of London had also been very active in promoting a wider concept of geographical studies in Britain and its efforts were given some encouragement when in 1895 the International Geographical Congress met in Britain for the first time. It was at this Conference that H.J. Mackinder made his famous speech on the scope of geography and its practitioners in England. He suggested that:

> speaking generally, and apart from exceptions, we have had in England good observers, poor cartographers, and teachers perhaps a shade worse than the cartographers. As a result, no small part of the raw material of Geography is English, while the expression and the interpretation are German.[21]

It is significant that he said 'England' and not 'Britain' as there were distinct differences in Scotland.

Cartography, at least, had advanced much further in Edinburgh than elsewhere. Dr J.G. Bartholomew of the publishing firm of Bartholomew and Sons was an outstanding cartographer and his achievements, when added to those of the Scottish Geological Survey, and the Bathymetrical Survey of the lochs and inland waters, gave Edinburgh a European standing in cartography.[22] It was in celebration of this that Geddes had designed the main exhibit in the Scottish room of the Outlook Tower as a huge map drawn to scale and properly orientated. A.J. Herbertson worked closely with Dr Bartholomew in the 1880s, the result being a pioneering *Atlas of Meteorology* published in 1889, one of a new series by Bartholomew mapping different geographical elements. The purpose of this work was educational rather than merely as a means of compiling information in cartographical form. Herbertson had been devoting much attention to the teaching of geography, and in 1898 he published an article, 'Report on the Teaching of Applied Geography', and a school textbook, *An Illustrated School Geography*.[23] He had been working on the manuscript of this during 1896 and 1897, developing an approach to geography in which the human and the physical took precedence over the more limited aspects of economic geography and the descriptive geography of exploration. He pays tribute to the help he received from H.R. Mill, an ex-student of Edinburgh, and presently working as the Librarian of the Royal Geographical Society in London, to J. Arthur Thomson, with whom he had worked at the Outlook Tower, to G.C. Cash of Edinburgh Academy, and J.G. Bartholomew who had produced the maps. Source material came from Reclus's *Geographie Universelle*, Sievers's *Landerkunde*, and Chambers *Encyclopaedia*. He does not mention Geddes but the whole work had a distinctly Geddesian orientation. Herbertson published his most famous book a year later, *Man and His Work: An Introduction to Human Geography* which he wrote in collaboration with his wife, who had just completed her own small but penetrating study of Le Play.[24]

The Royal Geographical Society of London recognised that there was much activity of the kind it wished to promote going on in Edinburgh in the 1890s, and in 1895 Geddes was invited to give an account of his Outlook Tower. As so often before, and subsequently, however, while Geddes' enthusiasm found a favourable response, his breadth of approach and his ambitious objectives of regenerating society did not elicit any financial support.[25] Instead the Royal Geographical Society decided to support the renewed efforts of H.J. Mackinder who had been trying since 1892 to arouse interest in setting up a Geographical Institute which he hoped would be in London. A Geographical School was established in 1899, though in Oxford and not in London. But through Herbertson, Geddes was to share in some of this success as

Herbertson was invited by Mackinder to come to Oxford to help him in the task of setting up the School.[26] Herbertson had spent the previous two years in Edinburgh, employed as a Lecturer in Industrial and Commercial Geography at Heriot-Watt College and giving his spare time to Geddes' work at the Outlook Tower, especially the Summer Schools.

Herbertson was to initiate the organisation of Summer Schools at Oxford at biennial intervals from 1902 and he modelled them on those of Geddes at Edinburgh. Herbertson and his wife, Dorothy, were both keen supporters of the Summer Schools which they knew greatly magnified their influence on the teaching of geography in Britain. Between 1899 and 1905 only nine students gained the Diploma in Geographical Studies and there were a further sixty-eight students between 1905 and 1915 when Herbertson had succeeded Mackinder as Director of the School, while over 850 students had attended the five Summer Schools that Herbertson and Dorothy had organised during his directorship.[27] Herbertson invited Geddes to give lectures at the Oxford Schools, which he did when his other engagements permitted him to do so. But the requirements of academic geography and Geddes' vision of the subject were steadily diverging in the later years.

For a moment, however, between 1896 and 1902, it seemed as if Geddes had had a chance to take geography in a new direction. He had begun modestly enough introducing geographical lectures at Dundee in 1898. But after the Paris Exposition of 1900 he conceived his new ambition for a National Geographical Institute, which was to be not only a museum but an active school, training students for a new kind of 'applied geography' by which was meant a Geddesian kind of environmental planning. The training of future geographers had become a live issue in the 1890s. In 1893 a Geographical Association had been formed to promote the teaching of geography to school teachers and it had gained the support of the London-based Royal Geographical Society. Herbertson became particularly interested in these developments as he had had direct experience of the differences between the approach of the economic and commercial geographers and the approach he favoured of human and physical geography. Subsequently he was to become a member of the Geographical Association and an editor of the influential journal the *Geographical Teacher*.

The extent to which geography was an academic discipline or a practical one requiring experiments and outdoor activity was a moot point at this time. Geddes very much favoured a new approach which included practical activity and outdoor work alongside academic study, and it was this which made the activities at the Outlook Tower as important to his mind as the exhibits in the museum. His ambitions for

[129]

his National Geographical Institute thus did not stop at the erection of a building and a vast number of galleries and displays. He also wished it to be a repository for the material collected by workers in the field whose work experience was a vital part of their training. There was no University Department or Institute in Great Britain at the time which remotely resembled this ambition. Victor Branford, who was to become one of Geddes' closest collaborators, decided that it would be best to adopt an American model for the purposes of publicity. It was in 1902 that he wrote his article in the *Scottish Geographical Magazine* on the Philadelphia Commercial Museum.[28]

Geddes' most influential support came from his old friend Dr J.G. Bartholomew, whose family firm had supported a small Geographical Institute on the premises of the firm for a number of years, and who would have dearly loved to see Edinburgh as a centre of geographical studies of national and international importance.[29] Bartholomew's support for Geddes stemmed from his appreciation that, as geographical studies multiplied and diversified, it became increasingly difficult to handle the material in an orderly fashion. He believed that some new method of classification was essential and that Geddes was the man to develop one. Such a classification had to be synthetic rather than analytical, and by correlating the biological with the physical, to pave the way to enabling the study of organic distributions including those of human societies. Two things helped to generate some support for Bartholomew's assessment of Geddes and his plan for a national museum at this time. One was Geddes' performance at the International Summer School at the Paris Exhibition of 1900 and his near success at establishing the exhibits as an Index Museum. The other was the work done by former students such as J. Arthur Thomson and A.J. Herbertson at the Outlook Tower and the current contribution being made by the young men Geddes had chosen to help him as his assistants in Dundee. Since the early 1890s Geddes had been sending his assistants to Montpellier to his friend Professor Charles Flahault. The purpose was to acquaint them with Flahault's work on botanical surveys.[30] By the turn of the century two of his assistants, Robert Smith and Dr Marcel Hardy, had begun to produce some outstanding work applying these ideas to Scotland.

Perhaps Robert Smith's own description of Professor Flahault's work will best illustrate why Geddes was so enthralled by the potential of this technique for the kind of work he wanted to do at the Geographical Institute. Smith wrote,

Professor Flahault's survey is essentially based on the distribution of tree vegetation. A number of the trees of Europe are what are called 'social' species, i.e. they tend to form forests where one species prevails, to the more or less complete exclusion of others. Each tree has its own particular requirements, which limit it to definite *regions* dependent upon the prevalent climatic, soil and other conditions. Thus, from a geographical point of view, forests of a particular tree indicate particular climatic and soil conditions; from a botanical point of view, the presence of particular associations of subordinate species; and from an economic point of view, regions suitably situated for the growth of certain cultivated species and for certain industries . . . . The principle upon which such a map of the vegetation is based will be seen to be quite different from that of 'floristic' maps where the object is to show the actual areas of distribution of the species in such a way that they may be easily compared . . . . Vegetation maps, on the other hand – such as those prepared by Professor Flahault . . . are not primarily concerned with the precise limits of the individual species, but show rather the distribution of a number of selected communities of plants which exist associated in nature. The species forming these communities are held together by their common requirements, and thus each association becomes indicative of a particular *ensemble* of environmental conditions.[31]

Here was a method with the practical potential of enabling synthetic work to be done in an orderly way in specific localities or regions. This botanical survey technique could serve to illustrate Geddes' intentions to a sceptical public. It gave promise of something which could be identified as a 'natural' region and a means for investigating it. But Geddes' plans for the National Geographical Institute went far beyond the prospect of developing new kinds of regional studies. He was still wanting to create an Index Museum of the world and educational facilities for the training of geographers and environmentalists.[32] If the London Royal Geographical Society was prepared to fund the Oxford Geographical Institute, he hoped that the Royal Scottish Geographical Society might support his schemes. He and his friends kept up a veritable bombardment of the Scottish Royal Geographical Society in 1902 through the pages of the *Scottish Geographical Magazine*. In the March issue there was a plea for an Institute of Geography in Edinburgh by Dr J.G. Bartholomew, followed by the outline plan and a note by Geddes. In May Branford published his piece on the Philadelphia Commercial Museum as an example of what had already been done in an American city. In June, Geddes published his key article, 'Edinburgh and its Region, Geographic and Historical' which was an attempt to put across the ideas he had developed at his Summer Meetings to a wider and more critical audience. He tried to show particularly how the work he had been encouraging at the

Outlook Tower could be made to lead on naturally to the development of a National Geographical Institute.

But the responses from eminent geographers both in London and in Edinburgh were not over-enthusiastic. Sir Archibald Geikie suggested that Geddes was being a little ambitious and warned that London's Imperial Institute was an object lesson in not embarking on grandiose schemes. Sir Clements Markham, President of the Royal Geographical Society, felt that the first need was to consolidate and build on the records and facilities of the London Royal Geographical Society to produce the kind of national facility that Geddes outlined.[33] Geddes' sense of a separate centre for Scottish geographers was lost very early along the way. This was not surprising because Geddes had chosen to emphasise, as the main objects of his new Institute, two giant globes, celestial and terrestrial, designed by M. Galeron and Elisée Reclus. Geddes wanted the vision of the Institute to be global not national, and he even tried to suggest that the regional surveys which could emanate from such an Institute would have universal rather than particular significance. What Geddes failed to convey in his outline were the practical applications he envisaged for his institution. He carefully planned exhibits on industry and commerce which were juxtaposed with exhibits of the educational system in an attempt to highlight past connections, present connections and possible future ones. There was a Peace Museum to educate visitors in world citizenship; there were exhibits of engineering, including railways and canals, which led on to what Geddes described as 'geotechnics' which encompassed all kinds of environmental planning from afforestation, irrigation, agriculture and hygiene. At the apex of all these exhibits there was a section of Comparative Civics in which all modern knowledge was drawn together and applied to the development and progress of representative cities. The connections between these exhibits though was obscured by the fact that Geddes still had at the back of his mind the task of charting the evolution of civilisation.

Why was Mackinder successful in establishing a Geographical Institute at Oxford where geographical studies had only a lowly status, while Geddes failed to get support for a Geographical Institute in Edinburgh where Scottish scientists and explorers had not only an honourable record in geographical exploration, research and cartography, but had also pioneered related subjects such as oceanography and sustained the Challenger Office for several decades? There are obvious reasons such as the soaring ambition, in conjunction with the lack of clarity, in Geddes' objectives for the National Geographical Institute; his 'outsider' status at Edinburgh University; and the reason his friends and disciples, especially V.V. Branford, felt was a strong contributory factor – Geddes' lack of influence south of the border. But beyond these reasons (and they could be multiplied) was a much more basic fundamental difference between Mackinder's view of geography and that of Geddes. Mackinder had been educated at Oxford with Curzon and Milner. He belonged to the small group of Liberal Imperialists (even standing for Parliament in 1900 at Warwick and Leamington though he was defeated on this occasion) who believed that the means to solving economic and social problems lay in tightening Britain's relationship with her Empire.[34] Unlike many of his generation, he did not get swept along in the tide of immediate, practical concern about social conditions in large cities in Britain.

His sights were set on the world implications for Britain of the process of industrialisation and the growth of the world economy, what the modern geographer could do to alert politicians to the growing vulnerability of Britain and her Empire and the best means of defending it. Geddes saw the potential of geographical studies in a completely different way. Geography was the means for studying people in their places or regions; finding out the history, culture, and traditions of each locality as the key factors in its evolutionary process. Such knowledge, in the Le Playist tradition, would facilitate the necessary adjustments to modern industrialisation, minimising the harm which could occur to the most favourable evolutionary trends. While Mackinder saw the importance of geography in political terms with an emphasis on the military defence of countries, Geddes concentrated on the interdependence of regions and the complementary nature of these regions in terms of their resources. In many ways the two men epitomised the paradoxical experience of European industrialisation in the nineteenth century.

On the one hand the process of industrialisation had drawn Europe away from the rest of the world and had made the industrialised countries more of a single entity in terms of the flow of capital, labour, markets and technological ideas in a historically unprecedented way.[35] The result had led to a higher degree of regional specialisation in manufacture than ever before and a great increase in world trade.[36] Geddes' view of the autonomy of city and region had some basis in this experience as economic survival had led each region to concentrate on those factors, industries or processes which gave it the competitive edge. On the other hand, Mackinder's views on the rise of Germany as the world's greatest industrial power after the United States, helped to shape his geographical perceptions. In the early 1900s, Geddes, Herbertson, and Mackinder were all to produce papers which were influential for a short time until some of their theoretical shortcomings were widely publicised. These were Geddes' publication in the *Scottish Geographical Magazine* of 1902: 'Edinburgh and its region, geographic and historical'; Mackinder's paper for the Royal Geographical Society of London in 1903, the famous 'The Geographical Pivot of History'; and Herbertson's paper for the Geographical Association in 1904 with the title 'The Major Natural Regions of the World'.

Geography and the region: the problems of definition

In all three papers the concept of the region was central. In Geddes' case it was the natural unit for evolutionary experiments in environmental planning; in Mackinder's case the region was merely a vague geographical concept used only to explain the implications of the location of scarce resources to the relationships between countries; in Herbertson's case it was used as a teaching tool to promote new approaches to geographical studies. The only paper which elicited criticisms on the question of defining the concept of 'region' was Herbertson's, mainly because he was the only one concerned with giving the concept academic viability.[37] For Geddes and his audience there was no question about Edinburgh and its region. Even the first vegetation map produced by Robert Smith, using Professor Flahault's techniques, was of Edinburgh and its region. In every way, the city and its surrounding hinterland was a 'natural' region. The point is of some importance, because whilst Herbertson spent the next nine years struggling to define the idea of the region under constant criticism, Geddes ignored the debate. His interest in 'geotechnics', large-scale environmental 'management', the nurturing of the best environmental trends, all required some kind of large geographical unit and the region seemed ideally suited for this purpose.

It was not the 'region' but the activity of 'regional survey' that Geddes wanted to promote. Such 'regional survey' work was seen at first as suitable for school children. Initially Geddes was prepared to start at that level hoping to build on the recognition of his work in encouraging nature studies. Herbertson's influential position since 1901 as editor of the *Geographical Teacher* provided a forum for the campaign. Geddes wrote a paper on 'Nature Study and Geographical Education' as further propaganda for his National Geographical Institute in 1902.[38] But he longed, as he was to write later to H.J. Fleure, to move out of the schoolroom into the university, to reach on a wider scale the kinds of students he attracted to his Summer Schools.[39] His sights were set on training young men like his assistant at Dundee, Dr Marcel Hardy, for a new kind of career. Hardy, who had been trained as a regional geographer in France before meeting Geddes, was to leave Europe in 1906 for Mexico to organise a 'colonisation land project'. He used Geddes' techniques, such as regional survey and the valley section as his development model, and his work on the

Tezonapa Valley was amongst the first successful land colonisation and reclamation schemes in Latin America, and has been frequently emulated. Hardy subsequently returned to Europe to work as a War Reparations Commissioner in Germany; he went to South America again in the 1920s as an agricultural planner in Uruguay; and in Scotland, on one of his many visits, he became the chief designer of Lord Leverhulme's model village at Leverburgh on Harris. In complete contrast, in 1926 he was in India as tutor to the son of the Maharaja of Indore. Hardy was frequently at the Edinburgh Summer Meetings offering lectures. Such a career as he enjoyed, acting as an environmental planner and adviser in a wide range of activities and contexts, was of a kind which Geddes foresaw as being of growing importance to the future. Such a new departure required a new kind of education. He wanted regional survey to become a training ultimately in the cause of a voluntarist, environmental planning.

Geddes' use of the 'region' in his 1902 article had a specific practical objective. He suggested that the regional portion of the National Geographic Institute should be the Outlook Tower which was already serving the purposes of a sociological institute. The work there displayed the two elements he wished to unite: the kind of environmental planning which seemed obvious to a natural scientist used to seeking and nurturing favourable environmental conditions; and his Le Playist inspired belief that favourable evolutionary change was at source a cultural matter. It was this combination which was to prove Geddes' Achilles' heel from this time forward. Many were to be dazzled by his insights or antagonised by his ambitions, yet friendly or hostile, there were few who were able to respond to one half of his message who could also respond equally to the other. In his 1902 paper he made one of the clearest statements of his position. It is worth quoting:

> What are the conditions of this sociological outlook? The natural environment − latitude and configuration of land and sea, climate, life and natural productions − not only conditions human population but determines its fundamental activities. These activities determine institutions, and these in turn determine customs and laws, ideas and ideals; yet ideals in turn profoundly modify, if not practically redetermine all. Hence our geographico-historic student should place himself if possible at some centre where the survey of the natural environment, and also of the influence of ideals is relatively complete, and also relatively simple and clear. Where shall we find such a centre?[40]

The answer was, of course, Edinburgh.

It was an unassailable position for a non-deterministic evolutionary scientist, since ideals were a product of culture, belief, emotion and religion, all factors depending in the last resort on the psychology of individuals. The latest trends in evolutionary theory emphasised the importance of biology and psychology in shaping the future. But Geddes' approach left the nature of any practical regional survey work so ill-defined that in the plethora of detail which such an activity produced, any grasp of an objective could be lost. The same seemed to be true of Geddes' National Geographical Institute. The basis of his Draft Outline Plan was in fact to retrieve or recreate some of the exhibits of the Parisian World Fair. While these had given credibility to his ideas of an Index Museum in Paris, starting from scratch in Edinburgh was another question. Sir John Murray, who was President of the Royal Scottish Geographical Society in 1902, was not impressed. He had known Geddes since at least the early 1880s and was well aware of his propensity for ambitious dreaming alongside his many and undoubted practical achievements. There was never to be a National Geographical Institute, and Geddes' contribution to geographical studies in Edinburgh was not to develop further. All that was achieved was that some further Summer Meetings at the Outlook Tower were devoted entirely to geography and the teaching of geography.

What Geddes took from this experience was a more clearly refined idea of what he meant by 'regional survey'. This was not a meaning that was accepted by all other geographers. Dr Marion Newbigin,[41] writing about regional surveys in 1913, did not mention Geddes at all. She suggests that the idea of a regional survey based on the ordnance survey map was first put forward by H.R. Hill in 1896. Hill was aware of the campaign since the 1870s by members of the Royal Geographical Society, including Francis Galton, to gain access for the public to the British ordnance survey maps. Hill recommended the use of the newly refined ordnance survey maps in his book for teachers of geography as a basis for local regional surveys in the late 1890s. Apart from this there had been a number of regional surveys undertaken at university level, one by Dr Newbigin herself, and others by geographical students at the University of Edinburgh since 1908, when a geographical lecturer was appointed. What she meant by regional survey had none of the critical juxtaposition of ideals and environment which was what Geddes meant by the term.[42] It is possible that Dr Newbigin's careful avoidance of Geddes' name in the very year of the founding of Geddes' national committee to promote regional surveys is another example of the mixed emotions that Geddes could arouse.

Another view of Scotland's contribution to geography in these years pays fulsome tribute to Geddes, suggesting that: 'Geddes' greatest contribution to geography was the idea and use of regional survey —

the stock-taking of an area — as a preliminary to any constructive work'. The writer was R.N. Rudnose Brown, who finished his botanical studies at Aberdeen University in 1899 with the newly arrived J. Arthur Thomson just appointed as Professor. Thomson sent him to Geddes to be demonstrator at Dundee in 1900–1901 after Geddes' successful Summer School in Paris. Rudnose Brown then followed his predecessors to Montpellier, but on his return was not so interested in mapping Scotland. There was not much left to do as Dr William Smith had taken over mapping the Lowlands after the sadly premature death of his brother Robert Smith, and Dr Marcel Hardy had covered the north-west and the Highlands. Instead Rudnose Brown followed another of Geddes' pedagogic syllogisms: 'Go and see for yourself and don't be content with the views of others' since 'The eye rather than the ear was the way to learn and understand'.[43] Rudnose Brown went off with W.S. Bruce on the *Scotia* to the Antarctic. Later he was to attend one of Herbertson's Summer Schools at Oxford as his only training in the teaching of geography prior to taking up his first teaching post in the subject at Sheffield University. The generation of lecturers who were introduced to the subject in this way tended to become 'regional' geographers. It was in this somewhat indirect way that Geddes' influence on geographical studies was mainly to be felt.[44]

The problem of defining 'the region' as an academic concept was left to A.J. Herbertson, who struggled with it for the entire period of his Directorship of the Oxford School. He made his last major statement at the British Association meeting in 1913 two years before his premature death. But a year before this he had published a paper which was quite out of character with his usual painstaking, scholarly attempts at definition. The purpose of the paper was to suggest on a general level the importance of geography to the sciences, and the title was significantly not 'Regions' but 'The Higher Units — A Geographical Essay'.[45] In this paper Herbertson extolled the virtues of the concept as a means for synthesising new knowledge on the relationship between man and his environment. The content, message, and choice of illustration were totally Geddesian and this article provides a clearer description of how Geddes viewed the region than anything Geddes wrote himself. As for Mackinder's use of the concept, he never had cause to take it further. He had already achieved the position of Director of the London School of Economics and he set his sights on a political career and a mission to save the British Empire.[46] The concept of the region had thus been given a central place in British geographical studies but at the same time there was total ambivalence over its use and meaning. It had to be rescued and recast by the next generation of geographers, H.J. Fleure and C. Fawcett who were both encouraged by Geddes and published their work just after the First

World War in the series he edited with Victor Branford, 'The Making of the Future'.[47]

Geddes' contribution to sociology in Britain

As the chances of achieving his National Geographical Institute began to fade, Geddes found his friends and contacts opening up new opportunities for him. From this time forward Geddes was to spend little time developing his Outlook Tower. He struggled to support it financially, still held his Summer Schools there, and encouraged others to become involved with its day-to-day administration. But he himself became more interested in promoting his ideas on civic museums, on regional survey techniques, and on 'geotechnics' to the widest possible audience. With J. Arthur Thomson in Aberdeen and Herbertson in Oxford, Geddes' closest collaborator became Victor Branford, the accountant and business entrepreneur who divided his time between his offices in London and Edinburgh. Branford tried but did not succeed in his efforts to put the Outlook Tower on a sound financial footing. After the failure of the attempt to create the National Geographical Institute, there was even less optimism than usual from Geddes about his ability to raise the necessary funds to support the Tower. By 1903 a financial crisis which would close the Tower seemed imminent. Even Mr Martin White, founder of Geddes' chair in Dundee, and a willing supporter of initiatives to promote sociological studies, could not be prevailed upon to give financial support. No one was quite sure whether the Tower was promoting geographical studies or social studies or, indeed, whether it was a museum at all.

For a while the activities there were kept going mostly by Victor Branford, who set up what he grandly described as the Edinburgh School of Sociology; and by Edward McGegan, yet another impecunious Edinburgh student who took over from Marr the day-to-day running of the place. When McGegan finally left to take up a more permanent post with the Village Trust in Bournville in 1904, Branford could only recommend to Geddes that he must sell the Tower. As such an outcome became imminent, a group of old Edinburgh friends and supporters of the Summer Schools banded together to form the Outlook Tower Association in 1905. With this help the Tower was refurbished and the exhibits renewed and rearranged, a number of small publications and postcards of the Tower produced and, while never wildly sucessful, the Tower survived quite well until 1914. In 1910 Geddes was able to raid the Tower exhibits to provide the material for his gallery in the RIBA first Town Planning Exhibition.[48]

But from 1903 Geddes had begun to concentrate on finding support for his projects in London, especially through the auspices of the newly-formed Sociological Society. Branford had initiated the idea of forming a British Sociological Society to act as a centre for disseminating information on developments taking place in social studies in Britain and elsewhere, and as a forum for promoting the ideas of Geddes. It was to be the beginning of a new era. At a meeting of the committee set up to discuss plans for the society, tribute was paid belatedly to Herbert Spencer who had recently died. His work was now considered out of date but he had been the last and the greatest of the British academic sociologists. The consensus at the time felt that the future demanded sociologists of action. Branford himself was greatly impressed by Geddes' outlook and activities. Although he earned his living in business,[49] he hankered after the academic life and relished the contacts he made giving lectures at the Edinburgh Summer Meetings and the Outlook Tower. He was a dreamer and idealist who was most interested in reinterpreting what he thought were the religious ideals which had inspired society in the past in some modern form. His publications, although covering a wide range of topics, were most often focused on the idea that there was a way of respiritualising modern science. His major work on this theme was published during the First World War with the title: *Science and Sanctity: being suggestions towards a theory of day-dreaming and visionary process*. His pursuit of a cosmic idealism was of the kind which appealed to many at the time struggling to establish value systems in the modern world.[50]

Branford, inspired by Geddes, felt that it was his personal mission to bring Geddes' ideas wider publicity. Since the mid 1890s he had been helping Geddes with his financial problems, funding the University halls, helping in the transition to the Town and Gown Association, setting up the Eastern and Colonial Association for Cyprus. He and his partner, Ross, remained Geddes' chartered accountants through all the vicissitudes of Patrick Geddes, Colleagues and Company. Branford also helped J. Arthur Thomson with the publishing side of Geddes' activities.[51] Above all, Branford valued his personal relationship with Geddes which, judging from the correspondence that has survived, was particularly intense between 1900 and 1905.[52] It was during this time that Geddes began writing to Branford that he believed he had at last found a way forward in his evolutionary work, and knew where he was going. He was stung perhaps to justifying himself against Branford's criticism that he could not see how Geddes' ideas worked in relationship to each other, for instance, the Summer Meetings, the International Assemblies and Exhibitions, the Outlook Tower, the Halls of Residence, and so on.

What impressed Branford was an extraordinary letter from Geddes explaining how he was trying to discover the essentials of religious life of past periods and to rediscover them in modern form. The contrasts Geddes described in this letter of the 'outer life' and the 'inner life' give an indication of the extent to which Geddes was trying to use the new science of psychology as a guide.[53] As Branford suggests in the letter he wrote in reply, Geddes had probably arrived at his main conclusions

> before you reflected on the method and path by which you reached them. . . . I am led to make this reflection by the idea that what is now engaging your thought is perhaps not so much the search for a more perfect sociological synthesis as for a more thorough consciousness of the method by which you have reached your existing sociological theory and practice.[54]

Branford's assessment was quite accurate, and from henceforward he took as his life's work, in the time he had from his business affairs, the task of helping Geddes to promote his methods. His efforts, from the founding of the British Sociological Society in 1903 to the establishment of Le Play House in 1920, and many other activities, were to have a pervasive influence on the development of the social sciences in Britain in the first two decades of the twentieth century.

The genesis of academic sociology in Britain is a subject which has received some attention from scholars. In his careful treatment of the subject, R.J. Halliday has tried to throw some light on the reason why certain elements such as civics, eugenics, philanthropy, and social work, were involved in the founding of the Sociological Society, while others such as statistical economics and social anthropology were not.[55] But the key question perhaps, is not who was included or excluded. Rather it is what kind of demand brought disparate elements together at this particular time and how did they seek to meet it? The demand above all was for informed social action but it came from many different sources. There was mounting public concern in the wake of such shocks as Rowntree's study of poverty in York, published in 1902, and the facts about the huge number of volunteers for military service in the Boer War who were rejected on grounds of ill health. These facts were publicised in the 1904 report of the Special Parliamentary Committee set up to enquire into the physical deterioration of Britain's masses. A campaign for National Efficiency was gaining political momentum.[56]

Eugenics and civics

In 1903 Francis Galton, at the great age of eighty-one, was prepared to come out of retirement and preach the message he had devoted his life to proving. He felt the time was ripe to persuade people of the connection between physical deterioration and racial degeneration. Biology was on his side as the latest studies of sex and variation had revealed the importance of chromosomes and genes. The condition of the people in Galton's view, thus had everything to do with the racial stock from which it was made up, and population growth and urbanisation had had the disastrous effect of encouraging the worst genetic stock to breed, while genetic stock of quality, 'the educated middle classes', had begun voluntarily to limit the size of their families. Galton used not only biology but population census returns to bolster his bald assumptions. Since 1870 the high fertility level amongst families in the upper classes had begun to decline, while amongst the lower orders it had not. He wanted to advocate policies which would reverse this trend. In the general uncertainty about politics, the role of the state, and alarm at the evidence of the social condition of the urban poor, Galton was able to strike a deep chord of response amongst some of the educated and upper classes.[57]

By a curious irony however, his pursuit of a questionable thesis helped to create a new methodology, mathematical statistics, with enormous implications for the gaining of new knowledge. Leading biologists wanted to refine their techniques in quantifying their subject and Galton's statistical methods were taken up. In the 1890s Galton's most able collaborator Karl Pearson, who had become Professor of Applied Mathematics at University College, London, had continued the work of advancing mathematical statistical methods.[58] From 1894 onwards Pearson had been particularly concerned with defending Darwin's theory of natural selection from the famous attack made upon it by Lord Salisbury at the British Association Meeting of that year. Salisbury had suggested that the process of natural selection could not be demonstrated. Pearson wanted to prove that the theory of evolution was likely to become a branch of the theory of chance, and a quantitative measure of the rate of natural selection could be found. Pearson worked alongside Professor Weldon, the zoologist, and together they built up a pioneering School of Biometrics. In 1901 the Journal of the Biometric School, *Biometrika* was launched.[59]

By 1904 the nucleus of a new discipline was coming into being and all that was needed was further financial funding and academic recognition. Galton himself was able to provide the former. In February 1905 he gave the University of London about £1,500 to establish a Eugenics Record Office, and then, until his death, an annual research grant of £500. On his death in 1911 he left the residue of his estate to the university to fund a Chair, the Galton Professorship of Eugenics, with the recommendation that the post should be offered to Karl Pearson. Further benefactors were found to fund the cost of a new building and laboratories, and Pearson gave up his Chair in Applied Mathematics to take the Galton Chair, and to be the first Head of a Department of Applied Statistics. Outside the universities Galton's work generated a eugenic movement without a vigorous intellectual base but boosted by a socially elitist enthusiasm. By 1907 it had broken away from the parent body of the Sociological Society to organise its own meetings and propaganda. From the chair, at the very first public meeting of the Sociological Society, when Francis Galton had given his paper on 'Eugenics: its definition, scope and aims', Karl Pearson had strongly argued that the founding of the Sociological Society itself was premature. He argued that English social studies needed some great figure to develop a sociological school and lead the way and he did not think that there was anyone of that stature in Britain.[60]

This left Geddes, when he came to give his paper, in an almost impossible position. It was well known that Victor Branford, who was the moving force behind the Sociological Society, was seeking to promote the ideas of Geddes and wished to suggest that he was the outstanding scholar whom Pearson called for. Yet others on the provisional committee did not share this view. There was a strong element representing the philanthropic world, such as C.S. Loch, Secretary of the Charity Organisation Society, and E.J. Urwick who was a Deputy Warden at Toynbee Hall. Their interest in the Sociological Society was to encourage a scientific approach to social problems and to find a means for training voluntary social workers.[61] Thus when Geddes stood up to give his first paper he did so from a position which was far weaker than Galton's and his eugenic message. Geddes wanted to transform the call for action into a programme of social and environmental planning and he believed his work at Edinburgh had shown him the way. But biological evolutionary theory was no longer totally supportive to his viewpoint,[62] he had no private income to back up his ideas, his subject which he called 'Civics: as applied sociology' was not a recognised academic discipline, and its closest related ally, geography, had no established chair in London, and was, in any case, dominated by H.J. Mackinder. Above all, Geddes really wanted to be the man of action, to inspire his audience to get up on their feet and go, since he was, at least, dependent on an emotional response to gain support. On all counts he was bound to lose against the Eugenists.

But he was not down-hearted by this as he believed that the eugenic movement was merely one part of the total equation Efo (Environment/Function/Organism) and he wanted to relate eugenics to the others. He was particularly anxious to emphasise that eugenics was complementary to civics and a crude brand of social Darwinism was biologically unsound. His message was to suggest that the movement of the masses in the course of the nineteenth century from the countryside to the cities had put the evolutionary equilibrium into jeopardy. The way to tackle social problems was not to see the end result of the disequilibrium, which was a social problem, but to tackle ways of encouraging a better equilibrium to re-establish itself. This meant gaining a new perspective on the urban environment and understanding the culture and traditions of a society and how these were worked out in a specific physical context. If no care was taken to safeguard this relationship the equation Efo = Ofe could not be met and the future became doomed to retrogression.

This was a complicated enough message to get across and Geddes did not help his cause with the methods he chose. He presented papers at the first three Annual Congresses, starting in 1904 at a high peak with a large audience and Charles Booth as his chairman.[63] Two years later the audience which turned up to hear his paper on the need for civic exhibitions was much smaller, and it was his brother-in-law from Edinburgh, James Oliphant, who was in the chair. At the first 1904 conference Geddes had tried to suggest that he was offering a new methodology to guide those who wished to solve urban problems. To begin with, the city had to be seen as a whole, not as an amalgam of disparate elements each requiring specific treatment (which at base was the Fabian approach). Seeing the city as a whole, however, was not straightforward; it required a special combination of science and art. Scientific facts, observations made in a systematic manner, combined with an artistic understanding based on cultural criteria, together made a new subject which Geddes called 'civics'. It was only possible to study this subject in a specific context and therefore the beginning of such a study had to be a practical social survey. Once the survey had been undertaken, then a new response to social problems could be generated which involved the control and nurture of both environment and people together. This made 'civics' a subject of 'applied sociology'.[64]

At the end of his first paper Geddes was heavily criticised for ignoring political, economic, and historical realities, but he was unrepentant. To him economics and politics as understood at the time were irrelevant to the task of improving social conditions. He was therefore beyond criticism and he bolstered his position by highlighting the weaknesses in the theories of others and enhancing his own by reference to the importance of science and technology in improving the lot of mankind. He was constantly encouraging others not to be afraid of science and to see what benefits chemistry had brought to agriculture, and biology to medicine. His objective in establishing 'civics' was to dispel fear of cities and mass urbanisation, and to release the creative responses of individuals towards solving modern urban

[143]

problems. Pursuing his ambition to study, at the same time, the physical structure of the city with the cultural ideals of urban life, he suggests that the best method for studying the city was to begin, on the one hand with its geographical location, on the other, with the evolution of its historical and cultural traditions.

The right way to look at cities then is as organisms subject to the natural processes of growth, blossom and decay. What he was striving for in his own words was 'The elemental and naturalist-like point of view even in the greatest cities'. Geddes always made much of the contrast between the 'urban' and 'rural' response to cities and he strongly identified himself with the latter. He believed that the countryman was able to observe the strange social and cultural milieu of the city with a keener eye, since his pre-conceived 'norms' were not shaped by the experience of urban life. He argued that, in spite of all the nineteenth century developments towards city-planning, 'we are still building from our inherited instincts like bees'.[65] To persuade people to shift from this to a rational scientific development of our cities, which would lead ultimately to the elimination of social problems, was the major difficulty. What was needed was a better understanding of the forces shaping the city.

His favourite illustration of this theme was Edinburgh, where he had developed his ideas of the regional survey, and which furnished him with the best examples, in buildings and layout, of the importance of the past in the present. Geddes, at the second conference of the British Sociological Society in the following year, attempted to define his subject more precisely. His arguments, however, proved to be even more unexpected. He tried to convince his audience that civics was a subject with the practical objective of leading students of civics towards accurate forecasting of the future. He really believed this was possible, though the chances of success depended on two factors. First an understanding of the present evolution of cities, and second, a keen perception of those factors in the present, the 'buds', discernible only to the trained eye, which were going to influence the future. To arrive at the former Geddes set out to establish a general approach for analysing the evolution of cities. He took the basic Le Play formula of society — place, work, family — made it place, work, folk — and construed from it a method for analysing the city. This method of notation was, however, merely a shorthand for the range of material relevant to the 'bud' hunter. The terminology place, work, folk, covered geography, economics, anthropology, nature — Geddes' argument being that all knowledge touching on the human condition needs to be encompassed by the student of evolution.

If this was the case, however, the problem was to utilise a wide range of knowledge in a specific context both to understand the evolution from the past to the present, and to determine the needs and direction for the future. Geddes provided an ingenious way out of this difficulty by utilising a 'biological' viewpoint and developing his notations. In the centre of the stage was man, an organism with physical and psychic properties. Modern man was a product of those elements of organic continuity built up between generations.[66] He had at once an inheritance which was organic and psychic, and a heritage which encompassed economic wealth and social and cultural traditions. Starting from here Geddes constructed a composite model, step by step, under general headings, referring to those factors he believed had most influence on man's inheritance and heritage. His steps were uncertain and certainly he managed to confuse his audience.

Yet the main thrust of his argument does emerge from all the subsections he used to make his case. Social problems in the city cannot be solved with reference to specific factors such as low wages or bad housing. Economic activity, occupations, and locations are only part of the picture. Of equal importance are biological units such as the family, and social, and cultural traditions which provide the basis from which man learns to adapt to technological change. The evils of the city such as disease, vice, and crime are not just problems, they are the result of this process of adaptability being allowed to go on without direction or control. To solve them the whole social life of the city has to be directed towards higher evolutionary goals. Yet self-conscious effort to achieve progressive evolutionary trends can only come about if all citizens are united. Students of civics could point in the right direction, but evolutionary progress could only come from within each and every organism.

What Geddes therefore wanted was to encourage in the present and future the Ruskinian aim of a society whose main concern was the nurture of its people. At some stage the economic and social system had to be fitted to the people and not vice versa. Geddes felt that the opposite trend had been taking place since the Industrial Revolution and, with a certain disregard for historical fact, he divided the subsequent period up into evolutionary stages. The Industrial Revolution period itself was the palaeotechnic age when the new science and technology was in its infancy and men became so intoxicated with their new powers that humanity was sacrificed on the altar of economic progress. Then he discerned a neotechnic age, when technology became more sophisticated and was harnessed, not only to produce goods, but also to lighten labour. What he looked for in the future was a geotechnic age, when science and technology could be used to improve the material and physical conditions of the people. And finally a 'eutechnic' age, when the whole system would be devoted to the nurture of people. In that age man would have become master of his own destiny and Utopia would be within reach.

This was the long-term objective which Geddes insisted was perfectly practical, in view of the rapid development of science and .technology. But it could only be realised if, in the short term, students of civics were recruited to work towards this end. A year later at the third conference of the Sociological Society, Geddes made a plea for encouraging civic exhibitions. The result of a regional survey was the collection of a mass of material which could be analysed most readily if hung as an exhibition. Such an exhibition would generate local interest and encourage citizens to participate in those kinds of activities which would lead to social and environmental betterment, the prerequisite for evolutionary progress. Geddes suggested that such exhibitions need not be local and regional only but national and international.[67]

Geddes felt that his subject 'civics' was built on the nascent social sciences but it offered a unique practical training. He was not alone in believing that many of the new social sciences should be studied in a practical as well as an academic way. Dr A.C. Haddon, the anthropologist, who was to become involved in the activities at Le Play House, spoke at the Sociological Society meeting in 1906 on 'Sociology as an Academic Subject'. He said:

> It seems to me, if we want to make Sociology a really vital subject we should teach it as far as possible from the points of view of actual observation and investigation; hence we should not make it a purely academic subject, or rigidly limit our area or scope. Of course, it is obvious, a teacher cannot convert every student into an original investigator, but there is an attitude of mind in teaching which encourages students in the belief that they can themselves contribute something of descriptive or theoretical interest or of practical importance, and I venture to think this is the way to breed real living students, whereas the formal method of teaching a subject is often apt to be dispiriting.[68]

Geddes and Haddon were amongst a very small minority at this time holding such a view, as other social scientists like H.J. Mackinder were far more concerned with gaining academic respectability for their subjects and the practical problems of recruiting students.[69] Geddes as usual, wanted to go further than Haddon, and to reach out, not just to students, but to anyone interested in cities and social problems, as evolutionary trends were no respectors of academic exclusiveness. From 1906 onwards he increased his commitments to give university extension lectures for the University of London.[70] His courses were all built around his personal evolutionary perspective. Whether the lectures were on 'The History and Principles of Biology', 'The Evolution of Life, Mind, Morals and Society', 'Nature Study and

Geography in Education', 'Contemporary Social Evolution', 'Evolution of Occupations', or 'The Evolution of Cities', Geddes always used his synthetic approach, his notations, and those elements of a social philosophy he had forged in his activities at Edinburgh. He constantly referred his students to the *Chambers* and *Encyclopaedia Britannica* articles he had written with Thomson on evolution and evolutionary theory in the 1880s, and the monograph *Evolution of Sex*.

The cell theory developed in *Evolution of Sex* remained one of the basic building blocks of his evolutionary perspective. Other 'building blocks' basic to his approach were his evolutionary approach to 'mind', and his neo-Romantic attachment to place, expressed usually in terms of symbols and Greek mythology.[71] The mortar which held this unusual edifice together was Geddes' belief that he was pioneering a movement towards synthetic studies which had educational and social bearings for the future. Such an education would develop a new generation more in tune with the ideal of creating the eugenic age devoted to the nurture of people, which could then be achieved through environmental planning. His students were constantly advised to read Geddes' papers on 'Civics: as Applied Sociology' that he had given to the first three conferences of the Sociological Society. By 1906, however, the Sociological Society had ceased to be dominated by Branford or to be interested in being a vehicle for Geddes' ideas. J. Martin White had duly offered his money to the University of London for sociological studies and L.T. Hobhouse and E. Westermarck were jointly appointed professors. In their inaugural lectures Hobhouse spoke of 'The Roots of Modern Sociology' in which he argued that sociology had grown from philosophy and history; whilst Westermarck addressed himself to the subject of 'Sociology as a University Study'. Neither Hobhouse nor Westermarck made any reference to Geddes and his ideas.[72] In 1907, Hobhouse became first editor of the *Sociological Review* which replaced the *Sociological Papers* which had been published annually since 1904. Geddes' last chance to come into the mainstream of British academic life had slipped away despite support from his disciples like V.V. Branford, and money from his former benefactor, J. Martin White.[73]

Social experiments in London: Crosby Hall and Chelsea

In fact Geddes' evolutionary stance had taken him too far from any recognised discipline within the university system, and he himself was probably happiest keeping it that way. His special contract at University College, Dundee, had given him freedom to construct his lectures and courses as he wished. He had always preferred seeking out individual promising students and trying to influence them, to building up his department and subject. As the initial interest in his ideas on 'civics' began to fade, he managed by a characteristic combination of initiative and luck, to establish his own base in London. He had been exploring the possibilities of developing student halls in London as he had done in Edinburgh, when controversy broke out over the demolition of the last mediaeval building of note in the City of London which had once been the residence of Sir Thomas More. Geddes' biographer, Philip Mairet, has recounted how Geddes entered the fray, marshalled influential support and while unable to prevent the demolition of the building, managed to ensure that it was done in such a way that it could be re-erected elsewhere.[74] Geddes then fought off commercial opposition, including a plan from Mr. Selfridge who wanted to incorporate the building as a show-piece in his new store. Although with few financial resources, he was able to persuade the London County Council to give him the fragments of the building, and after much effort found a benefactress of Crosby Hall, the student residence already built in Sir Thomas More's garden at his Chelsea residence, who was prepared to finance the re-erection. From the start Geddes had wanted the building re-erected on that site as it would give a magnificent hall and dining area to the student buildings and its historical associations fitted closely. As Mairet writes:

> Crosby Hall in Chelsea may almost be said to stand as London's memorial to Patrick Geddes, for it was not, as he had hoped, achieved as a gift of the City to the University, but primarily the result of his own initiative.[75]

The Hall was administered by the University of London but Geddes was given the position of warden, and Crosby Hall became the venue for many Geddesian activities from exhibitions to social evenings, from masques and pageants to meetings of the Utopia Society. This last little society had flourished since 1904 under the aegis of Miss Dorothea Hollins who had a small flat near Crosby Hall and was a fervent admirer of Sir Thomas More. The purpose of the Utopian Society was to foster Art, Learning, and Friendship, seeking inspiration from the example of Sir Thomas More. In 1908 the Utopia Group was transformed into the Chelsea Association, in order, Miss Hollins wrote, 'to extend Utopia to the civic and public sphere'.[76] A small group of contributors to the volume of *Utopian Papers* she edited, give a glimpse into the close circle of Geddes' friends at this time. They included, apart from Geddes himself, S.H. Swinny, Dr J.W. Slaughter, V.V. Branford, Dr Lionel Taylor, Sister Nivedita, F.W. Felkin and the Rev. Joseph Wood. As Geddes' bid, through the Sociological Society, to develop a new discipline 'civics', and to found a National Museum for the Social Sciences, signally failed to find support, so Geddes seemed to retreat more and more towards the Chelsea Association.

After the failure of his attempt to launch a 'civics' movement, there were few ripples of recognition of his work. Geddes was able to set up a separate Cities Committee of the Sociological Society to match the Eugenic Committee, but the latter had much greater support. The Eugenists actually withdrew from the Sociological Society forming their own Eugenics Education Society in 1907. Geddes continued to be very active, encouraging his small Cities Committee and engaging in his university extension work. But there was a perceptible move in his activities away from attempts to influence the social sciences in the universities, towards building on his contacts with teachers and with those interested, like the Utopia Society, on an amateur basis, with cities and social problems. With the backing of Victor Branford and A.J. Herbertson and the Summer Meetings at Edinburgh and Oxford, Geddes built up a small network of teachers who tried to use a 'regional survey' approach to education in a number of subjects. It was to encourage these people and make them feel they were part of a movement that the loosely federated Regional Survey Committee was finally formed in 1913.[77] As a propaganda machine, however, the Regional Survey Association was ineffective as it did not distinguish between the educational intention of teaching through observation and the rather pointless semi-geographical survey of the environment which brought the term 'regional survey' into disrepute.[78] Geddes himself saw no problem here as for him survey work of any kind was the essential interaction of organism and environment. This confusion, however, led a number of geographers to repudiate very strongly the idea of survey and region, and it was against this background that H.J. Fleure laboured to retrieve what he thought was important in Geddes' initiatives for the study of Human Geography.[79]

The 'Masques of Learning': propaganda for the concept of cultural evolution

Meanwhile, Geddes' interests in the last few years before the First World War had become more and more fully absorbed with his work in the town planning movement. As far as the social sciences were concerned, he left the academic world behind him and found a method of promoting his ideas which seemed much more fun to him. Inspired by the Great Pageant of Scottish History mounted in Glasgow in 1908, and building on his experiences of the Edinburgh Summer Schools since the 1890s when acting out scenes of history in costume had been very popular, Geddes came to construct and to write Masques. The vogue for this kind of activity seemed to reach its zenith in the Edwardian period.[80] But Geddes' *Masques of Learning* were unique in their scale, their social message and the enthusiasm which they engendered. A good description of one of Geddes' masques is to be found in Amelia Defries's book, *The Interpreter: Geddes, the man and his gospel.*[81] These affairs involved several hundred actors and actresses who rehearsed scenes in small groups often unaware of how their contribution would fit into the final performance. This Geddes had carefully worked out beforehand. His aim was to illustrate to players and to audience the origins of contemporary culture.

The first masque had been staged in Edinburgh to mark the twenty-fifth anniversary of the founding of University Hall. It was called the *Masque of Learning* and was performed by 650 participants as players, orchestra and choir, in the Synod Hall in Edinburgh in March 1912. So great was the success of the occasion that Geddes immediately formed an association, the Edinburgh Masquers of the Outlook Tower, to promote further masques. The original *Masque of Learning* was revised and extended into two parts, one, *The Masque of Ancient Learning* and the second, *The Masque of Mediaeval and Modern Learning.*[82] Using the device of the contents of a schoolboy's satchel, every major cultural change in civilisation was given a scene, and the conventional attitudes to education were portrayed as an amalgam of fragments, relics from former cultural traditions. Geddes loved writing these masques. Involving large numbers, putting across his educational views with unanswerable force, generating excitement, pleasure and emotion, he was in his element. After success in Edinburgh, Geddes brought his masques to London to Crosby Hall and was equally successful.

It was education with entertainment for a middle-class public. The masques themselves are a parody of what Geddes wanted to achieve in educational reform. His attempts to make geography and then sociology into new evolutionary studies requiring a fresh response at every level had not succeeded. But the masques put the message across to a wider public anxious to understand contemporary society and suffering from the fact that there was little in formal education that was relevant to helping them. For many who came across Geddes personally in these years his work was a revelation. It was at *The Masque of Learning* at Crosby Hall that Amelia Defries first came into contact with Geddes and became yet another convert and enquirer into the Gospel of Geddes. 1913 was not only the year of *The Masque of Learning* in London and the founding of the Regional Survey Association. It was also the year when the Town Planning Institute was first established and Geddes was to win the Gold Medal at Ghent for his exhibit on cities in evolution at the International Exposition on Cities and Social Progress held there. Miss Defries went to Ghent to help put up the exhibition and to learn at first hand the impact of Geddes' 'Applied Sociology' on the new world of modern town planning.

Notes

1. P. Abrams (1968) *The Origins of British Sociology: 1834–1914*, Chicago: The University Press, pp.77–100.
2. J.D.Y. Peel (1971) *Herbert Spencer: the evolution of a sociologist*, London: Heinemann, pp.15–20.
3. Abrams, op.cit., pp.107–12.
4. Much of the criticism that has been heaped on Geddes was that his premature synthesizing led to amateurishness and unclear thinking – Ruth Glass (1955) 'Urban Sociology in Great Britain: A trend report', *Current Sociology: La Sociologie Contemporaire* IV, 4: 12–14, and M. Hebbert (1980) 'Patrick Geddes Reconsidered', *Town and Country Planning* 49, pp.15–17.
5. See chapter 1, p.10.
6. J.F. Unstead (1949) 'H.J. Mackinder and the "New" Geography', *Geographical Journal* 113.
7. L. Febvre (1925) *A Geographical Introduction to History*, London: Kegan Paul, Trench, Trubner, pp.17–19, quotes A. Favre (1914) 'Les enseignements de la Guerre', *Grande Revue* September, p.439 – 'Geography comprehends all the sciences, opens all vistas, embraces all human knowledge'.
8. His contacts were particularly with the Reclus brothers, Elie, the anthropologist, and Elisée, the human geographer.
9. A. Silva White (1890) 'The Achievements of Scotsmen during the nineteenth century in the fields of geographical exploration and research' *Scottish Geographical Magazine* V.
10. The Chairman of the Council of the Royal Scottish Geographical Society denied the nationalistic interpretation of Scott's quest. He said: 'Geographers are aware that the recent expedition was not organised to make a dash for the Pole. It was based on a comprehensive plan, elaborately constructed, to obtain all available information . . . of the highest scientific value', *Scottish Geographical Magazine* XXIX: 152, (1913). W.S. Bruce wrote an appreciation of Scott in the same volume, pp.148–152.
11. An account of Bruce's career and the founding of the Scottish Oceanographical Laboratory is given by M. Newbigin (1913) 'Geography in Scotland since 1889', *Scottish Geographical Magazine* XXIX: 1913, 477–8. Bruce's reports to the Fishery Board of Scotland resulted in Scottish whalers going to the Antarctic in the early 1890s.
12. R.N. Rudnose Brown (1948) 'Scotland and Some Trends in Geography: John Murray, Patrick Geddes and Andrew Herbertson', *Geography* XXXIII: 108–9.
13. This is where A.J. Herbertson did his first work in meteorology which was an important step in his personal path towards geographical studies – E.W. Gilbert (1965) 'Andrew John Herbertson 1865–1915: an appreciation of his life and work', *Geography* L (229): 313–320.

14. Geddes used his work in his 1884 paper, *Principles of Economics*, to justify his biological approach. See chapter 3, p.59.
15. Sir John Murray (1900) 'A Bathymetrical Survey of the Fresh-water Lochs of Scotland', *Scottish Geographical Magazine* XVI (4): 193–235.
16. T.W. Freeman and Philippe Pinchemel (1977–84) *Geographers: Bibliographical Studies*, London: Mansell, vol.III, 'Archibald Geikie 1835–1924'.
17. L.J. Jay (1965) 'A.J. Herbertson: his services to school geography', *Geography* L (229): 350–61.
18. See chapter 2, pp.42.
19. See note 9.
20. E.W. Gilbert, op.cit., p.318.
21. Quoted from O.J.R. Howarth (1951–2) 'The Centenary of Section E (Geography)', *Advancement of Science* 30: 152.
22. Ibid, p.158.
23. See L.J. Jay (1965) 'The published works of A.J. Herbertson: a classified list', *Geography* L (229): 364–70.
24. See chapter 2, note 13.
25. Geddes addressed Section E of the British Association – O.J.R. Howarth, op.cit., p.161.
26. Herbertson had been offered a professorship at New York which he was about to accept when Mackinder rushed to Edinburgh to persuade him to go to Oxford – E.W. Gilbert, op.cit., p.319.
27. Ibid., pp.320–1.
28. See chapter 4.
29. Dr. Bartholomew was a brilliant cartographer who produced three outstanding atlases:
    1. *Atlas of Scotland* (1895), published by the Royal Scottish Geographical Society;
    2. *The Royal Physical Atlas* vol. I, *Atlas of Meteorology* (1899), J.G. Bartholomew and A.J. Herbertson; vol.II, *Atlas of Zoogeography* (1913),
    3. J.G. Bartholomew with Messrs. Eagle, Clarke, and Grimshaw. (C.B. Muriel Lock (1972 edn) *Geography: a reference handbook*, London: Clive Bingley, pp.59–60.)
30. Professor Flahault completed the first sheet of his vegetation map of France in 1897, and completed his survey of the distribution of tree vegetation between 1896 and 1900.
31. Robert Smith (1900) 'Botanical Survey of Scotland. I. Edinburgh District', *Scottish Geographical Magazine* XVI: 387. Smith mentions the work of three German surveys of vegetation by Drude (1890), Pax (1898), and Radde (1899).
32. See chapter 4, pp.110–13.
33. *Scottish Geographical Magazine* XVIII (1902): 218–19.
34. 'Mackinder joined the Tariff Reform League, worked for the Victoria League and in 1904 was a founder member of the Compatriots, an organisation designed to promote the wider patriotism of commonwealth' – quote from B.W. Blouet (1975) 'Sir Halford Mackinder 1861–1947: some new perspectives', *Research Paper* 13, pamphlet, School of Geography, University of Oxford.
35. J.Foreman-Peck (1983) *A History of the World Economy: international economic relations since 1850*, Totowa, New Jersey: Barnes & Noble, pp.94–155
36. W. Ashworth (1987) 4th edn, *A Short History of the International*

Economy since 1850 London: Longman, pp.192–225.

37. J.F. Unstead (1965) 'A.J. Herbertson and His Work' *Geography* L (229): pp.344–6.

38. Even after he had failed in his main plans, a School of Geography was held at the Outlook Tower in the autumn of 1904 entitled 'Introductory Course on Nature Study and Geography in Education'. It was held on Saturday mornings to attract teachers.

39. Patrick Geddes to H.J. Fleure, 7 June 1915, Geddes Papers MS10572, NLS.

40. P. Geddes (1902) 'Edinburgh and its Region, Geographic and Social' *Scottish Geographical Magazine* XVIII: 305.

41. Lecturer in Zoology in the Medical School (with J. Arthur Thomson), voluntary worker at Outlook Tower, editor of *Scottish Geographical Magazine* from 1902, lecturer in Geography at Edinburgh University.

42. M. Newbigin, op.cit., pp.473–4.

43. R.N. Rudnose Brown, op.cit, p.114.

44. T.W. Freeman (1961) *A Hundred Years of Geography* London: Gerald Duckworth, pp.86–8. B.T. Robson (1981) 'Geography and Social Science: the role of Patrick Geddes' in D.R. Stoddart (ed.) *Geography, Ideology and Social Concern*, London: Blackwell, 201–4.

45. Reprinted in the centenary volume (as a special issue devoted to A.J. Herbertson) *Geography* L (1965): 332–42.

46. W.H. Parker (1982) *Mackinder: Geography as an aid to statecraft*, Oxford: Clarendon Press.

47. *H.J. Fleure (1877–1969)* was born and brought up in Guernsey. He went to the University College of Wales at Aberystwyth for health reasons and because he won a scholarship there. He read zoology and geology as an undergraduate and undertook research in marine biology as a postgraduate, after a short time at Zurich University. He was appointed lecturer in zoology, geology and botany at Aberystwyth in 1904. He began teaching geography in 1907 and undertook an anthropological survey of the Welsh people. He gained the newly-endowed Chair of Geography at the college in 1917. In 1930 he gained the Chair of Geography at Manchester University – Alice Garnett (1970) 'Herbert John Fleure 1877–1969' *Biographical Memoirs of Fellows of the Royal Society*, 16: 253–9. *C.B. Fawcett (1883–1952)* studied geography at Oxford with Herbertson and Mackinder, and was very much influenced by the latter. He saw geography as an aid to the reform of government especially local government. He published *Provinces of England* in 'The Making of the Future' series in 1919.

48. McGegan published one major article on Geddes with Arthur Geddes and Frank Mears: 'The Life and Work of Sir Patrick Geddes', *Journal of the Town Planning Institute* 26 (1940): 189–95, and wrote of his memories of Patrick Geddes and the Outlook Tower at greater length in an unpublished manuscript of uneven quality in the Geddes Papers, NLS.

49. See chapter 1, fn.6, p.14.

50. The Indian nationalist poet, Rabindranath Tagore, wrote to Geddes that he had read Branford's book 'with a great deal of sympathy and interest. It is a highly suggestive book, very helpful for me as I am in full agreement with the idealism it represents' – 2 January 1924, Geddes Papers MS10576, NLS.

51. J. Arthur Thomson had been acting as Geddes' literary agent since the

early days of the Encyclopaedia articles. Thomson also helped in the publishing affairs of Patrick Geddes, Colleagues and Co. (usually bitterly complaining about inefficiency – J. Arthur Thomson to Patrick Geddes, 28 October 1896, Geddes Papers MS10555, NLS). J. Arthur Thomson was also one of the three editors of the Home University Library of Modern Knowledge published by Williams and Norgate. The other editors were Herbert Fisher and Gilbert Murray. Branford helped Patrick Geddes to amalgamate his 'Evolution' series with Bliss, Sands and Co.'s 'Progressive Science Series' – V.V. Branford to Patrick Geddes, 16 September 1897, Geddes Papers MS10556, NLS.

52.   Correspondence between V.V. Branford and Patrick Geddes, Geddes Papers MS10556, NLS.

53.   Patrick Geddes to V.V Branford, 19 January 1902, Geddes Papers MS10556, NLS.

54.   V.V. Branford to Patrick Geddes, 21 January 1902, Geddes Papers MS10556, NLS.

55.   R.J. Halliday (1968) 'The Sociological Movement, the Sociological Society and the Genesis of Academic Sociology in Britain', *Sociological Review*, 16: 377–398.

56.   G.R. Searle (1971) *The Quest for National Efficiency*, Oxford: Basil Blackwell, pp.107–41.

57.   Galton's eugenic thought was a celebration of the professional elite which he so admired – D.A. Mackenzie (1981) *Statistics in Britain 1865–1930: the Social Construction of Scientific Knowledge*, Edinburgh: Edinburgh University Press, p.51.

58.   B.J. Norton (1978) 'Karl Pearson and Statistics: the social origins of scientific innovation', *Social Studies of Science* 8: 3–34

59.   B.J. Norton (1973) 'The Biometric Defence of Darwinism' *Journal of the History of Biology* 6: 283–316.

60.   Karl Pearson (1905) in the discussion on Galton's 'Eugenics' paper *Sociological Papers* I, London: Macmillan, p.52.

61.   The Charity Organisation Society had already set up a School of Sociology for this purpose and Geddes was invited to lecture there – C.L. Mowat (1961) *The Charity Organisation Society, 1869–1913: its ideas and work*, London: Methuen, pp.112–113.

62.   Even he got around to admitting this eventually when he published his volume on *Evolution* with J. Arthur Thomson for the Home University Library in 1911, p.232.

63.   The first Congress offered eugenics, civics, and the training of social workers in the new 'sociology' of action to solve social problems. As such it attracted attention in the press and from politicians and social reformers.

64.   Geddes' first two papers 'Civics: as applied Sociology' Part I and Part II, are reprinted with an introduction in H.E. Meller (ed.) (1979) *The Ideal City*, Leicester: Leicester University Press, pp.75–183.

65.   H.E. Meller (ed.) ibid., pp.78–9.

66.   Geddes' pioneering attempts at socio-biology have gone unnoticed in the somewhat chequered history of socio-biology – see E.O. Wilson (1976) *Sociobiology: The New Synthesis*, Cambridge Mass.: Harvard University Press.

67.   P. Geddes (1907) 'A Suggested Plan for a Civic Museum (or Civic Exhibition) and its associated studies', *Sociological Papers* III, London: Macmillan, pp. 197–230.

68. *Sociological Papers*, ibid., III, p.295.
69. Especially in his work as Director of the London School of Economics — B.W. Blouet op.cit., pp.26—8.
70. Extramural lecturing brought Geddes much needed income and kept him financially afloat before 1910 and the formation of the Cities and Town Planning Exhibition.
71. See chapter 2, pp.18. See also Geddes (1913) 'Mythology and Life: an interpretation of Olympus', *Sociological Review* VI.
72. For a brief comparison of the different approaches of these two to sociology see J. Owen (1979) *L.T. Hobhouse: sociologist*, London: Nelson, pp.184—6.
73. Branford had tried to get White to finance a national Sociological Society based on the Outlook Tower before setting up the British Sociological Society in London — Halliday, op.cit., p.382.
74. P. Mairet (1957) *A Pioneer of Sociology: life and letters of Patrick Geddes*, London: Lund Humphries, p. 136.
75. Ibid., p.137.
76. D. Hollins (ed.) (1908) *Utopian Papers: being addresses to 'the Utopians'*, London: Masters, preface p.vii.
77. The Regional Survey Association was formed after the Easter vacation course on regional survey at the Outlook Tower in 1913, organised by Mabel Barker.
78. The International Visits Association of the Sociological Society (later transmuted into the Le Play House Educational Tours), undertook 'regional survey' activities which were little more than a method of educating participants — S.H.Beaver (1962) 'The Le Play Society and Field Work', *Geography* XLVII: 232—7.
79. A. Garnett, op.cit., pp.261—4.
80. Amateur masques were much performed at country house parties. Geddes took an upper class fashion and made it a tool for promoting his ideas amongst the widest possible audience.
81. A. Defries (1927) *The Interpreter: Geddes, the Man and his Gospel*, London: Routledge, pp.41—56.
82. *The Masque of Ancient Learning and its many meanings: a pageant of Education from Primitive to Celtic times devised and interpreted by Patrick Geddes*, Outlook Tower, Edinburgh: Patrick Geddes and Colleagues, (1913); *A Masque of Mediaeval and Modern Learning: A Pageant of Education devised and interpreted by Patrick Geddes*, Outlook Tower, Edinburgh: Patrick Geddes and Colleagues, (1913).

CHAPTER 6

The sociologist
of the town planning movement

THE PROBLEM OF ASSESSING GEDDES' IMPACT ON THE
nascent town planning movement is not at all straightforward.
Historically he remained outside the mainstream of the movement in
comparison with, for example, Thomas Adams and Raymond Unwin.[1]
Yet, subsequently, he has been claimed as the founding father of
planning methodology, though even here the extent of his influence
has not been made at all clear. In fact, Geddes attached himself to the
movement at a late stage in his career when his theories, objectives,
and methodology were already complete. Town planners turned to
him because he seemed to have some ready answers to vital questions.
What many of the pioneer practitioners of planning found was that
their training, often in design-related skills such as architecture, was
inadequate to help them with many problems that they faced. This
seemed less serious in the early days when town extension schemes
could draw on the ideas generated by the work in Port Sunlight,
Bournville, and New Earswick. But by the time of the 1909 Act,
although it was only devoted to suburban extension schemes,
municipal councillors in large cities had to be converted to the idea of
planning the urban environment. If planning was a serious commitment then planning practice needed to grow to encompass the whole
town and city, perhaps the city and region.[2] Thus leaders of the town-
planning movement, like Raymond Unwin and Patrick Abercrombie,

[156]

came to appreciate Geddes' perception of the city and region as an integrated whole. He became the guide and adviser to the small group of planners who were responsible for establishing a professional response to the theory and practice of planning.[3]

Geddes' influence was vital in sustaining their confidence in a critical stage in the professionalisation of their role, between 1909 and 1925. As Abercrombie wrote in 1927:

It is perhaps safe to say that the modern practice of town-planning in this country would have been a much simpler thing if it had not been for Geddes. There was a time when it seemed only necessary to shake up into a bottle the German town-extension plan, the Parisian Boulevard and Vista, and the English Garden Village, to produce a mechanical mixture which might be applied indiscriminately and beneficiently to every town in this country; thus would it be 'town-planned' according to the most up-to-date notions. Pleasing dream! First shattered by Geddes, emerging from his Outlook Tower in the frozen north, to produce that nightmare of complexity, the Edinburgh Room at the great Town Planning Exhibition of 1910.

He then significantly added:

Geddes' influence will never be known to the world at large — he works by his disciples — his teaching is of such sort that it does not get watered down in transmission: it is a sort of vital idea — a divine inoculation that goes on spreading its infusion without exhausting its original *élan*.[4]

Amongst this new and enthusiastic group of environmentalists, Geddes was convinced he had found the 'bud' hunters he had hoped to train at the Outlook Tower, or at the abortive National Geographical Institute.[5] This time potential disciples shared with Geddes, not just a concern for social reform and social peace, but a much greater practical concern for the physical condition of the city. Geddes' work in Edinburgh had given him experience of what was becoming a key planning problem: the need to renovate and rejuvenate historic city centres. His survey of Edinburgh and its region gave hope of an integrated approach to all aspects of planning. Furthermore, Geddes' particular expertise at organising conferences and exhibitions was especially valuable to the British town planners.[6] The exchange of ideas on planning on an international basis had been carried forward in this way yet, in the absence of direct municipal support, the collection of materials for this work had to be done on a voluntary basis. The material from the Outlook Tower was by far the largest collection in private hands in Britain.

The most significant departure of the modern town-planning

movement, which differentiated it from all the previous efforts at ordering the urban environment, was that it encompassed the possibility of redistributing resources for the benefit of the community as a whole. The quest for the 'good environment' created priorities which cut across the sacrosanct rights of property and the individual, and could thus only be enforced by elected bodies directed by legislation.[7] Such a revolutionary prospect was barely recognised even by planning practitioners themselves, and was certainly not politically feasible in Geddes' lifetime. In the quarter century since the 1875 Public Health Act and Cross Act, which had given sanitary authorities the right to put compulsory purchase orders on groups of buildings, the issue of redistribution had been fudged. Areas of slum buildings had been cleared but at the same time the poor had not benefited. But as the 'housing problem' remained unsolved there was mounting pressure on the government, and by the turn of the century the trade unions had a political programme which included housing.

It was, however, a big step from the Cross Act to the concept of municipally financed housing, and political objections to such a redistribution of resources were strong. The issue was debated at both local and national level and local authorities such as Birmingham, led by J.S. Nettlefold, chairman of the housing committee, fought hard to sustain the idea that housing was best regulated by the market. Nettlefold had evolved an ideal plan in which the role of the local authority was merely to regulate the market in such a way as to benefit landlords, tenants, and former slum dwellers together.[8]

But by the turn of the century the 'housing question' and concern over the quality of the urban environment had begun once again to get a wider press. The fear generated during the course of the Boer War about the physical deterioration of the urban masses of Britain, since so many of the volunteers for active service were rejected as unfit, created a demand for new measures. The campaign to promote national efficiency permeated social and political thinking at all levels. The ever-expanding scale of Britain's large cities in the 1890s and 1900s made the task seem daunting. Eighty per cent of Britain's population now lived in cities, and cities of less than 100,000 population had become insignificant. The leading provincial urban centres had become as large as London had been at the beginning of the nineteenth century.[9]

As numbers increased so did boundary extensions, and large cities appeared to be moving beyond any kind of personal influence of individuals. It was becoming obvious that no small elite, however rich, well organised and well intentioned, could dominate the economic, social, and political environment of major provincial cities.[10] A great wave of anti-urban feeling swept through those very families whose members had done so much over the previous quarter of a century to promote an ideal of civilisation amongst Britain's newly-urbanised masses.[11] The Chamberlains, Cadburys, and other Edgbaston dwellers moved beyond the city to country houses, as did the Wills and the Frys in Bristol, and many others elsewhere. In some ways the anti-urban feeling was a gesture of defeat. Voluntary workers trying to 'humanise' the industrial urban environment had been particularly creative over the past quarter century. Now this creativity seemed to disappear. The missions to the poor, undertaken mostly by socio-religious reformers, had helped to initiate a wide range of institutions and activities, from libraries, art galleries, and temperance coffee taverns, to working men's clubs, adult schools, the organised youth movement, the Kyrle Society, the People's Palace and many other such institutions. A belief in the concept of community in a defined place brought together by personal contact had underpinned this work. It was much harder to sustain this belief in the altered circumstances of the turn of the century.[12] Not only the scale of cities made the personal contact impossible: the poor and underprivileged were gaining a greater voice and the working-class sense of priorities did not coincide with that of the social reformers. The urban poor were no inert mass. The context of cultural evolution had been broadened by politics, propaganda and the widening range of options now available in a mass market. Social reformers trying to 'civilise' the city found themselves on the side lines of urban culture.

Yet those supporting the town-planning movement before the First World War were not so undermined by this. They had every confidence in the idea that it was the environment which shaped urban culture. Therefore they had within their power the chance to direct future social and cultural development.[13] The great planning pioneers of the twentieth century were all convinced of this regardless of the fact that they undertook little or no consultation with the people whose lives would be affected by their work.[14] While planning propagandists were reluctant to admit that their objectives were 'social engineering', the followers of Unwin and Parker, and others, had no doubts that their view of community development was 'right'. There was thus a certain ambivalence about planning activities and social objectives. The pioneers of the town-planning movement wanted to liberate citizens from the degenerating influences of the city, yet the freedom they offered was circumscribed by their own ideals of the 'good life' and the pursuit of liberal culture. They wanted legislation to sanction their work, yet in many respects the town-planning movement was a collection of voluntary organisations requiring personal commitment from their members. Aims and methods thus to some extent became incompatible, creating an unresolved tension between the planner and those for whom he planned, especially if they did not share his viewpoint.

In general, for all his anarchistic views and determinedly anti-centrist tendencies, Geddes was not able to resolve this dilemma. He drew most of his support, not from the people at large, but from those formed by the tradition of social reform in cities.[15] His emphasis on the importance of the voluntary social worker for undertaking the kind of sociological 'civic' survey which he believed promoted cultural evolution became an inspiration at the point when the work of the 'civic' social reformers suddenly seemed out of date.[16] What Geddes promised his small band of supporters was redemption in the big city if they were guided by his Le Playist approach. It was a profoundly conservative message for an activity, modern town planning, which itself was both new and progressive. Like Octavia Hill, who invented an idyllic rural past to justify her great innovations in the management of her urban properties and tenants, so Geddes' insistence on an idealised evolutionary historicism was the justification for large-scale environmental planning.

Geddes was to find much of his most committed support amongst the more intellectual wing of the social reform movement, from the men and women of the University Settlements (in Britain and America), from a small group of teachers dedicated to educational reform and teaching through the use of the survey method, and a small number of architect practitioners in the town-planning movement who found Geddes an inspiration and a guide. Geddes' relationship with the town-planning movement as a whole was more mixed. He irritated many with his 'Eutopian' dreams and he was never to succeed in gaining the strong support of two key figures in the movement, T.C. Horsfall and J.S. Nettlefold. The former was perhaps cast in too similar a mould to Geddes and would brook no competition to his own propaganda;[17] the latter, the housing pioneer and industrialist from Birmingham, was too involved in the politics of his own solution to the housing problem.[18] In later years, Geddes was never to take the Cities and Town-Planning Exhibition to Manchester or Birmingham. However, long before this, in 1903, Geddes was given a chance to do some practical planning which marked the initiation of his influence on the movement as a whole. He was commissioned, in competition with T.H. Mawson, the landscape architect, to advise the Carnegie Dunfermline Trust on how to lay out Pittencrief Park in Dunfermline for the benefit of all citizens.

Geddes' first commission for city development: the Dunfermline Report

In the years 1903–4, the Scottish American millionaire, Andrew Carnegie, had a particularly lavish burst of philanthropic activity. Already widely known for his support of the free library system in this country and for supporting the Scottish universities, he decided at this time to fund in the United States a scientific institute to study natural evolution and, in his native Scotland, to give a huge bequest of $2,500,000 to the city of his birth, Dunfermline.[19] Perhaps the latter bequest was to give the city a chance to achieve a different level of evolutionary development in the future. Since the size of the population of the city was only 25,000, this was munificence indeed, and the local council was exercised as to how best to spend the money. It was stipulated that the money should be invested to provide an annual income of at least £25,000. It was decided that the house and land of a local large landowner should be bought for the townspeople with some of the money and the land should be laid out as a park. The Carnegie Dunfermline Trust, set up to administer the fund and income, commissioned the landscape gardener, T.H. Mawson, for a plan for the park. With urgings from Henry Beveridge[20] and others, they also decided to ask Geddes as botanist, evolutionist, and Scotsman, to complete a plan to give themselves the widest number of options to choose from, especially those which might be in the forefront of current thinking on social reform.

Geddes was fifty years old in the year that he published his report, and his initiation into environmental planning in Dunfermline was thus not so much a new departure as a summation of his work over the previous quarter century. His brief was of the kind which exactly suited him. He was asked to advise on the laying out of Pittencrief Park and on suitable educational and cultural institutions to develop the social life of the town. As he wrote in the summer of 1904 to Mawson when both were commiserating with each other after being rejected (and Mawson was considering legal action to force the payment of what he thought was adequate compensation for his work):

> You were consulted more especially as a landscape-gardener and architect, and largely on account of your larger practice and wider experience in these matters than mine: whilst I largely owed my invitation to my having added a good deal of social and educational organising experience to my more limited practice as gardener.[21]

Yet his approach was not altogether built only on his Edinburgh experience. In the two trips he made to America in February–April 1899

[161]

and December 1899–March 1900 to raise money for the plans for the Paris Exhibition of 1900, Geddes had not only been fundraising and lecturing, he had also taken a keen interest in social reform techniques used in cities. He had brought himself up-to-date with current American practice where, in the absence of a 'social reform' element in the City Beautiful movement, parks were being created and run with reforming zeal, very much, as one historian has commented 'as if they were municipally run settlement houses'.[22] Geddes' plan for Dunfermline was a pioneering one in this genre of the 'reform' park, though he still wanted to have a foot in both camps — the park as pleasure ground, and the park as a means for promoting 'social reform'. The problem was that in Britain, municipal parks committees had neither the resources, imagination, nor political will to mount a campaign for 'reform' parks. Thus many of Geddes' most ingenious ideas for promoting activities and institutions in Pittencrief Park were not imitated or taken up elsewhere.

At the same time, the message and method of the work struck a deep chord amongst British social reformers and those concerned with the social and cultural problems of cities. Here Geddes was proclaiming with a passion reminiscent of Matthew Arnold's message in *Culture and Anarchy* (1859) that a torch of culture could be lit to lighten the darkness of provincial urban life. If a little town like Dunfermline could be made to produce a modern, progressive environment, alive, vigorous and independent, then there was hope for all other small provincial cities of Britain. Geddes was staking out his first claim in the town-planning movement at large in an area virtually untouched by others. He had found a lacuna between, on the one hand, the Garden City ideal of Howard which required totally new departures from existing settlements;[23] and on the other, the public health and housing response to town planning which was being promoted in the larger industrial cities.[24] His message was for the smaller provincial city.

It is worth quoting Geddes' enthusiastic denial that life in Dunfermline in the twentieth century need be second rate:

Here, as in all true progress, we must not only comprehend and transform the environment without but develop our life within. Our inevitable and permanent provincialism must be accepted as one of the facts of life. Dunfermline will and may enlarge and develop, but it cannot become Glasgow or Edinburgh. What is the vital element which must complement our provincialism? In a single word, it is Regionalism — an idea and movement which is already producing in other countries great and valuable effects. It begins by recognising that while centralisation to the great capitals was inevitable, and is in some measure permanent, this is no longer so completely necessary as when they practically alone possessed a monopoly of the resources of justice and of administration, a practical monopoly also of the resources of culture in almost all its higher forms. The increasing complexity of human affairs, with railway, telegraph, and business organisation, has enabled the great centres to increase and retain their control; yet their continued advance is also rendering de-centralisation, with local government of all kinds, increasingly possible. Similarly for culture institutions: the development of the local press has long been in progress; the history of the city library movement is in no small measure identified with that of this very town; while the adequate institution among us of other forms of higher culture is just what has been discussed in the preceding pages. We see, then, that the small city is thus in some measure escaping from the exclusive intellectual domination of the greater ones, and is tending to redevelop, not, indeed, independence, but culture individuality.[25]

Geddes had found his voice as the champion of a newly-inspired and potentially vigorous provincial urban life in twentieth century Britain.

The methods he used to bring this vision into reality in Dunfermline had three elements. First he tried to use the survey techniques pioneered by Professor Flahault; second, he exercised to the full his fine talent for observation; and third, he built on the ideas and institutions he had developed in Edinburgh. Geddes himself did not actually carry out a survey of Dunfermline. But he did persuade his former assistant and friend, Dr Marcel Hardy, to come to Dunfermline to undertake a full geographical and botanical survey of the park, the city, and the region. Hardy, using Flahault's technique, duly produced the basic data for nurturing the natural environment of Dunfermline. Geddes turned his own attention to the city and produced a rather idealised view of Dunfermline's historic past as a religious and administrative centre. In these ways he established the 'genus' to which Dunfermline belonged and with which he, of course, felt particular sympathy.[26] Geddes then had two problems: first, to preserve the town's historical fabric; second, to suggest what lines of development were possible within the constraints laid down by the circumstances of history, geography, and present conditions.

To carry out his practical work he called on the help of three experts: Mr Aitken, a draughtsman/architect, to produce drawings to scale; Mr Norval, a photographer, to produce photographs which could be touched up to provide 'before' and 'after' photographs; and John Duncan, his friend and artistic collaborator at the Outlook Tower in the 1890s, now in Chicago, who nevertheless designed the monuments for city and park which Geddes favoured as a symbolic means of conveying his ideas. Members of the Trust Committee, when confronted with the bill for paying for all this help, reacted with consternation as Geddes had not sought their prior approval.[27] In fact Geddes himself had no contract with the Trust. He had been introduced informally to Dr Ross by Beveridge in the summer of 1903. Ross, at the interview, asked Geddes to prepare a report for the park. It was agreed that Geddes would be easily able to produce such a report in a couple of months. In the event, it took Geddes eight months. He wildly exceeded his brief, as he was inspired by Carnegie's gift to think in terms of every possible cultural, recreational, and educational activity which might be introduced into Dunfermline, and only gradually limiting the scope of his response as he began to deal with specific locations in city and park. That he completed it in eight months was an extraordinary achievement.

Geddes sent letters to over 200 correspondents on an international basis, both seeking their ideas and publicising his own.[28] He, as well as his rival Mawson, was most anxious to use this opportunity to promote a new career for himself as a civic designer as well as a social and educational adviser. He was also well aware of Carnegie's interest in promoting favourable evolutionary trends. Since then there had been much publicity about Carnegie's latest gift to found a scientific institute in Washington to encourage 'the improvement of mankind'. The first three departments, set up in December 1903, were Experimental Evolution, Marine Biology and Historical Research.[29] If this was being done in Washington, why not in Dunfermline too? Geddes tried to put all his ideas together for the first time in a more than usually coherent manner. His international correspondence and the report itself were written as if they were all propaganda tracts. It was as if his first task was to take his critics and any sceptics with him. He also tried to involve the trustees in an evolutionary way with his work as it progressed (and possibly to pre-empt Mawson) by sending the report in sections in an unfinished state over a period of time. It was a disastrous mistake which totally alienated the Trust's chairman, Dr Ross, and many of the other members.[30] By the time Geddes had exceeded the time limit for his report by six months, the trustees had already begun work on the Park without it. In the end they proceeded without any plan as Mawson's report, too, seemed unacceptable and Mr Carnegie, on a visit to Dunfermline in May 1904, suggested that the Park should be left as it was![31]

Mawson's and Geddes' reports taken together, provide an astonishing contrast in their approaches to their brief. Carnegie's instructions to the Trust were that his money was

all to be used in attempts to bring into the monotonous lives of the toiling masses of Dunfermline more of sweetness and light; to give to them — especially the young — some charm, some happiness, some elevating conditions of life which residence elsewhere would have denied.[32]

It was enough to confirm Geddes that his unconventional approach might just be suitable for this occasion. Mawson on the other hand, had eschewed the rhetoric and provided what he thought was an up-to-date professional job. Mawson had made his living as a landscape gardener, and he profited by the expansion of the public park movement which called for men with his expertise.[33] His first such commission had been to plan the public park in Hanley, one of the Staffordshire five towns. He badly wanted to break into the market for urban design which he was shrewd enough to have spotted developing in the wake of the Garden City movement and planned urban estates generally.

He had never before had the opportunity of working on a long-term plan which was fully funded with an assured annual income. He was lifted by the prospect to envisage the creation over the years of a British 'City Beautiful'[34] which, while starting at the park, would gradually encompass the whole town, particularly any new suburban growth. With this prospect in mind, he suggested that the old and rather dilapidated houses on the property now owned by the Trust should be flattened (which included, unfortunately for Mawson the home of Mr Carnegie's Scottish factotum, the chairman of the Trust, Dr Ross).[35] The cleared space could then be used for creating a grand redevelopment scheme in the Beaux-Arts style. Mawson believed that the future expansion of the town had to be westward on the other side of the park, and he proposed a wide new boulevard running through one corner of the park to connect the old town with the new. Mawson hoped that his plan would be impressive enough not only to win him the contract at Dunfermline, but also to build up his professional reputation. He, like Geddes, was to suggest afterwards that his Dunfermline Report was the best work he did in Britain and he felt he owed his later commissions to plan towns in Canada and Greece to the publicity the report gave to his skills.[36] It brought him an indirect reward straight away in that he was to meet Mr Carnegie who promptly engaged him to work on the gardens and grounds of his Scottish castle home at Skibo.

Geddes' response on the other hand was on an altogether different plane. He knew that no ornate and expensive plan was ever going to make Dunfermline less of a provincial backwater. But he had ambitions to make it the pioneer of a new social movement. As he wrote to Dr Adler in the USA:

Again, while I have absolutely no official right to say this in any way, it does not seem wildly improbable that as the great Carnegie Library did originate in this city, so another larger movement may be beginning here, of which the experience and suggestiveness might before long be applicable in America — through others if not through Mr. Carnegie.[37]

Geddes hoped for future commissions but only to further his kind of social movement. When writing his report he made no concessions towards adopting a conventional approach. He began his report, as he was to do in his later Indian reports, in an almost ritual manner, as if he was a visitor to the city arriving at the railway station and moving on from there to explore the city street by street, district by district, so that his impressions of the town appear to develop in an evolutionary manner.[38] His use of a photographer, though a technique used by the Geological Survey since 1889,[39] was novel, and again emphasised his evolutionary approach. The result was to make his report quite striking in its visual impact. Apart from his cultural mission, his report had two major practical objectives: to save the old houses (threatened by Mawson) and the old mansion in the park, now all belonging to the Trust; and second, to purify the stream running through the park. On the first matter, he had a long battle with Dr Ross whom Geddes told one of his correspondents: 'hopes to live to see every stone of these old buildings carried away',[40] although not his own house, of course. Geddes made some impact with his insistence on Dunfermline gaining a list of its historic buildings. He was also successful in getting some work done on reducing the pollution in the stream.

Geddes' ambitions for Dunfermline were, like Mawson's, bold. Whilst he was not indifferent to the current fashion for a Beaux-Arts style 'City Beautiful' for new developments, in the old city he was more interested in a gentle, 'gardening' approach to clean and restore the ancient fabric. Then suitable places could be found to place statues and monuments to symbolise the cultural heritage of the past. This method would create a sense of Dunfermline's unique history and its potential as a cultural centre for the present and future.[41] As for his educational and social plans, they were meant to encompass men, women, and children, and he brought to his work his ideas on cultural transmission and on the importance of nature in the process of creative evolution. Thus, for the children, Geddes wanted to make the park into an adventure playground which might provide the means for refining instincts into intuition, the source of 'true' knowledge.[42] For adults there were a range of cultural institutions, some devoted to the liberal arts, drama, music, painting, some to special museums which aimed to transmit past qualities of Dunfermline citizens to the present

[166]

generation. Chief amongst the latter were craft museums devoted to the traditional skills of the region, an idea which he took directly from Ruskin.[43] For young adults there was, of course, an Outlook Tower and halls of residence for communal living to encourage them to devote their energies on a voluntary basis to promoting local and world citizenship.

He sent the outline of these plans to T.R. Marr for comment. Marr, since acquiring the job of warden of the Ancoats Settlement in Manchester in 1901, had been engaged upon a Geddes inspired survey of housing in Manchester. This, and his two years' experience of an industrial city, made him cautious in responding to Geddes' plans for reforming the social life of Dunfermline. He wrote:

> some years ago in Edinburgh I could have been dogmatic, but time has persuaded me that one can only build a useful plan for social work after careful and lengthy study of the problem at first hand. We have not yet (I refer to those of us who are engaged in practical social reform) a sufficiently large body of experience to generalise from, though we constantly forget this and generalise for all we are worth and feel amazed when what ought to be does not square with what is . . . I have felt constrained to say little or nothing of the *political* problems of our social work but I am profoundly convinced of their importance.[44]

Geddes, nevertheless, persuaded Marr to write in the section on cities and civic problems on the subject of 'Municipalities at Work'. Marr's list of the activities of local government though, was hardly path-breaking. His analysis did not go beyond that produced by Canon Barnett in 1883 in his article 'Town Councils and Social Reform'.[45] Yet as far as Geddes was concerned this was a matter of small importance. He was not particularly interested in any of the conventional or established methods of social reform. He was obsessed with what he called culture-development. As he wrote himself on the civic problem:

> Beyond and above discussion of traditional or current political problems, Liberal, Imperial, and Financial, is urgency of culture-problems and culture-policy; since expansion or development of institutions, of empire, and of wealth, are all dependent upon development of civilisation and race. In short, quantity depends upon quality, not conversely.[46]

The vitality of Geddes' 'Dunfermline Report' and its originality offset its failures of timing, presentation, and some of the more outlandish ideas it contained.[47] He believed it was the best thing he had ever written, and he only set out consciously to try to supersede it in his Indore Report in 1918. His friends and contacts ensured that it was widely reviewed, though the popular press tended to misunderstand the work. Only the *Morning Leader* managed the perspicacious remark of calling Geddes 'a revolutionary conservative'.[48] Geddes' insistence on his cultural mission to the people of Dunfermline however, alienated many who might otherwise have been sympathetic. T.C. Horsfall, for instance, despite his record of being a campaigner for bringing Art and Beauty to the city since the early 1880s, refused even to review it.[49] Ebenezer Howard on the other hand was highly appreciative though there may have been a barb in his comment that a copy of the report should be in every public library.[50] He presumably thought it was inspirational rather than practical; a criticism that Geddes was to encounter in all his subsequent planning work.

In the case of the 'Dunfermline Report' such a criticism was, perhaps, least justified. Most of his recommendations for renovating the old city were practical and cheap. It was his flights of fancy about the future which made the report appear unrealisable. His response to specific problems in specific places was firmly controlled by his skilled observation. By personally directing his helpers, particularly the photographer, he gained an intimate knowledge of the place which gave weight to his judgement. These insights were lost, though, overlaid by the propaganda for culture institutes of every kind which would have been found, if Geddes had had his way, on most of the available open space.[51] The cultural feast that was offered proved an indigestible menu for many even amongst his most ardent supporters. Lady Welby, who had offered Geddes the peace and quiet of her home in Harrow in the early months of 1904 so that he could complete the report whilst still honouring his lecturing commitments in London from January to March, found his terminology irritating. As she wrote to him:

> You know I am the last person in the world to complain of enriched terminology! But the more keenly one feels the necessity of this the more one winces at any approach to what I would like to call 'culture-jargon'. The word itself has become a sort of shibboleth. Culture institute, culture policy, culture resources and so on, make one long to go out and be just cow parsley under the hedge.[52]

In fact the Carnegie Dunfermline Trust put into operation a more modest plan for realising a new cultural environment for the town. The park was improved with attractive walks, new trees, the introduction of bird and animal life, and the provision of an open-air bandstand and tearoom. The town gained a School of Music and money was given to support local music societies and to subsidise concerts. In every district, institutes containing libraries, reading rooms and facilities for indoor games were established. The Carnegie Public Library was greatly extended. Other parks and open spaces for organised sport were acquired. A magnificent building containing a swimming-bath, as well as baths of every other description, was erected, and the Trust became involved in health education by founding a College of Hygiene and Physical Culture. Further educational work included the founding of a craft school, the encouragement of school excursions, and the provision of bursaries for educational purposes. A lady gardener was employed to teach the school children about horticulture. Prize shows for produce were organised. Poor children in weak health were given the chance of a spell in a convalescent home. Mr Carnegie was so delighted with these results that he supplemented his original gift with a further £250,000 in 1911.[53]

The schemes hardly matched Geddes' ambitious attempts to transform the cultural environment of Dunfermline. But, then, his projects would have required an understanding of his revolutionary ideas on the nature of knowledge and the adoption of totally new educational methods. From an evolutionist, this was, perhaps, rather a tall order. Geddes was unrepentant because his 'culture-jargon' was the means he used to try and persuade people of the vital connection between social processes and spatial form. Most of the support he had received over the previous twenty years in Edinburgh had come from people who, for a variety of reasons, had had some concern over the nature of the social environment of the city. But Geddes was alone in uniting the current ideas on the beneficial influence of nature as displayed, for example, in Howard's Garden City idea, with that of a park system in a well-ordered historic city. What he was aiming at, as he was to outline later in *Cities in Evolution* and elsewhere, was to do for the small provincial city what Sir Horace Plunkett's Irish agricultural movement had done to modernise and yet preserve the milieu of the Irish peasantry. Plunkett's watchwords: 'better farming, better business, better living' were matched by Geddes' appeal for: 'better housing, better living, better business'.[54]

Fallow years as practitioner and prophet of 'civics as applied sociology'

Geddes' desire to found a new voluntary movement, however, was to place him in an unfavourable position to influence the developing town-planning movement in the crucial years between 1904 and 1909, since the planners, for all their woolly objectives, were not interested in an old-fashioned social reform movement.[55] The most influential voice in 1904 was not Geddes but T.C. Horsfall. He had started in the 1880s like Geddes, inspired by Ruskin to bring Art and Beauty to the people of Ancoats, the slums of Manchester.[56] But by 1897, when Horsfall went on a study tour of Germany, his growing belief that the Germans had managed their urban growth a great deal better than the British, was confirmed. He began propaganda work to publicise this idea and it was Horsfall, not Geddes, who was asked to give evidence about urban problems and their solutions to the 1904 Committee on Physical Deterioration. By 1904 Geddes' disciple, T.R. Marr, warden of the University Settlement in Ancoats had produced his first report on housing conditions in the area. Horsfall then wrote a companion volume to this entitled *The Improvement of the Dwellings and Surroundings of the People: the example of Germany*. Its main thesis was to prove the superiority of German town-extension schemes. Horsfall drew on examples showing how local authorities had been empowered to purchase land on the outskirts of the towns and to lay them out in such a way that residential and industrial areas were separated. The initiative of the local authorities had created the right kind of environment for private investment in buildings which were of good quality, thus ensuring that the new suburbs would not exhibit all the bad features of the older, unplanned areas. Horsfall's book became the talisman for the campaign for town-planning legislation in Britain before the First World War.[57]

Geddes was left on the side-lines both as practitioner and prophet. As a practitioner of planning, there were few commissions. He sought to extend his experience in landscape gardening, not in Britain, but in the South of France where he had a number of contacts.[58] In this respect, Geddes was to place himself alongside his erstwhile rival, T.H. Mawson, in developing the landscape gardening skills which were to be the basis of their approach to urban design.[59] Mawson, however, had managed to acquire another influential personal client, William Lever, who wanted him to create the gardens for his new house, Hill House, in Hampstead. Mawson was also to do work for him on his other house, Roynton Cottage, Rivington, and through these personal contacts Mawson found himself well placed to benefit from his client's faith in his competence. When Lever funded the chair and the Department of Civic Design at the University of Liverpool at the end of 1908, Mawson was asked to give lectures in landscape design. Besides his practical work, he had also managed to publish a book on *The Art and Craft of Garden Making*. From here, he developed lectures on landscape architecture and civic art which he gave, not only in Liverpool, but also on a lecture tour in the USA in 1910. Subsequently he published his second book, *Civic Art: Studies in Town Planning, Parks, Boulevards and Open Spaces* which was to establish him firmly in the new field. Mawson, through prodigious hard work, was to gain for himself a niche in the academic development of town planning;

commissions to plan towns after the First World War abroad; and what he considered to be his greatest personal achievement, the role of founding father of the newly-self-conscious profession of landscape architects.[60]

Geddes, meanwhile, had been to some extent side-tracked from activities within the town-planning movement by his own and his disciples' attempt to sustain him in his role as instigator of a new civic movement. Branford lost the fight to keep Geddes as the central figure of the British Sociological Society, but he did not give up. He maintained contacts with Paul Otlet's International Bibliographical Institute in Brussels, and tried to arrange joint ventures with the Outlook Tower. He managed to get the London headquarters of the Sociological Society at 24, Buckingham Street, recognised as a University Extension Centre with some small financial support from the University of London.[61] He was continually optimistic that Geddes' educational methods and his evolutionary approach to sociology were going to become established within the mainstream of British academic life. By January 1907 he is suggesting that

> Is it not time we dropped the attempt to father on Le Play, a scheme and method which is widely different from anything employed in the research of the Le Play School? Instead we should call it the Le Play/Geddes method.[62]

He constantly pressed Geddes to write a review of what he calls the civic movement from his particular viewpoint, and when Geddes failed to do this, Branford sent Miss McGegan, Edward's sister, to attend Geddes' lectures and to report on them fully for the purposes of publication.

Geddes himself was trying both to pursue his self-appointed mission and to make a living.[63] His writing paper from these days is headed: 'P.G. and Colleagues, Landscape Architects, Park and Garden Designers, Museum planners etc. Special experience in city plans and improvements, parks and gardens, garden villages, type museums, educational appliances, school gardens'. He found his tie to Dundee increasingly irksome in these years. After the succession of botanist geographers, culminating in Dr Marcel Hardy, Geddes had persuaded one of his Edinburgh circle, the poetess, Rachel Annand Taylor, to collaborate with him in giving a literary significance to the planting in his Dundee garden. But by 1907 Mrs Taylor was in London trying to make a living as a University Extension lecturer and there were no more recruits of the old kind to act as Geddes' assistant in Dundee.[64] He began to turn more and more to his own three children, especially his eldest daughter, now in her twenties and his eldest son, Alasdair, now in his late teens. After encouraging them to have as wide and as varied an education as possible, Geddes was happy that Norah was interested in a career in landscape gardening and Alasdair was becoming old enough to start to help him in his work. Geddes was also in these years to acquire a young Edinburgh architect collaborator in the person of F.C. Mears.[65] Mears hoped, with Geddes' contacts, to get commissions for urban extension schemes, and he was also to become deeply interested in Geddes' survey approach.[66] Mears was eventually to become part of Geddes' family on his marriage to Norah in July 1915.

In London there was further cause for optimism with the developments taking place in the town-planning movement. In October 1907 the Garden City Association mounted another of its key conferences at the Guildhall in London to discuss the present state of affairs in the movement. Geddes was able this time to attend as the chairman of the Cities Committee of the Sociological Society. At this conference attended by representatives from over 100 local authorities and all the major professional institutions and reform bodies, the resolution was passed to press for legislation for town-extension planning.[67] Geddes' hopes began to rise that this was merely the first step towards a new response to the urban environment and the next would be a greater interest in his ideas.[68] He had little faith in parliamentary lobbying and state legislation, but the adherents of the new 'town-planning' movement were seeking enlightenment about cities and he thought they might become more interested in social survey and social service: civics as applied sociology.

T.R. Marr was not so sanguine. As he wrote to Geddes in October 1908:

Here and there one gets evidence that the idea of a definite Civics is catching hold. It is to my mind one of the most extraordinary phenomena of this extraordinary age that we have all sorts of politicians and social reformers pointing out how our towns have made the unemployed or unemployable, how they promote vice and crime and how necessary it is to have a change in our town conditions, yet nowhere does one find − or at any rate I have not noticed it − a proposal to set to work reconstructing these towns and absorbing the vast mass of unemployed labour in the task. We want the Ethico-social awakening you speak of to carry us forward to 'geo-technic tasks' (I wish you could devise some Saxon equivalents for these Greek terms). But just now those who are awake to social wrongs are too much awed by the horror of it all to do constructive thinking and this energy gets spent too much in vain bewailing. And those who are thinking and constructing are, alas! too frequently out of touch with the actual difficulties of the people. We want a bridge builder; a whole order of them, badly.[69]

A bridge builder was exactly how Geddes began to see his role in the town-planning movement. Only Ebenezer Howard had spent as much energy trying to relate beneficial social changes to changes in the urban environment. But Garden City principles could not be applied wholesale to existing towns and cities and the Garden Cities Association was in the process of sacrificing some of Howard's ideas in order to play a part in the political lobby for town-planning legislation.[70] Geddes very much wanted to step into the breach and provide a link between social reform and the urban environment. He was moving from his position as champion of the small, provincial city to encompass all cities, large and small. As opportunities appeared to be opening up, so did his ambitions.

The work involved in establishing Crosby Hall eased in the latter months of 1908 and Geddes threw himself into the task of refining the theoretical base between survey and action. He asked Marr to carry out a social survey of Manchester, but by now Marr's health had broken and he was quite unable to undertake the task.[71] Geddes then turned to his Cities Committee of the Sociological Society. Over the past year he had been trying to make his survey technique a more suitable tool for training his adherents how to become 'town planners'. His vision, as he wrote to Branford in 1907, was to encourage 'Practicable Progress in the opening future, the *Eutopia*, as I would call it, of the next generation — the Town-Planning and betterment of all kinds, like Edinburgh, Dunfermline or Chelsea work.'[72] But after working on many diagrams, using the technique of his 'thinking machines', he suddenly lost confidence in the results he had achieved so far. As was usual on these occasions, he poured out his anxieties to Branford.

The cause for his alarm, though, was not the critical question of how the survey could be used for promoting action. By now Geddes was confident that he had got beyond what he considered to be the diffidence of Booth and Rowntree who had merely described rather than prescribed.[73] Instead he felt that the system he had devised for the collection of data was artificial. It was based on a 'conventional' order of knowledge rather than on the life process which he believed was the critical core of city development. The survey had been divided into three main stages and was intended to be scientific. The first step was the exploration of the region and occupations (the Valley Section); the second, the historic context, past, present and possible future; and the third, the vital and developmental stage. The report using this material was also divided into three: first, it concentrated on the preservation of natural resources of the region and occupational developments; second, it sought amongst the historical and contemporary material (culled from newspapers) for the 'essential and desirable' by which Geddes meant the 'pure, lovely and useful' elements to be

[173]

found in the locality; third, and finally, the focus had to be on the facilities for promoting both individual and social culture with much emphasis on education.[74] This was the orderliness Geddes was bemoaning! He had reached the point of impossible conflict between his desire to order the environment and his belief in the importance of the evolutionary potential of the individual. It was a dilemma he never resolved. In this instance he tried to take refuge in constantly reordering his material to incorporate what he insisted was a more 'vital' sequence of survey and report. Taking this escape route tended to encourage his wilder flights of fancy and intellectual rigour was often lost in romantic fantasy. However, Geddes worked ceaselessly at publicizing his survey technique and the value of exhibitions to aid the full comprehension of the importance of traditions and cultural norms in creating the physical environment of cities.

Geddes' moments of self-doubt were few, and he made a number of important contacts in London. Raymond Unwin wrote the most influential book amongst incipient town-planners, *Town Planning in Practice*, published just before the 1909 Housing and Town-Planning Act, and in it he makes great use of both Geddes' historical approach to cities and his survey work.[75] Unwin was most receptive to Geddes' insistence on the need for surveys and exhibitions because in many respects they tied in with his own experience. Unwin had emerged as the leading exponent of town-planning in the United Kingdom at a time when there were very few experienced practitioners. He owed his eminence, not only to his talent and prestigious commissions, but also to his earlier training as an engineer as well as an architect. It was this combination which made him particularly responsive to the practical problems of town-planning. As he wrote himself:

> Drainage will not run uphill to suit the prettiest plan; nor will people, to please the most imperious designer, go where they do not want to go or abstain from going where they must needs go, and from taking generally the shortest route to get there.[76]

He was therefore very keen on the idea of survey before plan and, in the general dearth of information about surveys, he advises his readers to consult Geddes' papers on the subject. He was one of the first to promote a Geddesian technique and to draw inspiration from Geddes' work without necessarily going to the lengths of adopting Geddes' propaganda for his 'civics movement'.[77]

The ultimate educational and propaganda tool: the Cities and Town-Planning Exhibition

Unwin appreciated the kind of contribution Geddes could make to the town-planning movement, particularly the critical task of training professional planners. He was instrumental in securing for Geddes the role that came nearest to being his heart's desire. After the passing of the 1909 Act the Royal Institute of British Architects decided to mount a conference to publicise the event and to make a bid for putting British ideas to the forefront in the international town-planning movement. It was to be called the First International Conference of Town-Planning; it was designed to bring together both those representing the 'arts' and the 'sciences' in their work in the urban environment; it was to have an exhibition to complement the proceedings of the conference. Geddes was offered the post of director of the Exhibition, which was organised by a committee under the chairmanship of Unwin. Most of the United Kingdom material was to come from the Outlook Tower as no other institution had a comparable collection. The British contribution in this respect was to be greatly outshone by the exhibits from Germany and elsewhere. Daniel Burnham's 'master' plans for Chicago and its region, a prime example of large-scale environmental planning with the use of a Master Plan, filled a large gallery and created a considerable stir.[78]

Here Geddes had truly found his element. The aura of international citizenship which hung over the proceedings seemed a manifestation of his desire to promote world citizenship. Since the Paris Exhibition of 1900 and the collapse of his own International Association, Geddes had lacked a forum for his internationalism. When the chance came in 1910, his constant propaganda for civic exhibitions and his past experience admirably suited him for the job of exhibition organiser. Having found such a congenial vehicle for his personal work, Geddes was not prepared to let it go, and he managed to persuade leading members of the British town-planning movement to sponsor him to maintain the Exhibition on a peripatetic basis.[79] The aim was to urge the local authorities of cities to invite Geddes, now entitled Director of the Cities and Town-Planning exhibition to take the exhibition to their city. Geddes offered not only to mount the exhibition, but also to give a series of lectures, based very much on the format he used for his Summer Meetings to arouse interest in cities generally and planning in particular. Since his civic museum movement had failed to materialise, Geddes found in this way a means of continuing his propaganda single-handed and with few resources. From 1910 onwards, the Cities and Town-Planning Exhibition was to travel more or less everywhere Geddes went and it brought him his greatest successes.

The achievements with the exhibition have to be seen against the background of two major influences on the town-planning movement of the period: the growing international movements to co-ordinate the handling of legal, medical and social problems of cities in industrialising countries, and the political ambivalence about the resources necessary for making an impact in these areas.[80] The Le Playist pursuit of 'Social Peace' had become broadened and widened to include not only the peaceful and healthy transition of societies from a rural to an industrial base, but also the peaceful evolution and co-existence of nations experiencing these factors. These international organisations were perhaps the most tangible evidence of a hostility to power politics and the prospect of war. Geddes' passionate concern for peace, evidenced in his saving of the war and peace exhibit from the 1900 Paris Exhibition, and in all his propaganda for civic regeneration, was closely in tune with the ethos of the international movements. He was constantly ready to put his energies into any international activity for co-operation. In 1907 he became involved with the organisation of a Franco-British Exhibition for 1908, and he tried to organise a collection of civic exhibits from municipal museums in France and the UK to form a *Civic Gallery*.[81]

He had formed the idea of a World Congress of Cities which is how he saw the Garden Cities and Town Planning Exhibition in London in 1910. While British architects were trying to promote British expertise in promoting Garden City principles, Geddes was wanting to generate a World Civic Movement encouraging the pooling of ideas to promote happy and healthy environments and peaceful co-existence in the future. The international exchange of ideas had become a marked feature of the town-planning movement in Europe and America, especially since 1904, yet Geddes was the only one amongst the British town-planning pioneers with this vision of world development through direct contact between cities on an international basis. He had made one of his most important personal contacts in this respect at the Paris Exhibition of 1900 when he met Paul Otlet and Henri La Fontaine.[82] He had been aware since the mid-1890s of their attempts to organise a centre for the 'World's Culture-Resources' (to use a Geddesian phrase) in Brussels. The Office and the Institute of International Bibliography which Otlet had established in 1896 had continued to grow in the next decade and Otlet had worked assiduously to develop an institutionalised synthesis of modern knowledge. He intended it be a powerful tool for the standardisation, co-ordination, and co-operation of intellectual activities on the grandest scale, even if in practice the results were modest.[83]

Geddes had written to Otlet whilst preparing his Dunfermline Report, asking for help in organising the City Library. Rather perversely, in view of the fact that Mr Carnegie was the benefactor,

Geddes wanted the Dunfermline Library to remain small if a bibliography could be compiled of the books in the Dunfermline region. He believed Otlet's method could be used to do this. But Otlet failed to produce what was wanted for the Dunfermline plan, suggesting instead that a Scottish section of the International Institute of Bibliography should be set up which would house a complete duplicate set of cards from his Universal Bibliographic Repertory. There was, however, no way at that time that the RBU could have supplied the duplicate set even if the money to pay for it had been forthcoming. Otlet's aspirations quite outran his ability to realise them.[84] After some years of rather lack-lustre achievement and heavy ridicule of their lofty international idealism, the general response to Otlet and La Fontaine began suddenly to change. Growing international rivalry, overhung by the prospect of war, made their efforts seem more relevant, and money from a number of sources was forthcoming to support their work. The turning point was 1908 when they held an International Congress of Bibliography and Documentation in Brussels.

The interest this aroused generated enough income for them to move one step further in 1910 in realising their combined interest in making all knowledge universally available, and in promoting peace. They offered to set up an organisation which would act as an information centre for international organisations of all kinds. The idealism was again lofty. But in May 1910 they were able to mount a World Congress of International Associations, and this time they gained the full backing of the Belgian Government and massive co-operation from international associations all over the world. Although the aims and objectives of this congress were very vague, the spirit was one of affirming world peace and world co-operation, and by the end of the congress over 130 international organisations were affiliated to it. To the international organisations of health, education, social welfare, and communications were added international organisations of municipal bodies united by the common problem of adapting their administrations to dealing with the current problems of industry and expansion. In some ways this was exactly the kind of 'civic movement' that Geddes was seeking to establish actually being realised in Brussels. The outcome of the congress was the setting up of a permanent Central Office for International Institutions to foster and maintain international links. A World Museum was also projected.

A small beginning for the latter was made by the exhibits brought for the World Congress of 1910. The problem remained of finding adequate funds to develop a collection. Between 1910 and 1913 Otlet and La Fontaine were to enjoy their greatest success in raising money.[85] La Fontaine was not only the President of the Union of International Assocations in Brussels, he was also President of the

International Peace Bureau in Berne, and he successfully approached the Carnegie Endowment Trust, newly established by Mr Carnegie in 1911, to get support for both. He wanted to bring the Peace Organisation to Brussels, but the Council of the Peace Bureau resisted this, so that the funds offered by the Carnegie Trust were split. But the amount coming to Brussels was enough to make the possibility of a Second World Congress possible in 1913, and the start of the World Museum, to be known as the Palais Mondial, which would house all the different elements of the Central Office and more and, it was hoped, would grow into a great world centre. The Second World Congress was held in Ghent and Brussels in June 1913, and in that year La Fontaine was awarded the Nobel Peace Prize.[86] The scene was set for Geddes to bring the Cities and Town-Planning Exhibition to Ghent in the International Congress of Town Planning and Organisation of City Life, and for his aims of social peace within cities and between nations to be given their greatest recognition as he was awarded the Gold Medal.

Geddes' career between 1910 and 1913 mirrored that of the world movement of Otlet and La Fontaine with its lofty idealism and internationalist propaganda.[87] It was also the period of his greatest personal success. At the 1910 Exhibition, organised by the RIBA, he had been the lone voice advocating the connection between social reform and environmental improvement and the combination of the two in promoting social peace. What had impressed many visitors to that exhibition, and was to continue to win him support, was Geddes' confidence in explaining city development to all who would listen. Amongst roomfuls of town extension plans, which were generally rather dull fare for the uninitiated visitor, Geddes held forth personally amongst his own exhibits on the geographical and historical features which had produced the great cities of the world, and the importance of the cultural context in shaping the built environment. He offered a holistic approach to the city as a living organism explaining problems in terms of both the process of growth, blossom, decline, and decay of natural evolution and the levels of adaptability of the social organism, human society, to the demands of function and environment. He spoke with great confidence about understanding cities and controlling their degenerative tendencies, at a time when large cities had become more impersonal and seemed less under control than ever. But once again, these promising beginnings were not to lead on to generating the 'Civic Movement' which Geddes wanted so much, though for a brief while, Geddes really believed it was possible.

The Regional Survey as an educational method for planners

On the one hand he had secured support from the leaders of the British Town-Planning Movement for his peripatetic exhibition: but on the other the invitations from large cities were not forthcoming. He had hoped that the curators of municipal museums would invite him to come as he had addressed the national Museums Congress on a number of occasions, the last time being 1907, when he had elicited a favourable response. There was no move now though, to adopt Geddes' ideas of a civic museum in any major city. Undeterred, Geddes turned his attention to promoting his survey technique and its use for training town planners. Together with his Cities Committee of the Sociological Society, he produced a small circular entitled 'City Surveys for Town Planning'. He was asked to produce three short articles for the first of the newly named volumes of the journal: *Garden Cities and Town Planning* brought out in 1911. This was a chance to publicise the importance of survey work before planning, regardless of further propaganda on the need for a civic movement. Geddes was ready to seize the opportunity.

The brevity of his presentation, and his direct concern with practical details in these publications, were in sharp contrast to his former treatment of these ideas in the lectures for the Sociological Society in 1904, 1905 and 1906.[88] But his ultimate objective nevertheless remained the same. He allowed himself just one sentence on that:

> Our experience already shows that in this inspiring task of surveying, usually for the first time, the whole situation and life of the community in past and present, and of thus preparing for the planning scheme which is to forecast, indeed largely decide, its material future, we have the beginnings of a new movement – one already characterised by an arousal of civic feeling, and a corresponding awakening of more enlightened and more generous citizenship.[89]

The first article was to put across as emphatically as possible the need for all local authorities considering the 1909 Act to undertake a survey of their local areas before proceeding with any planning work. The second was to suggest that the best means of reviewing the data collected in the survey was through a civic exhibition; and the third and final paper dealt with the actual survey method which should be adopted. Geddes had to bring himself to admit that while he believed that each city was unique and that its cultural evolution (which was the only positive way forward) depended on its unique qualities, nevertheless a general survey method applicable to all cities had to be found. He therefore proposed a general outline of headings under which every city survey should be carried out since 'unity of method is necessary for clearness, indispensable for comparison'. The outline was as follows:

SITUATION, TOPOGRAPHY AND NATURAL ADVANTAGES:
  (a) Geology, Climate, Water Supply etc.
  (b) Soils, with Vegetation, Animal Life, etc.
  (c) River or Sea Fisheries.
  (d) Access to Nature (Sea Coast, etc., etc.).
MEANS OF COMMUNICATION, LAND AND WATER:
  (a) Natural and Historic.
  (b) Present State.
  (c) Anticipated Developments.
INDUSTRIES, MANUFACTURES AND COMMERCE:
  (a) Native Industries.
  (b) Manufactures.
  (c) Commerce, etc.
  (d) Anticipated Developments.
POPULATION:
  (a) Movement.
  (b) Occupations.
  (c) Health.
  (d) Density.
  (e) Distribution of Well-being (Family Conditions, etc.).
  (f) Education and Culture Agencies.
  (g) Anticipated Requirements.
TOWN CONDITIONS:
  (a) HISTORICAL: Phase by Phase, from Origins onwards. Material Survivals and Associations, etc.
  (b) RECENT: Particularly since 1832 Survey, thus indicating areas, lines of growth and expansion, and local changes under modern conditions, e.g., of streets, open spaces, amenity, etc.
  (c) LOCAL GOVERNMENT AREAS: Municipal, Parochial, etc.
  (d) PRESENT: Existing Town Plans, in general and detail. Streets and Boulevards. Open Spaces, Parks, etc. Internal Communications, etc. Water, Drainage, Lighting, Electricity, etc. Housing and Sanitation (of localities in detail). Existing activities towards Civic Betterment, both Municipal and Private.
TOWN PLANNING: SUGGESTIONS AND DESIGNS
  (A) Examples from other Towns and Cities, British and Foreign.
  (B) Contributions and Suggestions towards Town-Planning Scheme, as regards
      (a) Areas.
      (b) Possibilities of Town Expansion (Suburbs, etc.).

(c) Possibilities of City Improvement and Development.

(d) Suggested Treatments of these in detail (alternatives when possible).[90]

This was the technique which Geddes brought to the Town-Planning Movement. This was the way he introduced a breadth of vision into an activity which, in the hands of architects, engineers, and surveyors (who made up the total of the founder members of the Town-Planning Institute in 1913 apart from Geddes himself), threatened to be concerned exclusively with a simple ordering of the physical environment. Geddes' impact was greatest on the young professionals, either members of the Cities Committee of the Sociological Society or practising architect planners.

Geddes' ideas made a particular impact on Patrick Abercrombie, architect and lecturer in Architecture and Town Planning at the University of Liverpool and Editor, from 1910, of the *Town Planning Review*.[91] Another who became passionately interested in Geddes' ideas was H.V. Lanchester, architect and editor of *The Builder*, who had begun to take an interest in town planning since 1910. Another architect as yet relatively uninfluential but who was destined to play a master role in the development of British town planning after the First World War was George Pepler.[92] Geddes could also always rely on a sympathetic response from some of the older generation such as his friend, C.R. Ashbee of the Arts and Craft Movement, who was now, in the absence of commissions for houses, turning, like many others, to urban design.[93] Above all, Geddes enjoyed the support of Raymond Unwin, the doyen of practitioners of town planning in Britain, and soon to become town-planning adviser to the Local Government Board when Thomas Adams left for Canada. Geddes' influence on Unwin, Abercrombie, Lanchester, and Pepler was to ensure him an established place in British town planning.

What inspired these architects who came to their tasks as town planners from a training in the aesthetics of design, was Geddes' ability to supply them with a sociological rationale for their work. His 'Eutopia' was close to their own artistic ideal. As Unwin wrote in *Town Planning in Practice* in a chapter on 'Civic Art The Expression of Civic Life':

The artist is not content with the least that will do; his desire is for the best, the utmost he can achieve . . . In desiring powers for town planning our town communities are seeking to be able to express their needs, their life, and their aspirations in the outward form of their towns, seeking, as it were, freedom to become the artists of their own cities portraying on a gigantic canvas the expression of their life.[94]

But the actual 'expression' was in the hands of the planners and Geddes' survey and exhibition provided much needed cover between the stated ideal and the practice. Geddes' techniques promised insight into the life of the 'community' which the planner would then be 'free' to interpret. Geddes offered critical support to planning practitioners who perhaps, like C.R. Ashbee, had recently fled the city to restore their aesthetic inspiration in the calm of the countryside. Raymond Unwin's book, which drew heavily on Geddes' historical and visual approach to the built environment, was a best seller.

Thus Geddes' survey and exhibition supplied a technique, a perspective and a confidence to the early practitioners of planning. With the Cities and Town Planning Exhibition, he was also to acquire the patronage which would bring him commissions to undertake planning work in Ireland and India. One of the few cities prepared to invite him and his exhibition was his adopted city of Edinburgh. There he was able to mount a truly impressive exhibit of the city since the Edinburgh Survey had been in continuous process of development since the setting up of the Outlook Tower in 1892. There was also a number of Outlook Tower supporters who were able to help Geddes in his role as guide and interpreter of the exhibition with the result that the exhibition was a great success. One of those most impressed by it was Lord Pentland, then Secretary of State for Scotland, who had in his youth spent some months at Toynbee Hall with Canon Barnett, and shared Barnett's liberal political views and concern for social reform.[95] He was instrumental in introducing Geddes to Lord and Lady Aberdeen, friends of long standing to whom he had become even closer when he married their daughter in 1904. Lord Aberdeen was at that time Viceroy of Ireland and Lady Aberdeen was to be responsible for inviting Geddes to Ireland where he was to make his greatest impact in the years before the war.[96] During the war Pentland, now Governor of Madras State, invited Geddes to India, and was thus responsible for introducing him into the context where he was to gather his many commissions for town-planning work and to earn more than he had ever done before. Pentland was also prepared to put Geddes' name forward for a Knighthood in the Honours List of 1911, but Geddes was not prepared to accept the offer. He felt he could not sustain such a social position on his current income and, in any case, he did not care for that kind of recognition for his work.[97]

Geddes and the Irish experience: the professional 'trouble-shooter' as planner

The invitation to take the exhibition to Ireland in 1911 was most opportune for Geddes. It was a chance to preach his 'Civics Movement' in a country where Sir Horace Plunkett's Land Organisation Society had been so successful. Above all there was Dublin, a city Geddes had visited so often since the early 1880s, and which he felt was spiritually closer to his beloved Edinburgh than any other.[98] Geddes was invited by Lady Aberdeen in her capacity as President of the Women's National Health Association of Ireland, which was hosting the Annual Congress of the Institute of Public Health that year.[99] It proved to be not only an exciting opportunity for him, but also a key moment in the whole history of Ireland's health and housing problems when the need for taking action had never been greater, and the means for doing so fewer.[100] Over the next three years Geddes was to invest a great deal of time and energy most especially to the problems of Dublin, and he was to be joined in 1914 by Raymond Unwin. Together they were to fight for town-planning ideals and practices in the one city in the United Kingdom where environmental issues had become politicised.[101] It was one of those instances in which more was learnt by the practitioners than was actually achieved. The results were to leave Unwin struggling with the problem of clarifying the legal framework needed to carry out work in the city centre after the Easter Rising; whilst Geddes, now in India, constantly sought that combination of nationalism and passion for environmental reform which he thought he had found in Dublin.

The problems of Dublin which Geddes was willing to tackle in 1911 seemed overwhelming from any other viewpoint. The city, once the largest in the UK outside London at the beginning of the nineteenth century, had suffered an even more dramatic relative decline than Edinburgh, as England had climbed to world prominence and London, a world metropolis. Its population had increased from only 318,000 in 1851 to 382,000 in 1901. The industrial revolution had passed the city by, leaving only the remnants of its craft industries, mostly ruined by competition from industrialising Belfast. Unemployment was chronic and severe for men; for 60 per cent of women there was no work at all. Wages of both skilled and unskilled workers were lower than in England.[102] Many people emigrated from the city, but the pool of those living in abject misery and total want was constantly augmented by rural immigrants fleeing from the even greater poverty of the countryside. These made their way to the already overcrowded central areas of the city. Those classed in the 1913 census as being without regular employment numbered 104,000 people, and there was a substantial difference in the death rate between this group and the rest of Dublin: for the former 26.2 per 1,000 to 15.5 per 1,000 for the latter.

The misery of the poor, a disproportionate number in comparison with any other city in the UK, was compounded by the housing stock available in the city. By the beginning of the twentieth century more than half Dublin's families lived in tenements. In the 1911 census 21,113 households lived in one-roomed tenements and since only 8,914 were single or two-person households, the degree of overcrowding amongst the rest was very high. The tenements of the mediaeval areas of the city had been supplemented in the course of the nineteenth century by the multioccupation of the mainly eighteenth century town houses of the upper classes who had long since left the city for salubrious suburbs or the social life of London. These houses had often been laid out in squares with narrow gardens and stabling at the rear. Now the stables were also used as makeshift homes and shanties erected in any available open space. The façades of these once elegant homes hid from view chaotic and filthy slums. Public investigation after public investigation was held to try to determine what was to be done but there seemed to be no answer.[103] The Local Government Board inquiry of 1900 into the public health of Dublin at last spelt out the problem: it was abject poverty.[104]

In this instance it was impossible to argue that the answer to the overcrowding in the centre was suburbanisation. There was no way the poor could afford to live in suburbs. Nettlefold's ideas on housing reform within a free market system could operate in Birmingham where there was an upwardly mobile, skilled working class but not in Dublin. Geddes' fear of state intervention though, made him put forward this idea as a feasible one for Dublin in his 1913 report. But for Dublin's civic leaders the only hope of improvement seemed to be demolition and the subsidisation of new building by either private or public bodies to keep the rents of new housing low. Even here Dublin's problems were more extreme than those anywhere else in Britain. Private philanthropy was limited. Lord Iveagh, of the Guinness family, for instance, gave £250,000 for housing ventures, £200,000 for London, and £50,000 for Dublin. Other schemes were more generous to the Irish city, but clearly Dublin's housing problems completely outran the resources of private philanthropy. Dublin Corporation, however, was in no position to venture public capital into housing schemes. Its rateable value was far lower than that of any other city of its size in the United Kingdom, and it was in any case riven by political factions and disputes about its authority and intentions. By the time Geddes came on to the scene, politics and religion had left the corporation without leadership and resources, in dispute with the Lord Lieutenant, the Local Government Board, even the Port and Docks Board as well.

The Public Health Officer, Sir Charles Cameron, had worked in the city from the 1880s, walking a tightrope between the urgent need to remove the worst health hazards whilst leaving as much as possible untouched as the only 'improvement' was demolition, which always resulted in increasing overcrowding elsewhere.[105] But more and more tenements had to be closed and were often demolished. Between 1877 and 1914, 6,886 tenements were so treated. The result was that many streets and central areas became pock-marked by numerous waste grounds which was all that remained of former houses, creating an effect of even greater dilapidation and dereliction in the city. Some of the improvement schemes carried out under the Iveagh Bequest and by the corporation entailed the demolition of some of the oldest and historically most interesting parts of the city. Dublin was hardly the ideal city for launching a modern town-planning movement. Geddes was quick to perceive, however, that the very desperate nature of the health and housing problems made people more than usually willing to listen to him.[106]

He began by working with the organisation which originally invited him to Ireland, Lady Aberdeen's Women's National Health Association. The ladies who belonged to this were in the same mould as the ladies of the Edinburgh Social Union. They were willing to give their time and energies freely towards improving the condition of the poor, often working in co-operation with public health authorities, but providing the individual personal touch that no public authority could command. Geddes began as he had in Edinburgh with open spaces and waste ground in the city centre. He recruited teams of ladies to start clearing waste land and making gardens as the first step in the rehabilitation of the whole area. He converted Lady Aberdeen to his own ideas of gardens on waste grounds and the American idea of 'social reform' parks, and when she wrote her presidential report in 1912 she was very enthusiastic about these ideas.

> Whenever a garden occupies a vacant site, that site will make for the cheerfulness and health of the surrounding houses, instead of accumulating rubbish and encouraging disease. When something pretty is below their windows people stop throwing rubbish out of them, and take heart to clean their houses. These gardens serve many purposes . . . In these 'Garden Play-grounds' adequately staffed, the rough children become gentler, the idle children learn the happiness of active work and play, and all benefit from coming under the discipline of organised games.[107]

Voluntary workers gave their time to supervising the children in the afternoons.

This children's park movement made a considerable impact in Dublin with Lady Aberdeen's strong support. Geddes sent his daughter Norah, now trained as a landscape gardener, to oversee the garden building. Three such gardens were established in Dublin, and local branches of the Women's National Health Association had projects for seven more in smaller towns in the south. Another immediate result of the Cities and Town Planning Exhibition in Dublin and in Belfast in 1911 was the formation of the Housing and Town Planning Association for Ireland,[108] and again under Lady Aberdeen's influence, her Women's National Health Association formed a small collection of housing and town planning exhibits which was sent around the country with a peripatetic Health Exhibition. By starting in this simple way, Geddes began to win friends, and he was able to proceed on to the next stage by consulting civic dignitaries and local political and religious leaders about their aspirations for their city. He managed to interest the Royal Dublin Society so much in town planning that it sent invitations to W.H. Lever to come to lecture to them on Port Sunlight, and to C.H. Reilly, first Professor of Civic Design at the University of Liverpool, to tell them about the new art. But the time for talking had almost run out. Geddes' initial success in promoting town planning has to be seen against a background of growing political tension in the Home Rule movement, and the politicisation of the housing problem as just one more example of the wicked results of the Union.[109] Stronger action on housing became imperative.

The London Government set up a parliamentary departmental committee in 1913 to inquire into the housing of Dublin's working classes. Expert witnesses were called and Geddes was asked to give evidence as the representative of the Women's National Health Association. While accepting that poverty and bad housing were linked, Geddes wanted to make two points loud and clear. The first was that the poor were not an inert mass waiting to be dealt with by local authorities: 'we talk about the work people and the submerged tenth as if they were mere passive creatures to be housed like cattle. We must take them with us, and we must realise that we are working for the civic uplifting'.[110] The second point was his hostility to the provision of municipal housing and purpose-built, improved, tenement blocks. Even in the extreme case of Dublin, Geddes wished to argue for the retention of as much variety in building as possible, with special care for the surviving evidence of the historical past. The Committee's Report, whilst noting Geddes' evidence, kept its focus on what was believed to be the main problem: the urgent need for more working-class housing and the necessity of financial support from the state for a municipal housing programme. Parliament voted on this, and by 1914 large sums were allocated for housing projects in Ireland. Dublin had thirty-five schemes in hand.

The prospect of state finance brought the discussion of its best use to fever pitch. In March 1914 the Aberdeens invited Geddes to bring his Cities and Town Planning Exhibition (newly-decorated with the success of the Gold Medal in Ghent in 1913) back to Dublin. The heightened political tension galvanised Geddes to even greater efforts than before. He put all his energies into a great propaganda campaign to turn the prospect of mass municipal housing into a movement for civic renewal and modern town planning. Against him was the extent of the problem of working-class housing and those serving on the Corporation, such as John Clancy, who believed that housing and town planning were two completely different activities and that what Dublin needed was housing. On Geddes' side were the Aberdeens and the voluntary workers of the Women's National Health Association and the Dublin Housing and Town Planning Association which had been collecting material for what Geddes euphemistically called a 'Dublin survey' since his visit in 1911. Geddes not only mounted his exhibition in March, which was very successful, he also grasped the nettle by addressing himself directly to the two key issues, religion and politics, which were contributing to the growing social tensions. With Lord Aberdeen's permission, he approached the leader of the major dock strike then in progress, Jim Larkin, and discussed with him the problem of dockers' housing.[111] He located some corporation land a mile from the docks which he believed could be used, not just for housing, but for a 'garden village' for the dockers.

He had a grand plan to relieve the religious frustration of the Irish Catholics: nothing less than a new modern cathedral which he located in the city in proximity to the two existing mediaeval cathedrals and joined to them by what he called a 'via sacra'. This he felt would be 'the best monument for Home Rule . . . and for the cathedral of the Irish race'.[112] The Catholic Archbishop of Dublin was so impressed with the idea that the site was actually purchased from the Corporation, though there were no funds for the building. Finally, Geddes wanted to encourage a systematic development of the whole town, and he suggested to the Aberdeens that the continental practice of holding a town-planning competition might be both a propaganda tool and a way of producing new ideas. Since 1904, the German town-planning journal, founded by Camillo Sitte, *Der Städtebau*, had been publishing town-planning ideas produced in the course of competitions. As a way of publicising new ideas, Lord Aberdeen adopted Geddes' suggestion and he personally financed a first prize of £500 for a comprehensive plan for the extension of Dublin. Geddes at last began to believe he was having some impact and what he needed was reinforcements.

In April 1914 the Dublin Housing and Town Planning Association invited Raymond Unwin to the city. He gave them a lecture on 'How Town Planning may solve the Housing Problem' which was attended by the Aberdeens and other leading figures. Unwin's message was the need to relieve congestion in the city centre by suburban development which should be planned on garden suburb lines to prevent future problems. His pamphlet, *Nothing Gained by Overcrowding*, published by the Garden Cities and Town Planning Association in 1912, contained his views on low density development and design on garden city principles. The gaps between the problems of Dublin, the politics, the propaganda, and the prescriptions were obviously huge, but the urgent political need to be seen to be doing something helped to paper over the cracks. The Aberdeens persuaded Geddes to mount a Summer School of Civics in the last two weeks of July and the first week of August. He accepted delightedly, and made no concessions to the pressing problems of the present. The programme contained his usual lectures on the evolution of cities, on geography and nature study, on the importance of the region as a unit and regionalism as an educational method with the regional survey of cities. He persuaded Dublin's Medical Officer of Health, Sir Charles Cameron, to give a lecture, and brought the budding regional geographer, H.J. Fleure, from the University College, Aberystwyth, to supplement his own evolutionary and sociological lectures.

The Municipal Corporation had a much more direct anxiety about the advisability of some of its specific clearance and housing projects. It was decided in August to ask Geddes and Unwin to comment on some of the housing schemes drawn up by the city architect. This was Geddes' first chance of commenting on planning proposals in an official capacity; it was Unwin's first attempt to comment on plans for the redevelopment of an established historic city. Both men were determined to be guided in their judgement by what they considered to be their town-planning principles and to promote these clearly in their Joint Report. Geddes and Unwin inspected the city architect's proposals on 24 and 25 August. They met corporation officials including the architect, and visited all the sites due for development, both in the city centre and the suburbs. Geddes was learning the art of 'trouble-shooting' in practical planning with the greatest living British exponent of the art, Raymond Unwin, at his side. He was to use these techniques many times in the future in India.

The Joint Report welded together the approach of the two men. Geddes was concerned with preserving the historical core by 'conservative surgery', and working for the cultural evolution of all citizens; Unwin was anxious that town-extension schemes on garden city lines should be promoted, with high standards of house building in low density designs. They did not convince the city architect or other corporation officials with what appeared to be absurdly expensive schemes. But in any case, political events soon outstripped all proposals as, with the outbreak of the war, the promised state finance for housing suddenly disappeared. All projects were immediately halted. The competition for the city, however, had got underway and there were eight entries. The war delayed the judging of the competition as the chosen American assessor, John Nolan, was unable to reach Dublin until 1916. Once again events outstripped the planners as the Easter Rising and the destruction of the Central Post Office and part of the city centre created an immediate and completely different problem, overshadowing the impact of the competition.

Geddes and Nolan were to award first prize to the design submitted by Patrick Abercrombie, S. Kelly, and A. Kelly. Runner-up was the work of C.R. Ashbee. The brief had been to focus on three specific features influencing Dublin's future: (1) Communications, (2) Housing, and (3) Metropolitan Improvements. Abercrombie included both Geddes' cathedral and Unwin's garden suburbs as well as a vast new road scheme and underground railway for the city. Geddes suggested that the Abercrombie/Kelly entry deserved first prize as it gave the best idea of what town planning really was to the uninitiated, even though he knew that the plan was impractical. He suggested that Abercrombie did not know Dublin well enough and he needed to carry out a full survey.[113] This Abercrombie did after the war in 1922, publishing the material in 1925. In this way Dublin came to be one of the first cities where a Geddesian plan and survey were carried out though it is perhaps rather ironic that the plan came *before* the survey!

The Irish experience was crucial in establishing Geddes' position in the British town planning movement. Until Dublin he had been merely a propagandist. In Dublin he had worked with Raymond Unwin and been at the centre of one of the most concerted efforts to promote modern town-planning practice in a specific context. He had taken F.C. Mears with him and set him to work to draw up designs and house plans for the suburban developments and to provide architectural drawings for his other city centre schemes. He had been able to promote the career of his supporter Patrick Abercrombie, and to get the idea of survey and plan accepted, not only in theory, but also in practice. He was able to capitalise on the unique circumstances of Dublin in these years and make them work for him rather than against him; and the lack of any permanent result, especially the lack of any improvement in working-class housing, could be blamed on the war and on the Easter Rising, not on the failure of their planning policies. Indeed, both Geddes and Unwin were adamant that if only the money had not suddenly stopped in 1914, and some of their ideas had been carried out, there would not have been an Easter Rising.[114] The unfortunate city of Dublin provided the context for the initiation of all concerned in the relationship between politics and planning, and the planners were too new in their role to appreciate the limitations of their influence.

'Cities in Evolution': Geddes' influence on the concepts of town planning before 1914

Geddes' natural optimism, especially, had been unleashed by the activities of these years. After his first visit to Ireland in 1911, he managed to produce the manuscript of his long-awaited monograph on the growth of cities with the title *Cities in Evolution*. It was an extraordinary document displaying both the insights and the quirkiness of Geddes' evolutionary approach. Its uniqueness was compounded by its subject. There were very few books at that time on the general development of cities. It invites comparison with the American A.F. Weber's classic monograph on *The Growth of Cities in the Nineteenth Century*, published in 1899. The core of Weber's work was the collection and analysis of the statistical record of the growth of all cities in the world. It was a masterpiece of accuracy and scholarship which has never been superseded. But the analysis of the physical and social changes to be found in cities over the period was necessarily limited. Geddes on the other hand, eschewed statistical analysis entirely. He had not only been lecturing for so long on social evolution, he had also become totally ingrained in his own idiosyncratic evolutionary style. The book, therefore, lacks a balanced structure and an orderly survey of material, and the chapters read more like independent lectures, which many of them, in fact, were.

What emerges most clearly as the greatest contrast with Weber's work however, is the sense of place. Geddes' insights are all related to an astute perception of the physical environment and the changes, both technological and social, which were influencing cities and their regions. Geddes' experience in Dublin had confirmed his belief in the vital importance of teaching people about the historical evolution of the place in which they lived. He saw municipal housing schemes as a threat, not only to the fabric of the city, but to the people as well. What he wanted was the preservation of the best historical traditions of the past, the involvement of the people in their own betterment and the rediscovery of past traditions of city building which deliberately expressed the aesthetic ideals of the community. *Cities in Evolution* was meant to explain the urgency of these objectives because of the natural law of evolution and to persuade readers to take up survey work, the basis of his hoped-for 'Civic Movement'. Above all, Geddes wanted to change the perceptions of his contemporaries about city life. What he was writing was not a town-planning manual, a historical or geographical analysis of town growth, or a new educational theory. It was a rationale of town life for all those who lived in cities and for all nations where the majority of the population dwelt in cities. It was a polemic on civilisation and what cities had to offer for their citizens, which for Geddes was the chance for 'cultural evolution', in the way

Figure 6.1   Edinburgh: Upper High Street

'So bookish has been our past education, so strict our school drill of the "three Rs", and so well-nigh complete our life-long continuance of them, that nine people out of ten . . . understand print better than reality. Thus, even for the few surviving beautiful cities of the British Isles, their few marvellous streets – for choice . . . the High Street of Edinburgh – a few well-chosen picture postcards will produce more effect upon most people's minds than does the actual view of their monumental beauty . . .'

Source: P. Geddes (1915) *Cities in Evolution*, London: Williams & Norgate, p. 16. Illustrations from pages 17 and 11 respectively of the same book.

that he specially interpreted it.

It was this vision which had insulated Geddes from being overwhelmed by the problems of Dublin. Where others saw poverty, decay, and neglect, Geddes saw the former capital city of an independent Ireland, an elegant centre of eighteenth century culture, a noble city able to inspire its present and future citizens. But while his enthusiasm helped to sustain others, his interest in long-term, evolutionary trends appeared obtuse in the face of immediate problems. *Cities in Evolution* is a passionate book, full of energy, with paragraphs liable to go off at a tangent, with obvious emotional appeal. It was written at a time when planners were more aware of the social and political importance of their work than ever before. Geddes' place, work, folk, his valley section, his emphasis on regional survey, seemed like straws in the wind to be caught and used in the absence of any other input from the social sciences in the study of cities. In 1912, when the book was completed, Geddes' hopes of creating a civics movement were still strong after his success with his propaganda work in Ireland. He was to receive a rude shock when the editors of the

Home University Library, Herbert Fisher and Gilbert Murray, rejected the manuscript. It was the thankless task of J. Arthur Thomson (also an editor) to break the news to Geddes that his work did not fit into the pattern of the established series.[115] Thomson thought it should be published independently with a new title (Geddes had originally offered two: *The Study of Cities* or *An Introduction to Civics*), and Thomson did not care for Geddes' term 'Conurbation'. Geddes had invented the latter to highlight the process of urbanisation in industrial areas, and in the London region, which swallowed up the individual identity of former settlements. Thomson preferred 'town-group'. The book was not published until 1915 when the circumstances of war had created an entirely new environment for its reception.

The International Town Planning Movement was badly hit by the outbreak of hostilities. The conflict seemed particularly absurd to members of the town-planning fraternity in the belligerent countries who had met each other and worked together frequently over the past decade, united in their hope of creating a better future for all. Raymond Unwin was early converted to the hope of peaceful co-existence in the future, and became a founder member actively involved in the British League of Nations Society. He published a pamphlet, *The War and What After* in May 1915 which was when Geddes' monograph at last appeared. Even before this, in November 1914, the British Garden City and Town Planning Association had sprung into action to discuss the problems of Belgium. At this stage it was still widely believed that the war would soon be over and that Belgium had taken the brunt of the physical violence. To focus on the problems of Belgium seemed appropriate and challenging. The future of this small country with its ancient cities, set in beautiful countryside, starting from the low point of war to build again a modern nation, was an intoxicating prospect. It was also a morale-boosting exercise for the hundreds of thousands of Belgian refugees who had fled to France, England, and Holland. A Belgian Town Planning Committee was set up so that the Belgians could do their own planning and a context was provided by a conference, held in February 1915, organised by Raymond Unwin.[116] At these discussions, and many others that followed in Britain and France, culminating in the special Exposition de la Cité Reconstituée held in 1916 in Paris, Geddes' influence was to be found everywhere though he was personally away in India most of this time. In the special circumstances of war, the seeds he had scattered amongst the international fraternity of town planners on the need for civic surveys was given hothouse treatment and flowered unexpectedly.

Those concerned with promoting this activity all owed their inspiration and in some instances, their initial training, to Geddes. In Britain work was co-ordinated by Abercrombie and by H.V. Lanchester,

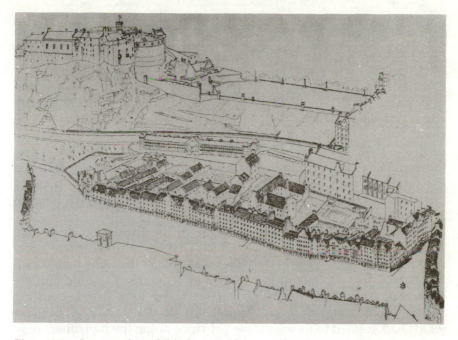

Figure 6.2 Grassmarket of Edinburgh: old agricultural centre and market place below Castle-town

and for the Belgians, by Raphael Verwilghen. Lanchester worked with Geddes in India in 1915 and 1916 where he undertook a survey of Madras; Verwilghen came across the idea of the civic survey in England during the early months of the war. However, the leading promoter of civic survey was to be Abercrombie. Abercrombie had chosen over the past five years or so to serve a kind of apprenticeship with Geddes. He was in close sympathy with Geddes' neo-Romantic approach to an ideal of city life. He had a deep appreciation for the Belgian poet, Emile Verhaeren. He published an article in 1912 in which he describes Verhaeren as the first modern poet who tells of the beauty of the modern town. As an architect, he approached his town-planning work as a designer, but his experiences, especially in Dublin, brought home to him the vital significance of knowing and understanding a place through survey before trying to plan. At the Ghent International Congress of Town Planning and the Organisation of City Life in 1913, Abercrombie had read a paper on 'The Modern Use of Great Monuments', which marked the high point of Geddes' influence on him with his ideas of cultural evolution and cities.[117]

Abercrombie, since 1915 Professor of Civic Design at Liverpool after the departure of S.D. Adshead, threw himself into the task of encouraging the development of civic surveys in Britain which would not only provide crucial data for post-war planning but would also provide work for out-of-work architects. Abercrombie himself became involved in setting up a regional survey for Lancashire, and a survey of Manchester. Verwilghen, in consultation with Abercrombie and H.V. Lanchester, set up a Civic Development Survey for the re-construction of Belgium. All these initiatives began to find problems in defining city and region, and balancing regional and national needs and interrelationships. Lanchester could offer only the example of his work in Madras, carried out in 1915–16, which was a sociological survey of a pre-industrial city. In fact, the Belgians soon found that the key factor in the future of their country was economic planning, which involved questions of central government finance and direction, which bore no relationship to the work of the civic survey.

This point was reinforced at the time of the Exposition de la Cité Reconstituée in Paris in 1916, for which Geddes returned from India. An Economic Conference of the Allies was held at the same time which highlighted the paramount importance of economic planning and co-operation. The Cities Exhibition was reduced to a morale-boosting exercise quite different from the debates on the future of Belgium which had seemed to be within the province of the town planners only eighteen months before. It served, however, to introduce the town-planning fraternity who were able to get there, to the lesser-known work of French town planners, especially that of Henard and Tony Garnier in Lyon, and to the French regionalist perspective.[118] Geddes' ideas roused some interest in France, but not at all on the same level as in Belgium three years earlier, and not at all as a moral movement. Geddes made a rather half-hearted attempt to intensify his propaganda by allowing Victor Branford to publish (as the first book in a new series, 'Making of the Future'), transcripts of the lectures he gave at the 1915 Summer School on the war in London. But commitments in India kept him in the East and he was to remain rather out of touch with the experience of total war in Europe.

Notes

1. M. Simpson (1985) *Thomas Adams and the Modern Planning Movement, Britain, Canada and the United States, 1900–1940*, London: Mansell; Frank Jackson (1985) *Sir Raymond Unwin: Architect, Planner and Visionary*, London, Zwemmer. Thomas Adams wanted to nominate Geddes as a Vice-President of the Garden Cities Association. Thomas Adams to Geddes, 24 March 1906, Papers of Sir Patrick Geddes, vol.I, 9/643, University of Strathclyde.
2. Gordon E. Cherry (1979) *The Evolution of British Town Planning*, Leighton Buzzard: Leonard Hill, pp.87–90.

3. See chapter 9.

4. Letter by Professor Patrick Abercrombie published in A. Defries (1927) *The Interpreter: Geddes, the Man and his Gospel*, London: George Routledge and Sons Ltd, pp.321–2.

5. See chapter 4, pp.106–9.

6. A. Sutcliffe (ed.) (1981) *Towards the Planned City: Germany, Britain, the United States and France, 1780–1914*, London: Blackwell, pp.163–79.

7. W. Ashworth (1954) *The Genesis of Modern British Town Planning: a study in the economic and social history of the nineteenth and twentieth centuries*, London: Routledge & Kegan Paul, pp.166–90.

8. J.S. Nettlefold (1914) *Practical Town Planning*, London: St. Catherine's Press, pp.98–102, offers proof of the efficacy of his policy by quoting the example of the development of Harborne Garden Suburb, Birmingham, undertaken under his direction. For an assessment of his work here see G.E. Cherry (1975) *Factors in the Origins of Town Planning in Britain: the example of Birmingham 1905–14*, Working Paper no.36, Centre for Urban and Regional Studies, University of Birmingham.

9. W. Ashworth, op.cit., pp.7–14; G.R. Searle (1971) *The Quest for National Efficiency: a study in British politics and political thought, 1899–1914*, Oxford: Blackwell, pp.60–68.

10. H.E. Meller (1976) *Leisure and the Changing City* London: Routledge & Kegan Paul, pp.237–52.

11. A. Lees (1985) *Cities Perceived: urban society in European and American thought, 1820–1940*, Manchester: Manchester University Press, pp.307–8.

12. H.E. Meller, op.cit., (1976) chapter 7.

13. R. Beevers (1988) *The Garden City Utopia: a critical biography of Ebenezer Howard*, London: Macmillan, pp.68–79.

14. R. Fishman (1977) *Urban Utopias in the Twentieth Century: Ebenezer Howard, Frank Lloyd Wright, Le Corbusier*, New York: Basic Books .

15. H.E.Meller (ed.) (1979) *The Ideal City*, Leicester: Leicester University Press, Introduction. In America Jane Addams was a keen supporter of Geddes' ideas.

16. J. Springhall (1977) *Youth, Empire and Society: British youth movements, 1883–1940*, London: Croom Helm, pp.53–64.

17. M. Harrison (1988) *Social Reform in late Victorian and Edwardian Manchester with special reference to Thomas Coglan Horsfall*, Unpublished PhD thesis, University of Manchester; and M. Harrison (1985) 'Art and Philanthrophy: T.C. Horsfall and the Manchester Art Museum', in A.J. Kidd and K.W. Roberts (eds) *City, Class and Culture*, Manchester: Manchester University Press, pp.120–147.

18. For Nettlefold, see note 8 above.

19. See *A Manual of the Public Benefactions of Andrew Carnegie*, compiled and published by the Carnegie Endowment for International Peace, Washington, 1919.

20. Henry Beveridge had been an admirer of Geddes since the early days of the Edinburgh Summer Meetings. A philanthropist and educator from Dunfermline, he was one of the first directors of the Town and Gown Association, 1896, and he was also appointed to the Carnegie Dunfermline Trust. When the Carnegie benefaction was announced, V.V. Branford alerted Geddes and told him to contact Beveridge, offering him his ideas for a Natural History Museum. (Branford to

Geddes, 8 May 1903, Geddes Papers MS10556, NLS). Beveridge responded by showing Geddes' scheme to Dr John Ross, chairman of the Trust and Carnegie's Scottish factotem.

21. Geddes to T. Mawson, 28 December 1904, Geddes Papers MS10536, NLS.
22. G. Cranz (1982) *The Politics of Park Design: a history of urban parks in America*, Cambridge, Mass.: MIT Press, p.62.
23. E. Howard (1898)*Tomorrow: a peaceful path to real reform*, London: Faber, 1965 edn.; R. Beevers (1988) *The Garden City Utopia: a critical biography of Ebenezer Howard* London: Macmillan, especially chapter 8.
24. W. Ashworth, op.cit., pp.81–117.
25. P. Geddes (1904) *City Development: a study of parks, gardens and culture institutes. A report to the Carnegie Dunfermline Trust*, Bournville: The Saint George Press, and Edinburgh: Geddes and Co., reprinted in facsimile by Irish University Press, Shannon, Ireland, 1973, with an introduction by P. Green, pp.215–16.
26. An ancient city (it had once been the capital of the kingdom of Fife), its physical structures contained many indications of its historical past.
27. H. Beveridge to Geddes 14 February 1904, Geddes Papers MS10536, NLS.
28. For copies of the letters he sent and list of correspondents see MS10612, NLS.
29. *A Manual of Public Benefactions*, op.cit., pp.79–105.
30. J.H. Whitehouse to Patrick Geddes, 10 February 1904, Geddes Papers MS10536, NLS.
31. H. Beveridge to Patrick Geddes, 21 May 1904, Geddes Papers MS10536, NLS.
32. *A Manual of Benefactions,*op.cit., letter from Carnegie to Dunfermline Trust, p.244.
33. T.H. Mawson (1927) *The Life and Work of an English Landscape Architect*, privately published, and for a brief review of his career see David Mawson (1979) 'Thomas H. Mawson 1861–1933', *Journal of the Landscape Institute*, no.127, pp.30–33.
34. See J.A. Peterson (1976) 'The City Beautiful movement: forgotten origins and lost meanings', *Journal of Urban History*, 2 August, pp.415–34 and W.H. Wilson (1980) 'The ideology, aesthetics and politics of the City Beautiful movement' in A. Sutcliffe (ed.),(1980) op.cit., pp.166–98.
35. Beveridge wrote to Geddes, 26 February 1904, that he had become somewhat rehabilitated amongst the Trust members:
    I am afraid considerably at Mr. Mawson's expense. It must not come to the public or indeed not to be mentioned that he has sent in a plan for driving new streets right through the town and destroying nearly half the buildings including the new baths and Dr. Ross's house! (Geddes Papers MS10536, NLS).
36. T.H. Mawson, op.cit., p.100.
37. Geddes to Adler, Geddes Papers MS10612, NLS.
38. *Dunfermline Report,*op.cit., p.21.
39. See below p.125.
40. Patrick Geddes to Mr Murray, 8 December 1904, Geddes Papers MS10612, NLS.
41. He divided his Report up into seven sections: Approaches to the Park and Adjacent Improvements, Parks and Gardens, Stream and Glen,

Nature Museums, Labour Museums, History and Art, Life and Citizenship.

42. See chapter 2, p.20.
43. The Walkley Museum, Sheffield — see chapter 3, p.78.
44. T.R. Marr to Patrick Geddes, 8 December 1903, Geddes Papers MS10566, NLS.
45. Published in *The Nineteenth Century* November 1883, and reprinted in Rev. and Mrs S.A. Barnett (1888) *Practicable Socialism: essays on social reform*, London: Longman, Green, pp.62—75.
46. *Dunfermline Report*, op.cit., p.124.
47. Amongst his 'Hygiene and Domestic Improvements' to produce a stronger, fitter people, he advocates that every house should have an 'out-hung sleeping cage' like the open-air sleeping box found in sanatoriums for consumptives despite Dunfermline's known climate! Ibid., p.32.
48. Geddes Papers MS10612, no.165, NLS.
49. October 1904, Geddes Papers MS10536, NLS.
50. Ibid., 2 August 1904.
51. See *Dunfermline Report*, op.cit., Book II, Sections E—F.
52. 2 August 1904, Geddes Papers MS10612, NLS.
53. *A Manual of Benefactions*, op.cit., pp.241—6.
54. *Cities in Evolution*, London: Williams & Norgate, 1915, p.382.
55. W. Ashworth, op.cit., pp.167—90.
56. See Michael Harrison (1988) op.cit.
57. A. Sutcliffe, op.cit., pp.70—72.
58. He went from France to Italy, both lecturing and working on gardens — Geddes to Norah September 1905, Geddes Papers MS10501, NLS.
59. T.H. Mawson (1927) op.cit.
60. David Mawson, op.cit., and G.E. Cherry (1987) 'Thomas Hayton Mawson (1861—1933), a biographical note in *Planning History Bulletin*, 9(2): 28—9.
61. V.V. Branford to Patrick Geddes, 17 January 1907, Geddes Papers MS10556, NLS.
62. Ibid., 7 January 1907.
63. Geddes asked F.C. Mears to investigate the possibilities of interesting the Japanese in his civic movement but without success. Letter to Patrick Geddes, 22 January 1908, University of Strathclyde, Geddes Papers — Mr N. Nakagana (Secretary of Home Development, Bureau of Local Administration, Tokyo) to Geddes, 25 January 1909, ibid.
64. Geddes tried to interest Dundee Town Council in his work. But he did not get much response. Walter Walsh (Minister), 31 May 1907, and J.H. Martin, 10 July 1907, to Patrick Geddes. Geddes Papers, University of Strathclyde.
65. Frank Mears first began working for Geddes' projects in 1907 and was made Secretary of the Open Spaces Committee of the Outlook Tower. It was in this capacity that he first worked closely with Norah Geddes in July 1915.
66. He worked on an urban extension scheme with Geddes in South Shields. Later he was to be co-author of the *Cities and Town Planning Exhibition Edinburgh. Guide Book and Outline Catalogue*, Edinburgh: Hutchinson, 1911.
67. A. Sutcliffe (1981) op.cit., pp.76—7.

68. A 'Cities Survey Committee' was set up at the Town Planning Conference, and the Sociological Society, under Geddes' influence, set up its own 'Cities Committee to promote the Survey and Investigation of Cities, and the study of Civics in the first place by promoting Civic Exhibitions', P. Geddes, *Cities in Evolution* (1915 edn) p.251.
69. T. Marr to Patrick Geddes, 15 October 1908, Geddes Papers MS10566, NLS.
70. A. Sutcliffe (1981) op.cit., p.78.
71. T. Marr to Patrick Geddes, 14 November 1908, Geddes Papers MS10566, NLS.
72. Patrick Geddes to V.V. Branford, 12 August 1907, Geddes Papers MS10556, NLS.
73. H.E. Meller (ed.) op.cit.,p.68.
74. Patrick Geddes to V.V. Branford, 20 September 1909, Geddes Papers MS10556, NLS.
75. R. Unwin (1909) *Town Planning in Practice: an introduction to the art of designing cities and suburbs*, London: Ernest Benn, especially chapters II, IV.
76. Ibid., p.140.
77. Ibid., p.141.
78. Royal Institute of British Architects (RIBA) Town-Planning Conference, London, 10–15 October 1910, *Transactions*, pp.407–26.
79. *Garden Cities and Town Planning*, vol.1, 1910.
80. See A. Sutcliffe (1981) op.cit., pp.163–70; and G.E. Cherry (1982) *The Politics of Planning*, London: Longman, chapters I, II.
81. Patrick Geddes to V.V. Branford, 12 August 1907, Geddes Papers MS10556, NLS.
82. P. Boardman(1978) *The Worlds of Patrick Geddes: biologist, town planner, re-educator, peace warrior*, London: Routledge & Kegan Paul, p.183.
83. W. Boyd Rayward (1975) *The Universe of Information: the work of Paul Otlet for Documentation and International Organisation*, published for the International Federation for Documentation by the All-Union Institute for Scientific and Technical Information, Moscow.
84. Ibid., pp.123–5.
85. The Carnegie Endowment for International Peace gave the Central Office of International Institutions (set up after the 1910 Congress) $7,500 for the first half of 1912 and $15,000 for the fiscal year 1912–13, $15,000 for 1913–14 and more money was intended for 1914–15 when the war interrupted activities – Ibid., p.192.
86. Ibid., p.196.
87. P. Uttenhove (1985) *Resurgam: La reconstruction en Belgique après 1914*, Bruxelles: Credit Communal de Belgique, pp.47–50.
88. See chapter 5.
89. P. Geddes (1911) 'The City Survey: a first step – I', *Garden Cities and Town Planning*, 1:18.
90. Ibid., p.56.
91. G. Dix (1981) 'Patrick Abercrombie 1879–1957', in G.E. Cherry (ed.) op.cit., p.104, and G. Dix (1978) 'Little Plans and Noble Diagrams', *Town Planning Review*, 49(3): p.332.
92. H. McCrae (1975) *George Pepler: knight of the planners*, Glasgow: Department of Urban and Regional Planning, University of Strathclyde, pp.21–34.
93. A. Crawford (1985) *C.R. Ashbee: architect, designer and romantic socialist*,

New Haven and London: Yale University Press, p.155.

94. Unwin, op.cit., p.9.
95. Lord Pentland (1860–1925) was born in Edinburgh, educated at Wellington and Sandhurst, and took a commission in the Irish Guards. He was, however, a 'liberal' with a social conscience, and he became a resident at Toynbee Hall with Canon Barnett, where he became concerned with promoting education and sport. He was the moving force behind the setting up of the London Playing Fields Society designed to secure open space for sports use by the London poor. He tried to go into politics in the 1880s as a Liberal and supporter of Home Rule for Ireland but was defeated. Instead he became interested in the Empire, travelling to India, Tasmania, Australia, and New Zealand with Lord and Lady Aberdeen. In 1904 he married one of their daughters, Marjorie. He was then 44 and she was 24. Pentland was to become Secretary of State for Scotland from 1905 until his appointment as Governor of Fort St George, Madras 1912–19.
96. Lady Aberdeen (1857–1939) was the youngest daughter of Lord Tweedmouth and married Lord Aberdeen in 1877. Her husband was a diplomat and this enabled Lady Aberdeen to use her considerable talents of organisation in many social, educational and philanthropic ventures. When he was Governor-General in Canada (1893–8), she founded the Victoria Order of Nurses and became interested in women's issues. In Ireland, when her husband was Lord Lieutenant (1906–15), she made a great impact, founding the Women's National Health Association, and campaigning for better conditions for Irish women. Her feminism, though, was very much in the Geddesian mould, and she always maintained that the role of wife and mother came first. She had five children and was happily married.
97. P. Boardman, op.cit., p.129.
98. The following account of Geddes' work in Ireland relies heavily on a number of papers published in 2 volumes: M.J. Bannon (ed.) (1985) *A Hundred Years of Irish Planning*, Dublin: Turoc Press.
99. M.J. Bannon (1985) 'The Genesis of Modern Irish Planning', vol.1, ibid., p.197.
100. M.E. Daly (1985) 'Housing Conditions and the Genesis of Housing Reform in Dublin 1880–1920', ibid., pp.77–130.
101. M. Miller (1985) 'Raymond Unwin and the Planning of Dublin', ibid., pp.263–306.
102. F.H.A. Aalen (1987) 'Public Housing in Ireland 1880–1921', *Planning Perspectives*, 2:186–7.
103. Report of the Royal Commission on the Housing of the Working Classes (1884–5) Cd. 4547; Report of the Departmental Committee to Inquire into the Housing Conditions of the Working Class in Dublin (1914) Cd. 7273.
104. Report of the Departmental Committee appointed by the Local Government Board of Ireland to Inquire into the Public Health of the City of Dublin (1900) Cd. 243.
105. See C.A. Cameron (1904) *How the Poor Live*,
106. M.J. Bannon (1978) 'The Making of Irish Geography III: Patrick Geddes and the emergence of modern town planning in Ireland' *Irish Geography*, 2:141–8.
107. Countess of Aberdeen (1911–12) 'Memorandum on the Work of the

Women's National Health Association of Ireland', *Annual Reports of the Women's National Health Association.*
108. The Housing and Town Planning Association of Ireland.
109. M.J. Bannon (1985), op.cit., p.241.
110. Ibid., p.212.
111. P. Kitchen (1975) *A Most Unsettling Person: an introduction to the ideas and life of Patrick Geddes*, London: Victor Gollancz, pp.248–9.
112. M. Bannon (1985), op.cit., p.200.
113. Patrick Geddes to Norah, 14 September 1922, Geddes Papers MS10502, NLS: 'I told A. [Abercrombie] privately his survey was less adequate than it should have been and thus than Ashbee's – but the latter was unable to finish, or . . .? (A. naturally did not like this!)' This hints of a rift between Geddes and Abercrombie.
114. A. Defries, op.cit., p.181.
115. Thomson to Geddes, 2 March 1919, Geddes Papers MS10555, NLS.
116. P. Uttenhove, op.cit., p.45.
117. Antonio Manno (1980) *Patrick Abercrombie: a chronological bibliography with annotations and biographical details*, paper 19, Planning Research Unit, School of Town Planning, Leeds Polytechnic, pp.4, 19.
118. A.S. Travis (1977) *The Evolution of Town Planning in France from 1900–1919: with special reference to Tony Garnier and planning in Lyons*, paper 48, Centre for Urban and Regional Studies, University of Birmingham.

CHAPTER 7

The challenge
of India and Palestine

BY 1914 GEDDES WAS ENJOYING CONSIDERABLE SUCCESS
with his Cities and Town Planning Exhibition. His work in Dublin had
brought him into the public eye and he contemplated taking the
Exhibition to America. He saw a chance of developing international
links and he had already been invited to take the Exhibition to South
Africa to the city of Pretoria.[1] Yet in the end, for both pecuniary and
personal reasons, he chose to go to India where he had been invited by
Lord Pentland, Governor of Madras. He was always searching for extra
income for his work and he assessed, quite rightly, that if he was a
success in Madras, other invitations would follow. The governor
network, especially the 'liberal' governors, friends of Lord Pentland,
could be utilised.[2] On personal grounds, the East was a challenge. All
his work, all his observations to date, had been in cities built by
Europeans. The alien cultural context of the East drew him like a
magnet.

Over the next decade Geddes was to find work, not only in India,
but also between 1919 and 1925 in Palestine. He practised 'the art of

living dangerously and abundantly for the sake of an endless future',[3] but for his considerable achievements he was to pay a heavy personal price. The first blow was the loss of his precious Cities and Town Planning Exhibition, the work of thirty-five years of effort, sunk by enemy action in the Indian Ocean off Minicoy. Lord Pentland wrote in his diary that 'Geddes has taken it like a sportsman'.[4] Friends in London, co-ordinated by H.V. Lanchester of the RIBA, co-operated to send out fresh material to build up a second Cities and Town Planning Exhibition. Much more serious were the losses Geddes was to endure in his family. His eldest son, Alasdair, who had become his ablest collaborator in his work, and had accompanied him to India in 1914, felt obliged to volunteer for military service. He was killed in May 1917.

For Geddes the loss was not only of a son, but also of the closest helper he had ever had in his lifelong search to find collaborators to carry out his ideas. Alasdair's life and education had, from his earliest years been dedicated to this end, and he had had the temperament and aptitude to fulfil his father's demands.[5] Very soon after the news of Alasdair's death reached him, Geddes suffered his greatest blow, the death of his wife Anna. She had gone with him to India where she contracted dysentery and fever. She was just 60 years old, and from the earliest days of her marriage she had given her husband total support, and worked ceaselessly at the administrative details which had kept his work and his family afloat, whilst leaving him free to pursue his ideas. It had been a very unequal division of labour and Anna suffered for her unselfishness, though always willingly, as she enjoyed a very warm relationship with her husband.[6] She was organising Geddes' first Summer Meeting in India at Darjeeling when she entered her final illness. Without her, Geddes increasingly lost his emotional equilibrium and grasp of reality. He indulged in romantic fantasising ever more frequently in his letters and published work.[7] He was, however, to dedicate his great Indore Report, the best work he believed he had done in India, to the memory of his wife and son.

When Geddes left for India in 1914, he had no premonition of impending catastrophe. What he had not anticipated was that his propaganda and educational work would become supplemented, indeed overshadowed, by commissions for practical planning work. As this process occurred during the next decade, Geddes welcomed it enthusiastically as he always enjoyed the involvement of specific schemes. This work was to build his reputation as a town planner over and above the mantle of the prophet of civic reconstruction which he had already gained. It was a vital extension of his image, since many who found his civics movement unpalatable or confusing were, nevertheless, impressed by his practical record. At the time, and even perhaps subsequently, few town planners have ever worked on so many cities as Geddes was to cover in both India and Palestine in the period 1914–24. It was an achievement that the Town Planning Movement could ill afford to ignore.

For a while in the 1920s and 1930s the library of the Calcutta Improvement Trust was a Mecca for aspiring young British planners, who came to study the complete set of Geddes' reports which used to be kept there. Visiting Indian cities and using them for comparison and contrast with cities at home, was a stimulating experience.[8] Yet the fact that Geddes' major practical work was in India has created problems for many seeking to understand his work. On the one hand his planning approach, conditioned by his civic reconstruction doctrine, was highly idiosyncratic; on the other, the context within which he worked, which profoundly influenced his achievements, was the product of the special historical circumstances of that decade in India. Geddes' concept of planning was to manipulate towards his particular perception of life-enhancement whatever economic, social and political circumstances he found. Under this double burden the reports have become, over time, increasingly obscure documents. Geddes' reputation as a planner has had to rest more on the assertion of his supporters than on an assessment of his work. Yet this need not be. For all the ephemeral nature of early town planning activities, it is still possible to piece together the context in which he worked and Geddes' planning response.

By going to India at the outbreak of the First World War Geddes had taken himself out of the mainstream of the development of modern town planning, which was concentrated on the phenomenon of rapid urbanisation in the west. What he was hoping for was a chance to apply his socio-biological approach in India. He thought that it might be possible to by-pass some of the stages of change suffered by western cities under the impact of industrialisation. Modern industrial-isation in India had barely begun and was, in any case, confined mostly to the textile and iron and steel industries. Geddes thought there was a chance in this mainly rural continent that his reconstruction doctrine could take root. In April 1917 Geddes wrote a long letter to his friend H.J. Fleure explaining how he saw this opportunity:

The transformation of India, from poverty to wealth, is becoming insistent. There is an Industries Commission on its rounds . . . There are others searching towards the transition, pointed out as necessary in *Cities in Evolution* before I had seen India. But I doubt if there is anyone here as yet, able to deal with this question. At present the Commission seems essentially paleotechnic, and missing entirely the agricultural and civic problem of Regional constructiveness, in which real hope lies. Still, I do not despair altogether of getting it before the public and government; though, as yet, I don't feel that the Cities and Town Planning Exhibition has elicited any *real* response, or discovered anyone of constructive power; all are in the stages of Lib-Lab Fabianism or its criticism as Radical Socialist alternative and Neotechnic and Geotechnic [approaches] are not yet apparently clear to any public man, Indian or European I can discover yet. Yet the progress has begun. The Foresters, the irrigators (though rude in their methods) are real; and real agriculture too is beginning. What magical change good seed makes! And it may be, by this new Pusa wheat and Five-rowed barley, and so on, that the transformation not only from poverty to wealth for the cultivators, but from policeman to peasant-helper and citizen in John Bull may come to pass: as that from lawyer to the same in the educated Indian.[9]

Geddes as usual, was optimistic in the face of the potential he saw, regardless of the difficulties he found which inhibited his work. His 'Reconstruction' message appeared to many in the British Administration to be superfluous and even dangerous. The Indian Civil Service, which provided the administrators for the municipalities, took little notice of his ideas and was generally hostile. The result was that, after Geddes' initial popularity amongst the 'liberal' governors, Pentland, and his friends Willoughby and Carmichael, he failed to get much further support from the British Administration. He remained all his time in India as an outsider, tolerated by the British but not encouraged. But Geddes was able to turn increasingly to the princely rulers of the native states to keep up his flow of commissions, helped in his search by his Indian friends, such as Sir Jagadis Bose.[10] He became a skilled entrepreneur selling his services, seeking commissions, and making more money in India than he had ever been able to do before. After his wife's death in 1917, when his daughter Norah tried to persuade him to come home, she received the following reply:

> You see, that just as I had constantly to leave home, since I earned nothing in Edinburgh, and as I now also earn nothing in London, I am compelled to work on here whether heartsick or not. I have pot to boil, past and dead horses to pay for, live white elephants (Tower etc) to provide for and provisions to make for the future.[11]

Geddes was able to capitalise on the stirrings of interest in economic and social change in India during and immediately after the First World War. Since he was rejected by the British Administration, he threw his energies into promoting his ideas amongst the leaders, particularly Indian leaders, of the cities he worked in. The drawback of this approach was that Geddes was left without any base from which to develop his ideas over a longer term. His role was mainly as a 'trouble-shooter', brought in to comment on the dilemmas faced by cities trying to develop a modern infrastructure whilst facing the ineptitude and inflexibility of their administrative structures. Indeed, the way Geddes expressed most of his planning ideas in his reports was designed to goad civic administrations into a new perception of their duties, and to avert some of the damage they were causing. Administrators of cities tended to turn to western-trained engineers to help with their urban problems, which usually meant one thing only: the total demolition of insanitary areas. Geddes felt that in India this response could be shown to be absurd, given the general levels of poverty and industrial development. In optimistic mood again, writing from Lahore in January 1917 to his daughter:

> What opportunities there are here — *and for the whole connection of us* ! — a whole continent with each province like one of the great European countries in population and with cities of the greatest variety and interest, and all needing human planning to cope with the devilries of the engineers, and their wastefulness and bungling and vandalism, for the greater part. Of course they may have their uses, but in cities, they are as yet mainly a curse, so far as I have seen.[12]

Urbanisation in India, 1900—1925

Geddes was not alone in perceiving that the problems of cities were demanding political and administrative expertise of a new kind. It was becoming obvious that the nature and the pace of Indian urbanisation generally had begun to change. While, numerically, the ratio of urban to rural population remained small for the period 1861—1941 as a whole (it was a mere 10 per cent) yet the figures show that for the first four decades of the twentieth century, cities were absorbing a larger share of the population increase.[13] Furthermore, the sheer scale of the Indian population as a whole meant that the proportion of the population living in cities actually amounted to as much as three-quarters of the entire British population in 1901. A.F. Weber, working at the end of the nineteenth century, highlighted three major features of the Indian experience of urbanisation which could be statistically determined with differing degrees of accuracy.[14] The first was that Indian cities had been very large in the pre-industrial era when modern technology was unknown. According to estimates made in 1823 Madras had a population of 817,000; Benares, 580,000; Delhi, 400,000; Calcutta, 900,000; and Surat 450,000. Thus the size of large cities in the early twentieth century was not unprecedented. But in earlier periods large cities were as likely to contract as to grow with shifts in their administrative, political or religious importance, or with the outbreaks of epidemics and plague.

A second significant point was that the statistics for growth in the decade 1881–91 showed a large increase in the case of industrial cities. Weber cites the examples of a 43.4 per cent increase of Hubli, 43.01 per cent of Karachi and 41.26 per cent of Ajmer. Rapid growth was now allied to economic development. Finally, the largest towns were gaining an increasing proportion of the total urban population. Weber, from his vantage point of 1899, was cautious about generalising on this latter tendency. He pointed out that the growth rates of individual large cities fluctuated according to the importance of the industrial, religious or administrative functions of the city, and to the ravages of epidemics and diseases. But, at least in this respect, time was to show that the concentrations of urban population in the larger cities was an established trend of Indian urbanisation. Large towns of 100,000 inhabitants or more thus began to gain an importance out of all proportion to the general level of urbanisation as a whole. Weber suggested that a city of this size or more had a vital economic, social and cultural role to play, not only locally, but for its rural hinterland, and as an outpost of national civilisation.

The Indian urban sociologist, G.S. Ghurye, seized on this idea when he suggested that one of the most significant trends in Indian urbanisation was the development in the period 1881–1941 of such large cities in practically every major region of the country.[15] In 1881 there were only 18 great cities in the entire subcontinent. By 1921 there were 35, and by 1941, 47. The distribution of these cities was not very even with some territories, such as Bombay State and the United Provinces, far outdistancing the rest. But behind these large cities was another group of smaller but still significant cities in virtually every region, province, and native state. If urbanisation is accepted as a vital factor in the modernising process, then the period 1891–1931 was the transitional stage for India. A new response to urbanisation had been on the political agenda since the 1880s with moves to encourage new methods of local government, and even more crucially, to raise local taxes to pay for municipal services.[16] Both were sensitive issues in the context of a colonial country.

While the western democratic principle of 'no taxation without representation' could not apply in this context, it still became essential to find the means of showing that local taxes were being used to pay for the economic and social needs of the local community. This was a complicated matter. The administration of most cities was closely controlled by the Indian Civil Service, and there was less necessity, politically, to consider the needs of different groups of citizens. It is true that the extension of services in municipalities, particularly sanitation, education, and medical care since the late 1870s, made greater demands on local taxation, and gradually the British, under Mayo and Ripon, had extended the principle of local self-government to all municipalities under British rule. But with a few exceptions, such as the great Presidency cities, this was a dead letter in political terms, since the chairman, the municipal commissioner, was usually a British official with a concern for efficiency, but none for innovation.[17]

However, some Indian nationalists had begun to see, in local self-government, a chance to exercise some power despite the British. In the period 1909–25, whilst Geddes was in India, local self-government in the municipalities was taken more seriously, and it became the exception rather than the norm to have a British official as chairman.[18] In these circumstances several prominent Indian leaders and members of Congress, such as Surendra Nath Banerjee, Pherozeshah Mehta, Lala Lajput Rai, G.K. Gokhale, Vallabhai Patel, C.R. Das, Subhash Chandra Bose, and Jawaharlal Nehru became associated with municipal work.[19] Their hopes, however, were soon to be disappointed. As Jawaharlal Nehru put it:

> The whole steel frame of municipal administration as erected by government, prevented radical growth or innovation. The financial policy was such that the municipality was always dependent on the government. Most radical schemes of taxation or social development were not permissible under the existing municipal laws. Even such schemes as were legally permissible had to be sanctioned by government, and only the optimists, with a long stretch of years before them, could confidently ask for and await this sanction.[20]

Nehru resigned before his three-year term of office was completed as the limits of his capacity to expand the range and impact of municipal work became apparent. Urbanisation generally was to aid the nationalist movement, but the leaders were not able to develop power bases through the exercise of municipal administration at local levels.

Thus the interaction between the growth of municipal responsibilities and the recruitment of local leaders for municipal service which had provided the context for the implementation of town planning legislation in Britain, was missing in India.[21] Yet the problems of Indian cities continued to get more pressing. Although growth rates did not approach those achieved in the post-independence period, those cities stimulated by contact with the British increased very rapidly. In 1838, Bombay for instance, had a population of 236,000, which made it already larger than the cities of Birmingham and Leeds at that time, though not quite as large as Manchester and Liverpool. By 1872 this number had multiplied to 644,000. By 1891 it was 821,764 and by 1921, 1,175,914. This was roughly comparable to the size of London in the second quarter of the nineteenth century, when ideas on the need for a public health movement were first mooted. Certainly by the end of the nineteenth century Bombay's problems were large enough to stimulate a similar response. Between 1896 and 1910 plague was endemic in the city, each fresh visitation carrying away many victims, weakened by the periodic famines endured by the city's poor.[22]

A report of 1906 estimated that over 200,000 migrants had come to the city in the last five years. Immigrants were desperately poor and ignorant, and since they stayed in the city for only part of the year, they accepted appalling conditions. Plague, famine, and tuberculosis were rife. A religious worker, working amongst the poor, wrote in 1908: 'The vocation of the Bombay Mission seems to be especially to prepare people for death. . . After being with us for a short time, death carries them off'.[23] Poverty was not the concern of government. But the mortality rates and the speed of urban growth demanded action. What in fact was done was the setting up of the Bombay Improvement Trust in 1898, the first in India. Such an organisation was deemed eminently suitable in the circumstances. Its major function was to control the development of new areas especially planning new roads and controlling public health nuisances. It came into existence 'entrusted under a Special Act. . . With the work of making new streets, opening out crowded localities, reclaiming land from the sea to provide room for the expansion of the city, and constructing sanitary dwellings for the poor and the police'.[24] In the years before the First World War, Improvement Trusts were set up in other large cities, both in British territory and the Native States. Hyderabad, for example, gained an Improvement Board in 1914, and Calcutta, where the scale and extent of India's urban problems reached a new extreme, also in 1914.[25]

Improving cities in the Indian context

Geddes, since his days with the Edinburgh Improvement Trust, was aware of their shortcomings. Their biggest drawback, in his view, was their need to be financially self-supporting. Instead of working in the interests of the people, especially the poor, Geddes wrote: 'The "Improvement" methods, derived their advantage, even their survival, from the opposite view-point and interest, that of the propertied and land-speculating classes and their economists; by making site space and working class dwellings permanently and increasingly dear'.[26] Housing for the poor in India though, was not the same kind of problem that Geddes had tackled in Edinburgh. There he had been renovating tenements which had formerly been palaces or the homes of the middle classes.[27] In Indian cities, while central areas had deteriorated because of severe overcrowding, the new slum areas were actually located on the periphery of the city wherever open space was available. It was usually a stretch of land that really was unfit for habitation and on this land huts of kutcha sprang up, serving as the nucleus for the growth of a new slum. The huts were scattered in a haphazard manner, and these areas often had

> no proper access other than a narrow twisted lane which was mire and slush in rainy weather and a dusty beaten track in the dry season. Drainage was totally absent and protected water unavailable. The huts themselves were little hovels built of the flimsiest material, walls were built of mud and stone or bamboo matting, and the roofs of kerosene tin sheets, rags, gunny sack, canvas, bits of wood, reeds and hay. It was rarely possible to stand up inside one of these structures.[28]

The Bombay Improvement Trust had tried to overcome the problem by building huge purpose-built tenement buildings (chawls), containing rooms 10 feet by 10 feet, with one bathing space, and a latrine for every six to eight tenements. These chawls were built of low-quality building materials and received no maintenance. They were also multi-storey and had little light or air around them. Geddes likened them to prisons without access to any kind of natural facility. In the densely crowded suburbs of Bombay in the climatic conditions of the city these improved tenement dwellings became stifling. Geddes thus became convinced that, in the Indian context, Improvement Trusts were doing more harm than good in their activities in Indian cities.[29] In Madras, on his first commission with the Cities and Town Planning Exhibition, he advised the municipality to appoint a town planning officer rather than set up an Improvement Trust. The former could have a wider brief, working for similar objectives to an Improvement Trust without its financial restraints.[30]

His success with this recommendation brought him a rich personal reward. He was able to secure the services of an architect-planner to work with him in India. At Geddes' instigation, Lord Pentland wrote to Raymond Unwin at the Local Government Board in London, asking him to suggest the name of a suitable person for the post of the first town planning officer in Madras. He recommended H.V. Lanchester. Lanchester was Vice-President of the Royal Institute of British Architects in 1913, and a founder member of the British Town Planning Institute. He had already visited India a couple of times, and had come as a member of the New Delhi Development Committee in 1912 (he was very disappointed not to get any work himself on this project). He had been commissioned to do some planning work in Indore and in Gwalior. Since his time as editor of *The Builder*, he had become far more deeply interested in town planning, and he was seeking ·openings to develop his skills in this respect. After India, he pursued a career as a colonial town planner in Africa. After the 1910 RIBA Exhibition he had become a fervent admirer of Geddes.[31]

When he took up his appointment as Town Planning Officer in Madras in 1915, he tried hard to carry out the kind of survey work which Geddes insisted was the essential preliminary to proper planning. His report on Madras contained material from a major survey of local conditions, including maps illustrating the occupational structure of the city, population densities, plague black spots, rates of mortality, and infant mortality, and much laboriously collected material of this kind.[32] This work was important to Geddes since he never carried out any detailed survey work in India himself. Instead he was able to quote Lanchester's Report as an example of how it should be done. Lanchester's architectural plans for Madras, however, still retain elements of the grand design with Beaux-Arts vistas and impressive buildings in the classical style, rather reminiscent of his 'City Beautiful' designs for the municipal buildings of Cardiff in Cathay Park. His architectural proclivities and Geddesian sociology do not seem quite to match up with each other. However, Lanchester found his contact with Geddes very stimulating, and he worked with him on many plans between 1915 and early 1917. He was particularly important in helping him in Lucknow where Lanchester had set up an architectural office to supplement his earnings and where he could most easily produce the detailed plans and drawings to back up Geddes' ideas. The two men worked as a team and also independently, on occasions superseding each others' work. Geddes followed Lanchester to Indore and Lanchester followed Geddes to Cawnpore (Kanpur).

Lanchester wrote a general text book on the *The Art of Town Planning* (published in 1925) in which he drew on his experience in India to contrast the major differences he perceived between town planning in the west and the east.[33] The crucial difference he suggested was that town planning in the east 'arose out of health measures dealing with insanitary and overcrowded areas', not as in the west, as a movement leading towards new developments in urban living.[34] This was an important point because the dominance of the public health hazards of Indian cities conditioned the nature of the response to their problems. The need for drastic measures seemed overwhelming. At one conference of the All India Sanitary Association just before the war, Dr Kailas Chundar Bose had produced statistics to suggest that health standards were deteriorating rapidly in large cities. The incidence of tuberculosis in Calcutta had increased fourfold between 1880 and

1911. Infant mortality rates in most Indian cities were appalling. In 1911 in Calcutta it was 362.1:1000; in Bombay it was 379.8:1000; in Bangalore, 267.8:1000. In Britain in 1911 the rates in London were 91:1000; for Birmingham, 111:1000; and for Liverpool, 125:1000. Slums in the centres of large cities were different in kind as well as in quality.[35] Calcutta had probably the worst slums in the world. The slum area in the city centre consisted of huge 'streetless blocks' of building, which covered areas ranging in extent from 20 to 270 acres, but most commonly extending over 100 acres. Infant mortality rates in these slums were more than one in two.[36] These were the kinds of facts which Geddes had to contend with in his attacks on the Improvement Trusts and the activities of the civil engineers and sanitarians. Just how original he was in this respect can be highlighted by a brief examination of the far more conventional approach of the first chairman of the Calcutta Improvement Trust, E.P. Richards, an engineer by training.

Western responses to Indian urban problems

E.P. Richards's report, *On the Condition, Improvement and Town Planning of the City of Calcutta and Contiguous Areas*, was published privately in England in 1914, where he was convalescing after a complete breakdown in his health.[37] In it he displayed the three most common western responses to Indian urban problems. First, a root and branch approach to health black spots, fully utilising sanitary legislation to enable the compulsory purchase and demolition of all building in these areas. Second, he tried to assess Calcutta's needs by comparing the city with other cities of comparable size, examples being drawn mostly from the west. Calcutta's position seemed all the more hopeless when it was compared with London, Paris, Berlin, Rome, Venice, Hamburg, Dresden, Budapest, and Chicago. Finally, for finding ways of doing the little that could be done, he turned to the Town Planning manuals and legislation of the west. He was totally pessimistic that Calcutta's problems could ever be tackled even using such legislation. Only a completely authoritarian regime with huge resources and a vigorous policy of demolition would make any impact whatsoever. His gloom was compounded by the limitations under which the Calcutta Improvement Trust had to function. Unlike its counterparts in Bombay and elsewhere, the CIT was set up under an Act based on the 1890 English Housing Act, aimed at the problems of working-class housing and slum clearance. Faced with slum clearance in central areas, its powers were in no way adequate for it to undertake this work.

Richards's pessimism was shared by British members of the ICS with responsibility for urban problems. The problem of inadequate legislative powers could be overcome. The Hon. Mr E.G. Turner ICS, of Bombay, recommended those seeking practical advice on improvement schemes to consult the 1910 Finance Act, the Kingsway Improvement Scheme Act, and the German Laws of 1893, 1911, and 1913.[38] Further shortcomings could be remedied by the state legislature. In fact, in 1915 Bombay State pioneered the first Town Planning Act in India. But the major problem was that legislation was powerless in the face of the practical obstacles to its implementation. In some cities the demolition activities of the engineers in central areas had sparked off communal riots. Tensions between Moslem and Hindu, between caste and caste, were brought to the surface by the destruction of a temple or sacred place or the location of an abattoir. A minor improvement scheme in Cawnpore had resulted in serious rioting in 1914.[39]

In Bombay the administrators tried to adopt a more positive approach. Unfortunately this usually cost money. Mr J.P. Orr CSI, ICS, Chairman of the Bombay Improvement Trust, gave a lecture entitled 'Social Reform and Slum Reform' in which he suggested that the way forward was to offer the poor better housing. Yet the few chawls built by the Trust hardly gave substance to his hopes. Another ICS officer in Bombay, Mr A.E. Mirams, Surveyor to the government, was particularly enthusiastic in the cause of town planning.[40] He, like Lanchester, was to further his career in town planning in Africa. In India, however, he turned his back on the problems of Bombay and concentrated his efforts on the small towns and villages in the state. He spent his time on propaganda work, travelling ceaselessly from Sind in the north to Belgaum in the south, giving lectures on the 1909 Town Planning Act. It was an isolated effort and for all his enthusiasm, his activities were rather amateurish. His lectures were illustrated by slides which he had sent from England from The Garden Cities and Town Planning Association. There was generally a total absence of ideas on how to improve the physical environment of cities.

The task of introducing modern town planning in India was thus one of such magnitude that only the most enthusiastic and optimistic propagandist of the movement would have been tempted to respond to it.[41] In 1914, as Europe plunged into war, perhaps only one man could have taken up the challenge so wholeheartedly. Geddes felt that both time and place were ready, not just for the transmission of modern town planning ideas, but for his entire doctrine of Civic Reconstruction.[42] In some respects, he was right that the time was critical. It was as if the whole context of Indian life and culture hung in the balance between the diverging forces of old established customs and the impact of western ideas. The process of modernisation on the western model, which had been gaining momentum in the course of the nineteenth century, had been greatly boosted by the First World War.[43] People and government were actively seeking change. The integration of India into the world economy had been gradually taking place for some time, but by the turn of the century the cumulative impact of these changes was beginning to be felt. The British government became uneasily aware of the need for planned economic development. In the wake of the Indian contribution to the war effort, the British government felt an obligation to offer help and advice in this matter. An Industrial Commission was sent to India in 1918. World economic integration, however, had wider implications. Closer ties brought cultural as well as economic links.

The cultural context for environmental planning in India

A new global culture, produced initially in the west, but now incorporating all nations east and west, was in the making, the First World War generally creating an awareness of the changes that were slowly taking place. Dr Anthony King has written persuasively about the connections between economic development, cultural domination, and urban form.[44] He argues that the built environment is a most sensitive indicator of the cultural environment which produces it, and that India provides an extraordinary example of how this works.[45] The model was not the simple one of the two-way process of mutual influence between subject nation and imperial power. With the growth of world economic activity, India was being drawn into an international culture, based on modern technology, which was not simply synonymous with all that was British. The political tensions inherent in developing any model of India's future could be seen reflected in the culture, life-style, and building projects of the elite of Indian society. There was a dual influence at work. British cultural propaganda on the one hand, and on the other, the more diffuse but all pervasive process of modernisation which accompanied the growth of closer international economic relations and new technologies.

Sons of the Indian social elite had been coming to England for their education in small but increasing numbers since the 1880s, giving substance to Macaulay's famous dictum of the 1830s of

> imparting to the Native population knowledge of English literature and science through the medium of the English language to form a class who may be interpreters between us and the millions we govern; a class of persons, Indian in blood or colour, but English in taste, opinions, in morals and intellect.

But an education in England and chances to study and travel in Europe

[213]

exposed these young men to a wider context, the international culture of modern life. As agents of cultural transmission they experienced both personal and national conflict. While wishing India to take her place in the modern world, adopting many western institutions and activities, they were also anxious to reject the prospect that such modernisation was a sign of subservience to the British.[46] The difficulties of this elite group are illustrated by the vicissitudes in the fortunes of the nationalist movement.[47] Both within and without the movement, in the period 1890–1918, questions of education and religion were keenly debated. These issues, the keystones in the cultural life of the nation, seemed suddenly of paramount importance. It was a critical debate about India's future. Amongst the British in India there were a small number of enlightened administrators and missionaries who involved themselves closely with these issues.

One such was C.F. Andrews, both Christian missionary, and close supporter of Rabindranath Tagore in his work at Santiniketan, which brought him into contact with Geddes, and even more closely with Geddes' younger son, Arthur. Andrews, as a friend of many Indians, Christian and non-Christian, was particularly sensitive to the issue of cultural identity. He wrote prolifically on the subject. For example, he wrote in 1912,

> This age in which we live is the renaissance for India. There is a tide of new learning surging in, destroying ancient faith and practice, undermining the old foundations of morality and of Indian society, producing an eager, restless, throbbing mass of student life, pushing onward amid a ferment of new ideas, and 'the moral unsettlement of a period of transition'.[48]

The fact that this 'ferment' was to be found amongst only a tiny minority did not deter him. The bald statistics were that India at that time had a population of 313.5 million, of whom 18.5 million were literate, and 1.6 million literate in English; whilst there were only 29,187 men in colleges of higher education in India and, even more significant, only 342 women.[49] He wrote rhetorically:

> only a few thousand students of no particular importance! No, these are the precursors of a new age, these are the first fruits of a renaissance, these are the future leaders of a nation that has been dumb for centuries and is being disillusioned.

Andrews's expression of the responsibility of the British in the face of this challenge could have been written for Geddes.

> It is the God-given task of this great empire under whose government we live to mould this power, to shape it so that it does

Figure 7.1   Narrow housing lane in India

'In housing areas there is no need of wide dusty streets. Indian tradition is far wiser with its use of narrow lanes, opening into pleasant squares, each containing a shade-bearing tree. The narrowness of the lanes makes for shade and quietness, and leaves building sites large enough to enclose courtyards and gardens.' Teppakalam, Trichinopoly, 1945.

Figure 7.2   A courtyard house, India

'A courtyard, bright with colour-wash and gay with old wall-pictures, adorned with flowers and blessed by its shrine.' Benares, Uttar Pradesh, 1944.

not fail to give to the world the contribution which lies hidden away in the centuries of India's priceless history and in its ages of solitary evolution'.[50]

What drew Geddes to India was the hope that he might study this 'solitary evolution' and reinterpret it in the terms of the modern world. From his youth he had been fascinated by the east. In the 1880s members of the Fellowship of the New Life had looked with a romantic longing to India for spiritual enlightenment in their distress at the results of western industrialisation. Edward Carpenter, for instance, read the Bhagavad Gita from the great Hindu epic, the Mahabharata, and went on a visit to India.[51] The romantic response was a reaction to the hardening of British cultural attitudes against the artefacts of Indian civilisation in the wake of ever tighter political domination.[52] Indian literature, art, and architecture were denigrated and described as debased and vulgar in comparison with the achievements of the civilisations of the white races. In architecture particularly, the imperialists built in the classical style of ancient Greece and Rome, creating an architecture that was a symbol of European domination. When Geddes got to India, he was not interested in the European imports.[53] He wanted examples of indigenous architecture and urban form. On his arrival in India he toured the subcontinent from north to south. What he was seeking he found in the south, in the great temple cities of the ancient Dravidian culture. He wrote a euphoric article about the temple cities as examples of the integration of culture, history, and urban form at its best, and in his enthusiasm the deeply romantic vein with which he viewed India was given full rein.[54]

Geddes' exploration of indigenous culture and traditions

The forum in which Geddes initially gave his eulogy was the Madras Literary Society, an imitation of the Royal Asiatic Society, which Lord Pentland had resuscitated from its formerly moribund state.[55] Pentland's activities were matched in the other Presidency cities, and Indians and British met together to hear scholarly papers on Indian culture and customs. It was part of the 'renaissance' defined by Andrews.[56] In Madras, a member of the society was so impressed with Geddes' enthusiasm for the temple cities that he began the research on their history which he subsequently published, dedicating his work to Geddes.[57] In such circles Geddes found a great deal of support. There was a growing consciousness of the need to repair the damage inflicted by British cultural domination. Neglect of historical knowledge had been paralleled by neglect of the actual fabric of Indian historic buildings. These had been allowed to fall into ruin. This disintegration became ever more serious as the skilled craftsmen and masons able to do the necessary repair work were themselves a dwindling number. In 1912 the Government of India had commissioned a report on Indian architecture which highlighted the decline in standards of craftsmanship and knowledge. A South Indian, Mr A.V. Ramachandra Ayyar, took the initiative to press for an All-India Sthapathya Vedic conference to be called to discuss the problems of Indian architecture and town planning. A conference was held in 1918, a small indication in itself of changing attitudes.[58]

There was, however, the conflict inherent in all these activities between preserving India's heritage, and pioneering her path into the modern world. Sympathetic missionaries like C.F. Andrews understood the dilemma very well. It was a similar problem to trying to convert Indians to Christianity. To follow Christian teaching left the Indian convert culturally stranded in his own country, alienated from his own society. Andrews escaped by giving his services to Rabindranath Tagore, helping him in his educational work in the hope that, through a basis of goodwill, east could meet west and both retain the best of their separate identities.[59] He liked to suggest that he was following in the noble tradition established by the great religious leaders of India in the nineteenth century, who had sought to reform religious practices in order to bring Indian society into closer contact with western ideas and ideals. There was, however, an ambivalence in his position which was also to be shared by Geddes.

For the religious leaders of the late nineteenth century, however, the work became more complicated. Geddes had come into contact with the mission of Swami Vivekananda to the west. Vivekananda had had a western education, and had been exposed to western ideas at a formative age. His attempt to build bridges between east and west, while still serious, had a sharpness that had not been there before. Vivekananda had become famous overnight after his address at the Parliament of Religions at the Chicago World Fair 1893. He had burst upon the scene, a glamorous, oriental figure (he was only thirty at the time), and the interest he roused was perpetuated by the biography of his life written by Romain Rolland, one of France's most influential writers before the First World War.[60] Geddes met the Swami at the 1900 Paris Exhibition. But far more important to him personally, he met a European disciple, Margaret Noble (known as Sister Nivedita), who stayed with the Geddes' for some months in Paris and became a close friend.[61]

Sister Nivedita had been fascinated by Geddes' socio-biological approach to social analysis, and his belief in sociology as social religion which inspired the individual to social service. She hoped she would be able to write a book about his approach, but she found herself often confused and she did not succeed. She did, however, write a book about the social life and customs of India, *The Web of Indian Life*, which she published in 1904.[62] It revealed her sympathetic response to Geddes' ideas and he, in turn, was to use her book as a guide in his search for Indian social customs. She was a vital source to him, especially in discussions about the importance of household gods, and the sacredness of the home, and in her descriptions of the role of women in Indian society. The subject of women and their emancipation was an extremely delicate one. For centuries women had been subservient to their men, whether Hindu or Moslem, and Geddes found considerable difficulty in trying to transpose his evolutionary views of women as the agents of cultural transmission in this context. Since he believed this role was vital to higher evolutionary development, he found the domestic ideals emphasised by Sister Nivedita a crucial starting point for his socio-biological approach.

He followed this up by selecting those parts of the holy books which exalted the sacredness of women's lives as wives and mothers; whilst ignoring other sections which justified their servility. He had allies to hand to help him in this task. Mrs C.M. Villiers-Stuart's book on Moghul gardening combined for him a fascinating account of the history of the subject with a strongly feminist bias.[63] Mrs Villiers-Stuart suggested that the preservation of the garden, and the vital cultural traditions which it served, was the work of women. Even the woman in purdah, with no external role, still had a civilising role to play:

> India is no exception to the rule that it is women who preserve intact the old religious observances; there as elsewhere, it is they who keep old memories fragrant — so the Indian garden is above all, the purdah woman's province. The day begins with the housewife's reverence, the pradakshina about the sacred Tulsi bush, which is generally planted in an altar built for the purpose in the centre of the forecourt.[64]

Through ancient religious observances, the housewife's dedication to her role was constantly renewed. Geddes was willing to see in this a way of revitalising the nurturing traditions of women in the towns and the cities where he worked. In almost all the reports he wrote, he mentions the need for every household to have its Tulsi bush to encourage the spiritual, and thus renew the energy for the practical achievements of the Indian housewife.[65]

The third major influence on Geddes before he arrived in India was a book by a public hygienist, Dr Turner, on sanitation in India.[66] Aware that his planning work would most probably be concerned intimately with these matters, Geddes was eager to absorb useful advice. Dr

Turner wrote that water-borne sewage disposal in India was both exorbitantly expensive and impractical. Instead, he suggested that the waste matter should be collected and used for gardening, and in this way eliminate the need for building an infrastructure of drainage pipes. It was a solution that had been tried in England in the 1840's before water borne sewage disposal had been widely adopted. It was direct, simple, cheap, depended on arousing new social and civic consciousness, and it resulted in the enhancement of gardens. It was the socio-biological answer. With Dr Turner's help, Geddes thought he had found a solution which could effectively clean up areas, and at the same time revive and nurture the customs and traditions described by Sister Nivedita and Mrs Villiers-Stuart. Above all, if public health could be treated in a socio-biological way, indigenous patterns of urban form could be saved, and thus the cultural heritage of the past preserved as the strongest guarantee against future deterioration in the environment.[67]

In the pursuit of his socio-biological approach, Geddes was constantly searching for indigenous customs and traditions which could be revived and made to serve modern purposes. He became convinced, after his discovery of the temple cities of the south, that he would find there the traditions of a civic consciousness which could be the model for his work in India. Geddes' predilections for the south were picked up by one of the first Indian professors of sociology and economics, a Bengali, Professor Radha Kamal Mukerjee, who was the first to be appointed to a chair in these subjects at Lucknow University in 1921.[68] He wrote of the domestic and communal traditions of South India:

> There almost every house has an orchard, which receives the sewage of the house that is the main stay of a profitable vegetable garden. Every street is lined with shady trees and its width guaranteed by the periodical car procession. Every village has its central park, tank and temple. The tanks are sacred, the trees are sacred, and the temple; it is covered by dense rich foliage, which perhaps gives the name and sacred distinction to the village. It is from the temple that there radiates the impulse which uplifts every house so that each may become itself the temple of God.[69]

Geddes and the Indian nationalist movement

For Geddes this was the example of 'Social Religion' realised. He was impatient with the fact that in a colonial country, in the political circumstances of the time, it was also a message with political implications.[70] Geddes had an early initiation into the activities of the nationalist movement as the Annual Conference of the National Indian Congress met in Madras in 1914. Initially the governor, Lord Pentland, had hoped that Geddes' exhibition and lectures would be available for the delegates. The loss of the first exhibition meant that Geddes was not able to make his usual impact, but an effort was made to mount a small exhibition using material sent out from England. His message to delegates, however, was firmly apolitical. His lectures and exhibition were designed, in Lady Pentland's words, 'to give the Indian public a new idea of the meaning and possibilities of town planning and of the opportunities of local authorities'.[71]

But even in cultural terms, Geddes was out-of-step with Congress leaders. While Geddes was eulogising about ancient Indian urban forms, and the domestic arrangements, for example, of courtyard houses (usually the first target for demolition by British sanitary engineers),[72] leaders of the Indian National Congress were taking their own families from traditional homes to the new-style bungalows. In Allahabad, the Nehru family and their relatives and cousins were all building themselves new bungalows outside the old town.[73] As public health hazards in the towns and cities increased, the elites of other towns and cities followed their example. The move was accompanied by the adoption of a more modern and westernised life-style.[74] The Mahatma Gandhi, in retaining his traditional Indian life-style in a self-conscious contrast to these developments, showed also his political astuteness. He provided a reference point for the elite to keep them in touch with their Indian background. Geddes was deeply impressed by Gandhi's personal stature as a religious leader, and his devotion to the Indian way of life.[75]

By chance the two men were both in Indore in 1917, Geddes working on his town planning report, and Gandhi attending the 8th Annual Hindi Language Conference, of which he had just become president, held there that year. A few months later, Geddes sent Gandhi a copy of his Indore Report, and they exchanged letters. They expressed a sense of common purpose in their respective missions, agreeing that one has but to introduce religion into social and political life to succeed. Gandhi writes though, that the purpose of doing so is that, while working for change, it gives you 'a perfect organisation in working order to fall back upon'.[76] Geddes, on the other hand, saw religion as an essential ingredient to social evolution. Geddes criticised Gandhi for the westernised form of his conference, calling it 'really perfectly English. Hindi apart, and as a conference, it might have been in London or Manchester. . . but surely your real problem. . . was to revive, enrich, ennoble your language and literature'. He offers him the example of the Welsh Eistedfodd as a better model. Geddes asked

Gandhi to work with him for Civic Reconstruction. The Mahatma gently declined, pleading an already overfull timetable.

Geddes wanted an Indian collaborator very much as a means for transmitting his doctrine, and he tried to develop his friendship with Rabindranath Tagore for this purpose. Yet once again, while Tagore was sympathetic, he was not interested in becoming involved with Geddes.[77] He managed to gain far more in terms of practical help from Geddes and his son Arthur in return for his friendship than Geddes ever received from him. Geddes had been hopeful because of their common ground in wishing to promote educational changes. Tagore's primary concern in these years was in setting up his school at the remote ashram founded by his father at Santiniketan in West Bengal, defining his educational ideals, and raising funds in both the east and west to support his educational ambitions. Tagore worked to create reform of higher education in touch with Indian culture and tradition. Geddes wished to save the imminent reform of higher education in India from following the British pattern. He thought the moment might be opportune to put forward the pattern he offered to the Royal Commission on the higher education of Scotland all those years ago in 1890.

Tagore wrote a satirical short story on current methods of education which he called *The Parrot's Training*, and he dedicated it to Geddes. It is a savage little story about a raja who comes across a bird, singing and hopping about, but otherwise completely ignorant and since 'ignorance is costly in the long run', the raja decides that money must be spent on educating it.[78] The pundits when asked for advice suggest first that the bird must not be allowed to live freely in his nest. He must be put in a cage: a gilded cage, of course, which required constant scrubbing and cleaning. Scribes were sent for and copies upon copies upon copies of books were made which was all very costly. By this time the raja had acquired a whole department of education concerned with the education of the little bird. When the system was complete all, including the raja, were very pleased, but when someone remembered to look, it was found that the little bird was dead.

The two men continued to correspond over the next few years, though in the early 1920s Geddes became exasperated at Tagore's constant demands for help with Santiniketan and at the same time his constant unreliability about details, meetings, and plans. The brunt of this clash of forceful personalities was born by Arthur, Geddes' second son, who worked with Leonard Elmhirst and C.F. Andrews at Santiniketan, both teaching students and supervising the building and the layout of the college.[79] Geddes and Tagore remained united by their dreams and visions. They shared a belief that educational activities must be designed to nurture creativity instead of killing it. As

Tagore wrote in a letter 'education should never be disassociated from life' and the key to creative education was 'freedom and spontaneity'.[80] Geddes, signalling his total acceptance of this view, wrote in return: 'The difference between us is while I work out (an equivalent of) musical notations *prosody* of thought, you make say six poems!'[81] By this time Tagore had plans for an international university dedicated to the cause of human welfare; whilst throughout his time in India, but particularly in 1918, Geddes hoped he would be able to influence the development of higher education in India.

Geddes and higher education in India: the Chair of Sociology and Civics at Bombay

It was a hope that went back a long way. In 1901, when Mr Carnegie was promising to fund a new-style scientific institute,[82] Mr Tata, the Indian iron and steel magnate, made it known he would fund such an institute in India. He asked Sister Nivedita for advice from educational experts in the west and she wrote to Geddes. He responded at length, writing two reports which were published in the Indian-based magazines *The Pioneer* and *East and West*.[83] But Mr Tata's choice was not for Geddes' plans and the institute set up at Bangalore was on more conventional lines.[84] By the time Geddes reached India in 1914, what direction the development of higher education should take had become a burning political issue. Both the control and content of university education were under attack, and matters came to a head in the years immediately preceding and during the First World War.[85] Since Curzon's Universities of India Act of 1904, strains within the system had become increasingly intolerable. More students were being recruited, yet the cultural framework for their studies, based on the British system, became ever more at odds with the current ideas and aspirations of both British and Indian. The defects of the system were compounded by inadequate educational provisions at lower stages in the primary and secondary schools, and imbalances between boys' and girls' education, as well as the difficulties of developing schools in communities of mixed religions and culture.

In 1918, in the wake of considerable unrest, the British set up a Royal Commission to investigate the most prestigious of India's universities, the University of Calcutta. Men of the calibre of Sir Philip Hartog and Sir Michael Sadler were appointed to it.[86] The commission was the focus of Geddes' propaganda efforts. Sir Michael Sadler was an old acquaintance from London days and Geddes had high hopes of having some impact.[87] It was for this reason he devoted most of the second volume of his Indore Report to the projective plan of a Central

University for India at Indore, which he wanted organised on socio-biologial lines.[88] Geddes' chances of wielding some influence seemed fair. The only university where modern scientific studies were carried on on any scale was Calcutta. Bombay University was an examining body only, although since 1914 the state government had offered the university government funds to start some teaching by setting up a school of research in economics and sociology.[89] There were numbers of colleges and university colleges set up in cities and small towns, some by enlightened maharajas such as Baroda, some by state governments such as the universities of Allahabad and Lucknow. Most were devoted to technical training or to agriculture. One or two of these gained national fame such as the agricultural college at Pusa. Geddes wanted to build, particularly on the latter more practical foundations, colleges which would serve the local communities, not only in practical matters, but in cultural pursuits as well, to bring together both place and people in a favourable evolutionary pattern.[90]

His concern for establishing his particular viewpoint led him here, as often as his planning work, to sidestep political and religious issues. The split between Hindu and Moslem, fostered by the British and widening as a nationalist movement progressed, was reflected in the demands for making the Hindu college at Benares a Hindu university. The Moslem movement countered with the demand for upgrading the Moslem college at Aligarh in a similar fashion. A leading figure in promoting the Hindu university scheme was Annie Besant, now a Theosophist and residing permanently in India.[91] Geddes had met Annie Besant again in Madras to renew an old acquaintance made when he was a student in London in the 1870s, and had given her and Charles Bradlaugh's daughter lessons in the natural sciences forbidden to them by London colleges.[92] He also met members of the Royal Commission in hill stations where they retreated for recreational breaks. But Geddes was out of touch with both British intentions and Indian aspirations, and his informal contacts did not bring him any influence. Measures were put in hand to give Indians more power over their universities.[93]

Rather perversely, he failed largely to capitalise on the one chance he really had for making his mark on Indian higher education. This was the invitation he received from Bombay University in 1918 to fill the Chair of Sociology which had at last been established. Here was Geddes' chance to introduce the social sciences to India. Over the last decade or two Indian academics had been aware of developments in this area in the west.[94] Sri Brajendranath Seal, Professor of Philosophy at Calcutta was the most outstanding example. He wanted to initiate studies in comparative sociology and to study the social institutions of India in a context of race, religion and culture. Calcutta had appointed its first lecturer in anthropology in 1919, and a Department of Anthropology was founded in 1921.[95] Geddes at Bombay was thus given a golden opportunity to map out his own subject and to get university backing for his activities. It was an opportunity, however, that in many ways came too late. Geddes had already had four strenuous years of work in an often hostile climate. He had suffered the severe personal blows of the death of his wife and eldest son, both of whom had the power to restrain and direct his activities more constructively, and he was still making considerable sums of money with his exhibitions and town planning schemes. He needed the money to sustain his projects at home, particularly his beloved Outlook Tower in Edinburgh.

He therefore succeeded in negotiating with the university terms similar to those he enjoyed in Dundee; mainly that he should be resident in the university only four months of any year and that an assistant should be appointed for the rest of the year.[96] There was, however, an enormous difference between a Chair in Botany at Dundee and one in a new subject such as sociology at an Indian university without a teaching tradition. Geddes tended to give the few postgraduate students a rather raw deal. Ever hopeful that he might find an Indian collaborator, he sent the best of them to England for further training. Two leading Indian social scientists, Professors G.S. Ghurye and N.A. Toothi, began their academic careers in this way.[97] The very different reactions of these two men to Geddes was, however, an indication of his mixed impact in Bombay. Geddes had brought his Cities and Town Planning Exhibition to Bombay and set it up on permanent display as a teaching vehicle for his students. He had changed the designation of his chair, adding the title civics to sociology. He then proceeded to offer his students indoctrination in civic reconstruction. Ghurye never forgave him. Geddes had put great pressure on him, as he was to do later on to the young Lewis Mumford, to become his collaborator and assistant. Ghurye escaped after his visit to England and his subsequent discovery of the work of American and French sociologists. In contrast, Toothi found Geddes stimulating and promoted civic surveys and the Le Playist approach to sociology after his return to India. One of Professor Seal's students from Calcutta, Radha Kamal Mukerjee, also took up Geddes' ideas enthusiastically and even wrote a couple of books inspired by them, though Ghurye's comment on this was that Mukerjee did this with every current fashion.[98]

Between thirteen and eighteen students (the number was disputed) enrolled in 1919 for a three-year course in what they believed was the new subject of Sociology. Geddes, in his first year, picked out the ablest to train as his assistant, Mr Pherwani, to work with him until his son Arthur, who came to India in 1920, was ready to help him. The emphasis of the course was on practical work, and Geddes sent his students to old friends: one to Lucknow to Mr Botting, Chief Executive now of the municipality; one to H.V. Lanchester, who had returned to pursue his private practice in Lucknow and Cawnpore; a couple more to Indore where Geddes remained on good terms with the Ministers of Home Affairs and Commerce; and he took one himself to Jamshedpur, when he went to comment on the new plans after the strikes and riots there.[99] When Geddes left in April 1920 for Palestine,[100] the students were left without guidance, and the University Senate tried to retrieve the situation by controlling the appointment of Geddes' assistant for the following year. Geddes was exasperated and completely unrepentant about his conduct. He wrote to the Senate that he was conducting not only a new course in India, but an experimental one which had to be allowed to run for three years without interference; that he was training his students in 'pure sociology' which they learned through observation, hence field-work was absolutely essential; that the academic components of such an education were an encyclopaedic grasp of the scope, methods and interrelatedness of biology and sociology; an ability to understand the evolutionary methods of organising material and books; and finally, a grasp of Geddes' graphic methods which could be used as tools of analysis when they faced practical problems.

In the next two years of the preliminary course, Geddes did make an effort to give at least two major series of lectures, one on 'The Essentials of Sociology in Relation to Economics', which was subsequently published in the *Indian Journal of Economics*; and another on 'Civilisation: A Challenge', which was taken down by an amanuensis and much later, in 1938, was edited by F.J. Adkins, with what proved to be an abortive idea that the manuscript might be published.[101] The work contained a constant reiteration of Geddes' old ideas at the same simple level as had already been published in the volumes in the 'Making of the Future' series which seemed even more out of date in 1938. The students at Bombay in the early 1920s were given no choice but to submit to this indoctrination of Civic Reconstruction. Many, especially in view of the long periods when the course failed to run owing to Geddes' absence, voted with their feet, and the initial healthy recruitment figures seriously dwindled. Geddes tried to fill the gap by opening his class to women. He had always found the staunchest practical support from women for all his initiatives. But this time it was not altogether successful. By the time Geddes reached the last year of his five-year contract with the university in 1924, his health had become seriously undermined and, with it, his ability to enthuse his students with his unconventional course.

There is little direct evidence as to whether Geddes' approach and ideas made any impact. From time to time Geddesian influences seem to resurface in, for example, one book written by Aloo Dastur, *Man and Environment* (1954), which does cite Geddes and elaborates on his concept of the Valley Section. However, the *Indian Journal of Sociology*, established and edited from Baroda in 1920, by Alban G. Widgery, was anxious to define from the beginning the limitations of civics in comparison with the range of studies encompassed by sociology. The journal made a great effort to put Indian scholars in touch with what was happening in Europe and America, listing western journals and periodicals of interest to sociologists and offering a comprehensive review section of the latest work.[102] Geddes' warmest support came, not from sociologists, but from one of India's most outstanding natural scientists, Sir Jagadis Bose. Bose was an immensely warm and affectionate person, and he cared for Geddes, especially after the death of Anna.[103] Geddes tried to repay him by writing his biography which was published in 1920.[104]

Bose not only gave emotional and moral support to Geddes, he was also important in getting him work. He invited him to come and give lectures at his institute; he helped him set up the Summer Schools in Darjeeling; and he got commissions for him in the early 1920s, when Geddes had more or less fallen out with the British administrators, and the commissions from the maharajas were drying up. It was Bose who managed to get him the contract to plan Osmania university's campus at Hyderabad, and Geddes was also to work in his later years on plans for Lucknow Zoo. Bose had originally met Geddes at the 1900 Paris Exhibition through an introduction by Sister Nivedita. When the two men met again, Bose was on the point of achieving his life's ambition, his own scientific institute. He had left the University of Calcutta to strike out on his own even though he had no personal resources. However, his work was so outstanding that he was able to attract funds, and his cherished institute was opened in 1917. The speech he made and the opening ceremony were masterminded for him by Geddes. He spoke of the challenge of the future for India and the two ideals. On the one hand was the ideal of efficiency so that India could take her place in the modern world; on the other, the cultural legacy of the past which must be maintained, especially the ideal of renunciation of worldly gain, which could be India's greatest gift to the present. The ceremony was limited to Indians only, except of course, for Geddes himself, who wore Indian dress for the occasion.

Geddes' response to the First World War

These stirring personal events of Geddes' life in 1917 and 1918 were taking place against a background of world conflict. Geddes, by spending most of the war in India, was remote from the Eurocentred struggle. Yet he was keenly aware that the breakdown of the old order signified by the war gave a chance for change and new ideas to be accepted. During the early years of the war, Geddes had tried to keep in touch with what was going on by returning to Europe every year. He put a considerable amount of effort into turning his usual propaganda machine, the Cities and Town Planning Exhibition, his lectures, and his Summer School, towards the subject of post-war civic reconstruction. It was in these years that his espousal of civics hardened from being an educational subject into a doctrine. In the summer of 1915 Geddes returned to London and set up a Summer School with Dr Gilbert Slater, formerly Principal of Ruskin College Oxford, and now just about to take up his new appointment as Professor of Economics at the University of Madras.

Victor Branford negotiated with the publishers Williams and Norgate for a new series, the 'Making of the Future', with himself and Geddes as editors. Notes were taken at the 1915 Summer School, and when these were deciphered (the process was extremely difficult since the lectures were disjointed and neither Geddes nor Slater was available for comment as both had gone to India), the result was published in the first of this series with the title of *Ideas at War*. Victor Branford then collaborated with Geddes on a second volume, *The Coming Polity: a Study in Reconstruction*. This did not prove to be any better in quality but it made civic reconstruction as proposed by Geddes into an article of faith.

> The great deficiency of the Mechanical Age is its sacrifice of Life to Things — therefore the first effort of the new age must be to make it eutechnic: not only must physical health bulk more largely in our minds than the possession of commodities endowed with exchange values, but also physical health and well-being must be regarded as important and valuable mainly as a condition of the inner life of the soul.[105]

Apart from these propaganda activities, at the end of the war Geddes had found another outlet for his energies. This was to work in Jerusalem and take part in activities which were intended to lead to a future Israel in Palestine. The world might not be ready for Civic Reconstruction, the 'New Jerusalem', however, was an extraordinary challenge. Since the Balfour Declaration of 1917, the hopes of the Zionists had mounted, and Geddes went with Chaim Weizmann to

Jerusalem in the summer of 1919.[106] He became determined to work there regardless of the delicate political situation. It was his activities in Jerusalem which made him late in taking up his duties at Bombay, so that he arrived after term had started. It was his interest in Palestine over the next two or three years which dominated his activities and relegated his academic role in Bombay in some respects to a secondary position.

Geddes viewed Jerusalem much in the same way as he had seen the challenge of India in 1914. The city offered him a new cultural context with exciting possibilities in which to practise his approach to problem solving.[107] His Civic Reconstruction Doctrine was now a formula; his concepts of the city, of civilisation, of the good life, of social progress, had become inflexible. But at the same time he was constantly seeking new environments in which to work to test his powers of observation and his hypotheses. In the true evolutionist fashion, he saw himself as an organism, constantly day and night engaged in problem solving. Problem solving was the primal activity of all organisms undergoing the process of evolution. India, and subsequently Palestine, not only offered him a completely different environment to that in which he had worked in Europe, they also offered an ancient and rich cultural heritage. Geddes' addiction to symbolism as a way of communicating his views on cities and civilisation was fed afresh by selected ancient social customs and religious rituals; at the same time he had the challenge of problem solving in an alien environment, demanding the keenest powers of observation from a foreigner.[108] These were the qualities which he brought to his Indian and Palestinian reports. The former tended to confuse readers of the reports unfamiliar with Civic Reconstruction doctrine. But amongst his writings were many valuable perceptions resulting from his powers of observation and determination to think freshly on urban problems.

In all the reports, however, Geddes takes it upon himself to voice the social and economic aspirations of the people. If they do not understand or object to his views, that is because they have had a deficient education or do not understand 'civic reconstruction'.[109] A town planner with such convictions would only gain a large number of commissions and work in exceptional circumstances. In India, Geddes was lucky since he arrived at a period of comparatively rapid change in the relationship between British administrators and Indian society. On the one hand the 'liberal' governors, backed by the authoritarian framework of British colonial rule, could encourage Geddes in his work, aware as they were that the immediate cause of most unrest was often the local conditions as, for example, as late as 1919 in the Satyagraha in Delhi. On the other hand, Geddes was able to establish a strong rapport with educated Indians willing to listen to a Britisher

[229]

sympathetic to their cause of separate national identity and development. It was a combination which brought Geddes, as far as work was concerned, more work, more money and more enjoyment than he had ever experienced before.

Notes

1. J.A. Thomson to P. Geddes, 16 August 1913, Geddes Papers MS10555 NLS.
2. Pentland was a great friend of Lord Carmichael, Governor of Bengal, and Lord Willingdon, Governor of Bombay. Lady Pentland (1928) *The Rt. Hon. John Sinclair, Lord Pentland GCSI: A Memoir*, London: Methuen, p.140.
3. One of Geddes' pedagogic aphorisms to complement his 'by living we learn'.
4. Lady Pentland, op.cit., p.214.
5. Victor Branford wrote a memoir of Alasdair – *A Citizen Soldier: his education for war and peace – being a memoir of Alasdair Geddes*, London: Headley Brothers, in 1917.
6. See chapter 1, p.7.
7. He began to write letters and an autobiographical memoir in verse in the early 1920s. See P. Boardman (1978) *The Worlds of Patrick Geddes: biologist, town planner, re-educator, peace warrior*, London: Routledge & Kegan Paul, pp.43–5.
8. Students of architecture and civic design at Liverpool University who studied with Professor Abercrombie were encouraged to go to India to complete their education. One such, J. Linton Bogle, became the Planning Officer in Lucknow, and wrote a monograph on Indian planning incorporating Geddesian ideas, though without direct acknowledgement to Geddes himself. (J.M. Linton Bogle (1929) *Town Planning in India*, London, Bombay, Calcutta, Madras: Oxford University Press).
9. Patrick Geddes to H.J. Fleure, 4 April 1917, Geddes Papers MS10572, NLS.
10. He showed his gratitude for Bose's friendship and the help he had given him in securing work on his behalf by writing a biography of his friend *An Indian Pioneer: the life and work of Sir Jagadis Chandra Bose*, London: Longmans, 1920.
11. Patrick Geddes to Norah, 23 September 1917, Geddes Papers MS10501, NLS.
12. Patrick Geddes to Norah, 2 January 1917, Geddes Papers MS10501, NLS.
13. K. Davis (1951) *The Population of India and Pakistan*, Princeton, NJ: Princeton University Press, p.128.
14. A.F. Weber (1899) *The Growth of Cities in the Nineteenth Century* reprinted Ithaca NY: Cornell University Press, 1965, pp.123–8.
15. G.S. Ghurye (1953) 'Cities of India', *Sociological Bulletin* 1–2: 305.
16. H. Tinker (1954) *The Foundations of Local Self-Government in India, Pakistan and Burma*, London: Athlone Press.
17. Ibid., p.333.
18. H.T.S. Forrest (1909) *The Indian Municipality and some practical hints on its*

everyday work, Calcutta and Simla: Thacker, Spink & Co., 2nd edn, 1925.

19. A. Bose (1970) 'Administration of Urban Areas', Survey of Research in Social Sciences (Major Group III, ICSSR), Delhi: unpublished paper, p.12.

20. J. Nehru (1936) *An Autobiography: with musings on recent events in India* London: Bodley Head, reprinted 1958, p.142.

21. See E.P. Hennock (1973) *Fit and Proper Persons: ideal and reality in nineteenth century urban government*, London: Edward Arnold. Geddes believed it was vital to generate this link. He wrote in *The Reports on the Towns in the Madras Presidency*, Madras: Government Press, 1915, 'A new level of civic statesmanship must therefore be reached, in which each individual or group interest no longer acts for self interest alone, but in subordination to the City's plan, which is more and more destined to become its veritable charter', p.88.

22. E.G.K. Hewat (1950) *Christ and Western India: A study of the growth of the Indian Church in Bombay City from 1913*, Bombay: p.313.

23. Ibid., p.318.

24. Ibid., p.312.

25. By the end of the war all regions in India were considering Improvement Trusts – see, for example: 'Improvement Trusts: Question of provision from the grants to be made by Government under the scheme for the constitution of —— for larger cities in the United Provinces', *Proceedings of the Department of Education*, Government of India Municipalities Act, January 1919.

26. P. Geddes (1922) *Town Planning in Patiala State and City: a report to H.H. the Maharaja of Patiala*, Lucknow: p.20.

27. See chapter 3.

28. *Slum Clearance in India*, Publications Division, Ministry of Information and Broadcasting, Government of India, 1958.

29. With the honourable exception of the Lucknow Improvement Trust which Geddes holds up to his other clients as a model – see, for example, P. Geddes (1922) op.cit., p.91.

30. 'Recording G.O. No. 536M sanctioning the deputation of an officer of the I.C.S. Sub-collector's standing for a period of 6 months for the purpose of advising local boards and municipal councils in connection with ..... operations', Public Works Department, Government of India, State of Madras. Municipalities 6 April 1915 Town Planning.

31. Information of Lanchester from the unpublished material kept on file in the library of the Royal Institute of British Architects.

32. H.V. Lanchester (1916) *Town-Planning in Madras*, Madras:.

33. H.V. Lanchester (1925) *The Art of Town Planning*, London: Chapman & Hall.

34. Ibid., p.201.

35. Quoted in E.P. Richards (1914) *Report on the Request of the Improvement Trust, on the Condition, Improvement and Town Planning of the City of Calcutta and Contiguous Areas*, published privately, p.246.

36. Ibid., section IV.

37. Geddes refers to Richard's report as a 'stately volume' in *Cities in Evolution*, 1915 edn, p.240 which he wrote before setting foot in India.

38. E.G. Turner, ICS (1914) *Notes on the Operation of the Housing and Town Planning Etc. Act 1909 and Allied Matters*, Bombay: State Government records.

39. Cawnpore (Kanpur) had been largely destroyed during the Indian Mutiny. It was subsequently chosen to be a major centre of industrialisation, and was thus growing very rapidly in the early twentieth century.

40. He gave a lecture before the London Town Planning Institute on 5 December 1919, on 'Town Planning in Bombay under the Bombay Town Planning Act 1915'.

41. H.E. Meller (1979) 'Urbanisation and the Introduction of Modern Town Planning Ideas in India 1900–1925', in K.N. Chaudhuri and C.J. Dewey (eds) *Economy and Society: essays in Indian economic and social history*, New Delhi: Oxford University Press, pp. 330–50.

42. He was involved with Victor Branford at this time in producing the 'Making of the Future' series. *The Coming Polity: a study in reconstruction*, London: Williams & Norgate, 1917, was an exposition of the Le Play/Geddes method which Geddes wanted to use in India.

43. Though the relationship between population growth, industrialisation, and urbanisation was to be markedly different from the earlier European experience – see A. Bose (1970) *Urbanisation in India*, New Delhi: Academic Books, p.63. See also B.F. Hoselitz (1954) 'Generative and Parasitic Cities', *Economic Development and Cultural Change*, 3: 278–94.

44. A.D. King (1976) *Colonial Urban Development: culture, social power and environment*, London: Routledge & Kegan Paul.

45. Ibid., chapter 10, New Delhi.

46. For a discussion of the important distinction between 'modernisation' and 'westernisation' see M.N. Srinivas (1976) *Social Change in Modern India*, Cambridge: Cambridge University Press, pp.53–4.

47. Anil Seal (1968) *The Emergence of Indian Nationalism: competition and collaboration in the later nineteenth century*, Cambridge: Cambridge University Press.

48. C.F. Andrews (1912) *The Renaissance in India: its missionary aspect*, London: Church Missionary Society, p.49.

49. Ibid., p.295.

50. Ibid., p.50.

51. There were a number of new translations of these works, for example, 'Bhagavadgita', trans. by K.T. Telang (1882) *Sacred Books of the East* viii. 34.

52. J. Fergusson (1862) *History of the Modern Styles of Architecture*, London.

53. S. Nilsson (1968) *European Architecture in India 1750–1850*, London: Faber & Faber.

54. P. Geddes (1919) 'The Temple Cities', *Modern Review* (India) 25:. He also wrote a eulogy on temple cities in a letter to his family, 7 January 1915, Geddes Papers MS10515, NLS.

55. Lady Pentland (1928) op.cit.

56. C.F. Andrews, op.cit., p.23.

57. C.P. Venkatarama Ayyar (1916) *Town Planning in Ancient Dekkan*, Madras: Law Printing House, with an introduction by Professor Patrick Geddes.

58. Records of the Department of Education, Government of India Municipalities, serial no. 14, 1 March 1917.

59. Rabindranath Tagore (1861–1941), Bengali poet, mystic, and educator. He was awarded the Nobel Prize for literature in 1913, and knighted in 1915. He renounced his knighthood in protest at the massacre at Amritsar in 1919. In 1901 he founded a school in Santiniketan in West

Bengal. In 1921 the school was extended to become the Visva-Bharati University. Geddes, and more directly, his son Arthur, helped Tagore to plan and to build the first buildings. See correspondence between Tagore and Geddes, Geddes Papers MS10576, NLS, and Leonard Elmhirst (1958) *Rabindranath Tagore and Sriniketan*, quarterly booklet reprinted from *Visva-Bharata Quarterly* Autumn, Santiniketan, West Bengal).

60. Romain Rolland (1929) *Essai sur la Mystique et l'Action de l'Inde Vivante: I. La Vie de Ramakrishna. II. La Vie de Vivekananda*, Paris: Delamain & Bontelleau.

61. Margaret Noble (1867–1911), was of Irish parentage and birth, the daughter of a Congregational minister. Trained as a teacher, she was one of a group of educationalists who founded the Sesame Club in the early 1890s. She met Vivekananda in 1895 (when he was 32 and she was 28) in London, where he was lecturing on Indian religion and philosophy. She went to India in 1898 to found a school for Hindu girls in connection with the Ramakrishna Mission in Calcutta, of which Vivekananda was the head. In India she began to wield considerable influence by her writings as well as her teaching. Her home was visited by leading Indians and visitors from Europe and the USA. She met Geddes in 1900 in Paris and, under his influence, wrote her most famous work, *The Web of Indian Life*, in 1904. She died of dysentery in Darjeeling at the age of 44.

62. She dedicated her book to Geddes. P. Boardman, op.cit., p.183.

63. C.M. Villiers-Stuart (1913) *Gardens of the Great Mughals*, London: Adam & Charles Black.

64. Ibid., p.248.

65. See plate 9, opp. p.33, in J. Tyrwhitt (ed.) (1947) *Patrick Geddes in India*, London: Lund Humphries.

66. Dr J.A. Turner, with contributions by B.K. Goldsmith, S.C. Hormusji, K.B. Shroff and L. Godinho (1914) *Sanitation in India*, Bombay: *Times of India*.

67. Jacqueline Tyrwhitt has collected extracts from Geddes' Indian Reports illustrating these views. For his public health approach see J. Tyrwhitt (ed.), op.cit., pp.66–83.

68. There was a review of *Cities in Evolution* in the first volume of the *Indian Journal of Sociology* (1920) edited by Albion G. Widgery from the University of Baroda. Widgery was also the reviewer, and he called for a book 'similar in purpose to Professor Geddes' but prepared in relation to Indian cities'. This challenge was taken up by Mukerjee who published a book on *Civics*, London: Longmans, Green, 1926, and *Regional Sociology*, New York: The Century Co., 1926.

69. Radha Kamal Mukerjee, in his foreword to the book by J.M. Linton Bogle (1929) *Town Planning in India*, London, Bombay, Calcutta, Madras: Oxford University Press.

70. The end of the First World War was to see the implementation of the Montagu/Chelmsford reform of the Indian Constitution, allowing greater levels of self-government in nine major provinces.

71. Lady Pentland, op.cit., p.214.

72. See for example, his appendix on the subject of courtyard houses in P. Geddes (1919) *A Report to the Corporation of Calcutta: Barra Bazar Improvement*, Calcutta Corporation Press, pp.28–9.

73. B.R. Nanda (1965) *The Nehrus. Motilal and Jawaharlal*, London: Allen &

Unwin, pp.30–1.
74. A.D. King (1977) 'The Westernisation of Domestic Architecture in India', in Dalu Jones and George Michell (eds) *AARP: Art and Archaeology Research Papers*, June (102 St Pauls Road, London N1).
75. See correspondence between Geddes and Gandhi in Geddes Papers MS10515 NLS.
76. Ibid., letters between January and April 1918.
77. Exchange of letters between Tagore and Patrick Geddes, Geddes Papers MS10576 NLS.
78. Rabindranath Tagore (1918) *The Parrot's Training*, translated by the author from the original Bengali, Calcutta and Simla: Thacker, Spink & Co., p.1.
79. *Rabindranath Tagore: Pioneer in Education: essays and exchanges between Rabindranath Tagore and Leonard Elmhirst*, London: John Murray/Visva-Bharati, 1961.
80. Tagore to Patrick Geddes, 9 May 1922, Geddes Papers MS10576, NLS.
81. Patrick Geddes to Tagore, 17 May 1922, ibid.
82. The Carnegie Institution of Washington was formally set up in 1902 – *A Manual of the Public Benefactions of Andrew Carnegie*, Washington, DC: Carnegie Endowment for International Peace, 1919, p.77.
83. P. Geddes, *On Universities in Europe and in India: and a needed type of research institute, geographical and social. Five letters to an Indian friend*, Madras: National Press, (reprinted from *The Pioneer* 14 August 1901 and *East and West* September 1903).
84. Although the Tata Foundation was to pioneer a new approach to industrial relations and the social environment in an industrial context. See article by J. Seabrook 'Company becomes cosmos: a business conglomerate that shares the Gandhian view of wealth', *The Guardian*, 6 February 1986.
85. Aparna Basu (1974) *The Growth of Education and Political Development in India 1898–1920*, Delhi: Oxford University Press.
86. Sir Philip Hartog (1939) *Some Aspects of Indian Education, Past and Present*, University of London Institute of Education, Studies and Reports no. VII, London: Oxford University Press, p.59. Also, M. Hartog (1949) *P.J. Hartog, A Memoir by his Wife*, London: Constable, p.77–85.
87. See chapter 5, p.96.
88. P. Geddes (1918) *Town Planning Towards City Development: a report to the Durbar of Indore*, Part II, Indore: Holkar State Printing Press, pp.1–18.
89. S.R. Dongerkery (1957) *A History of the University of Bombay*, Bombay: p.59.
90. Geddes attracted the support of two Britons who were pursuing similar ideas. One was H.H. Mann, of the Agricultural Institute in Poona, who undertook agricultural surveys in Deccan villages in 1917 and 1921. He was responsible for drafting the scheme adopted by the Syndicate of the University of Bombay for a 'Proposed School of Economics and Sociology' for which Geddes was later appointed to the first Chair in Sociology – H.H. Mann to Patrick Geddes, 6 May 1917, general correspondence no. 1384, Geddes Collection, University of Strathclyde. The other was H.W. Lyons, Professor of Economics at Indore Christian University who was responsible for suggesting Geddes' name to the Maharaja and thus gaining him the commission to work in Indore.
91. See Aparna Basu, op.cit., chapters 4, 5, and 6.

92. Patrick Geddes to Anna, 9 February 1915, Geddes Papers MS10502 NLS; P. Boardman, op.cit., pp.74, 263.
93. P. Hartog, op.cit., pp.48–68.
94. M.N. Srinivas and M.N. Panini (1972) 'The Role of the Social Sciences in India: Sociology and Social Anthropology', *Indian Institute of Advanced Study, Delhi, Indian Council for Social Science Research Meeting, 7–13 October 1972* (unpublished paper).
95. See also editorial on need to develop postgraduate work in these disciplines written by A.G. Widgery (1920) *Indian Journal of Sociology* 1(3).
96. Evidence of Geddes' relationship with the University of Bombay is to be found in correspondence from H.H. Mann to Patrick Geddes, MS10515 (NLS), and between Patrick Geddes and the Registrar of the University, MS10516 (NLS), and unpublished papers at the University of Bombay, Registrar's Office.
97. Geddes sent Toothi to Aberdeen to study with J. Arthur Thomson, an experience which made Toothi a life-long supporter of the Geddes approach. See Toothi to Patrick Geddes 1919, general correspondence no.1425, Geddes Collection, University of Strathclyde. Ghurye went to London and studied with L.T. Hobhouse and W.H.R. Rivers – see K.M. Kapadia (ed.) (1956) *Professor Ghurye Felicitations Volume*, Bombay: Popular Book Depot, Introduction.
98. Ghurye became violently hostile to Geddes and his methods, as recounted in an interview with the author, 2 January 1973, Bombay.
99. Geddes went to comment on the work of F.C. Temple (1919) *Report on Town Planning of Jamshedpur*, .
100. See chapter 8.
101. The unpublished manuscript is in the Geddes Papers MS10618 NLS.
102. See, for example, 1(1920):339.
103. See correspondence J. Bose to Patrick Geddes, Geddes Papers MS10576 NLS.
104. P. Geddes (1920) op.cit.
105. P. Geddes to G. Slater (1917) *Ideas at War*, London: Williams & Norgate, p.188.
106. Bernard Wasserstein (1978) *The British in Palestine: the Mandatory Government and the Arab-Jewish Conflict 1917–1929*, London: Royal Historical Society, chapter 1; C. Weizmann (1918) *Zionism and the Future Problem*, London.
107. P. Geddes (1921) 'Palestine in Renewal', *Contemporary Review* CXX.
108. See chapter 3.
109. In the letter in which he gave his consent to his son's decision to fight in the war, he also reiterated his view of the importance of their work in India.

   It is also the fact that I increasingly feel the value of our own exhibitions in India and that of my conservative yet constructive attitude and influence in cities and towns to be of direct political as well as social value – and not merely the chance of life in the ways it has been growing towards so long – but of an unexpectedly direct bearing on order and stability – even of the Empire – not only by economy etc. but by tending to check the revolutionary spirit by the Eutopian one – and cast out devils by ideals, so rendering a very direct form of service even in and for these times of war'. Patrick Geddes to Alasdair, 17 June 1915, Geddes Papers MS10501, NLS.

CHAPTER 8

Social reconstruction
in India and Palestine

INDIA HAD ALREADY EXPERIENCED HER INITIATION IN
twentieth century urban form with two major developments both
dating from 1911. The new capital, New Delhi, had begun to take
shape; and the city which was to be the symbol of India's progress in
industrialisation, Jamshedpur, was being built by the steel magnate
and large-scale entrepreneur, Mr Tata.[1] New Delhi was the personification of British Imperial Power, the last word in 'civil lines', built on
extravagantly luxurious Garden City principles. The scale of the road
layout and the size of the plots for each bungalow far exceeded any
English prototype. Lutyens's designs for the Viceroy's Palace and the
ceremonial grounds before it created a symbolic impression of the
strength of British rule. Geddes' views on the layout of the land and
especially the parks of New Delhi were not favourable. He was asked
to write a brief report for the New Delhi Commission but this report
has, unfortunately, been lost. He and Lutyens had an antipathy for
each other which extended to their work and their views on the social
evolution of the future.[2]

In 1920 Geddes was also asked to comment on town extension
schemes for Jamshedpur in the wake of industrial unrest. His advice
was particularly sought on the provision of workers' housing. The
management wanted to adopt a paternalistic line on this as a

[236]

demonstration of goodwill to placate workers who were striking for higher wages. In an Indian context, Jamshedpur was quite extraordinary. It had been laid out initially by a Pittsburg engineer on a gridiron street pattern which was a direct imitation of an American industrial city. When war demands led to increased production, there was a need for more labour and the city began to expand rapidly. An English sanitary engineer, officially employed by the Government of Bihar and Orissa, a Mr F.C. Temple, was commissioned to produce town extension schemes. He produced a model example of an English Garden suburb.[3] The Indian workers however, wanted more money, not English-style suburbs to live in. Their response convinced Geddes, if he needed more evidence, that ideas on urban development produced by western culture were totally inappropriate for modernising Indian society. What was needed was a reinterpretation and evaluation of indigenous customs and the physical urban forms they produced in order to create a new modern environment which was nevertheless rooted in the eastern culture in which it had to thrive.[4]

The early Indian Reports: explorations in the biological viewpoint

Geddes was given his first chance to work out his ideas in Madras soon after his arrival. While he was assembling the substitute Cities and Town Planning Exhibition sent from England, he offered a course of lectures to Borough Surveyors of the Madras Presidency towns and municipalities. He then wrote reports for twelve little towns and one suburb of Madras for which they were responsible and which he tried to visit. His work was mostly at the level of propaganda. He wanted to encourage the surveyors to map their towns accurately and to learn by first-hand observation how to 'read' the environment, that is, how to become aware of the historical, cultural and social factors which had created the present. He also set himself to solve the key Indian urban problems: plague, pestilence, and overcrowding, and the perpetual problem of water – its scarcity for much of the year and its superabundance during the monsoon. The solutions he put forward were brilliant adaptations of the techniques he had developed in his work in Edinburgh. Jacqueline Tyrwhitt's volume *Patrick Geddes in India*, has given them wide publicity: the themes of 'diagnostic survey', 'conservative surgery', 'the socio-biological approach', and the importance of trees, gardens, and open spaces, have all been illustrated by selected extracts, many of them from the Madras Reports.[5]

The Madras Reports lend themselves to this kind of editing because of the original intention of putting across Geddes' particular message.

His personal knowledge of the towns was quite scanty, often the result of a morning or afternoon visit, and one, Nellore, he never even visited at all. It did not matter since his major purpose was to change the perspective of these borough surveyors on their work. The tiny towns of the Madras State, their population numbered in thousands rather than tens of thousands, seemed ideally suited for his purposes. The key problem of public health he could in this context look at, with justification, from the viewpoint of the 'scientific' gardener. His major point was that disease and health hazards were the product of dirt and neglect. Public health problems were thus social problems. The people themselves had to be involved in the elimination of the causes of plague, fever, and dysentery. He used case studies to prove his point, and the little port and town of Cocanada (Kakinada) provided him with a good example. Here was

> a town in germ, even a city in its infancy, and thus a living infancy, growing from within, as village and city should ever do; needing help and guidance no doubt, but not the present too common alternatives; which are a) to be let alone till its disease, overcrowding and deterioration compel attention, with b) sweeping demolition followed by c) the imposition of a new plan from authorities above and without; a plan which, whatever its elements of European merit is, as we shall see, little related to village life with its characteristic Indian customs and Indian requirements.[6]

The way to keep Cocanada free from plague and pestilence was to encourage the people to revive and cherish those customs and traditions which were directed towards encouraging cleanliness. At the same time, modern knowledge of bacteriology must be utilised by town planners to ensure that the result of these endeavours would be successful.

Geddes found the ancient custom for encouraging cleanliness in the annual festival of Diwali, or Pongal, as it was known in parts of South India. In his advice on how to clean up Coconada, he suggests a full-scale revival of this festival.

> For what European has ever seen at home anything which fully corresponds to Pongal? To realise this we must combine at least three of our high festivals. Of these the first is the temporal one, known as 'spring-cleaning'; but this is nowadays merely of material and sanitary endeavour: it has lapsed from continuity with the spiritual spring festival of Easter renewal and inward purification, of which it was originally the outward part. Were this renewal of material sanitation with moral arousal again accomplished, the *public health* department of our medical profession and the *public holiness*

[238]

department of our clerical one would again become one as with every priesthood of the past.

The European may probably answer that this is Utopian. Be it so, it is however historical: and we are not here concerned with western cities, but with the simpler improvement of the Indian village, the checking and mending of its incipient slum; and the important point where the European sanitationist fails (and with him his educated Indian supporters, municipal or other) is that he forgets that for the surviving old-world communities of India as for our own Western origins, health and cleanliness have been traditionally approached, not directly as we now do, but from the side of religion, that is to say of public festival and ritual and of personal participation to suit. 'Pongal' may thus be a name to conjure with, more potent to simple ears than 'Microbes', 'Rats', 'Drains' or other contemporary western slogans, too humbly sanitary as these are.[7]

When Geddes writes about Bellary, a town decimated by plague (he picked up three fleas himself on an afternoon walk through the town), he advocated a general cleansing by sanitarian and inhabitant together in rather stronger terms.

Now for the different treatment, really a bacteriological one. It is first mapping the entire areas of inferior health conditions: and this house by house, lane by lane. Then – instead of spending all one's resources upon a heroic frontal attack for the destruction of some one clamantly overcrowded and fetid area, on a scale proportionately costly and therefore disproportionately small when all is done, we go along all the thoroughfares and lanes without exception, cleansing, paving, draining, whitewashing as we go. The whole neighbourhood is thus improved and heartened up and its aggregate valuation is improved, instead of diminished. Its population above all is inspirited, not annoyed, alarmed, embittered or depressed. Note here that this psychological factor is constantly overlooked by those sanitationists whose working theory keeps down their health-consciousness and health-conscience, their health efforts correspondingly to the level of the rectum, which forgets the regulation of the whole being by the nervous system, above all by the mind and mood, which re-determine health through brain.[8]

His gardener's instinct and Dr Turner's *Sanitation in India* alerted him to the natural and very cheap method of waste disposal. In Conjeeveran he noted:

With peculiar approval, the practical way of dealing with drainage which I observed at least at two points, and which I believe to be capable of imitation, and thus at a thousand points throughout the cities of India; not simply the guiding of drainage (of course not in the undue quantity of a sewage farm) into a garden, but actually the formation of new gardens, for the express purpose of receiving such impure waters, and thus converting what would have been in each case a fetid and poisonous nuisance into a scene of order and beauty.[9]

Since a water-borne sewage system was impossible to engineer and underground pipes, in any case, cost a great deal of money, Geddes was particularly pleased with this solution which both involved active collaboration from the people and the making of something of value and beauty out of what might have been a health hazard and nuisance.

His confidence appears to know no bounds as his socio-biological approach seemed to suit Indian conditions so well. He was delighted with the solutions he found to the water problems of Indian cities. The traditional method of saving water had been to dig tanks, but gradually tanks became neglected and became breeding grounds for the mosquito. In the wake of the then relatively new knowledge of the causes of malaria, the British had adopted a policy of filling in the tanks. Geddes wrote in his Bellary report:

This filling up of tanks, created by ancient foresight, labour and sacrifice, seems often too lightly suggested. That the edge be regularised, and the slope of the bottom also, so as to avoid the irregular development of stagnant and mosquito-breeding pools and to admit of the steady retreat of the water towards a central and deeper portion – that surely is the more practical policy, and an enormously cheaper one. By thus arranging a pool of retreat, loss by evaporation would be diminished, as well as mosquitoes abated. And if annual drying up prevent keeping up the supply of fish to keep down larvae we may efficiently replace fish by ducks, whose incessant searchings in mud and water, on bottom and on the surface are so peculiarly thorough and efficacious; while even if complete drought supervenes, the ducks survive. Is it not part of that curious apathy to minor agricultural interests which is so common amongst the educated classes, Indian not less than European, that such simple aid against one of the gravest scourges should not be provided and maintained?[10]

Tanks were a social amenity, for water supply, for washing, for food if stocked with fish. During monsoons they could absorb excess water and the great temple tanks were both aesthetically beautiful and symbolic of spiritual values as they were used for ritual bathing. Geddes made his case to preserve tanks seem unanswerable.

He rounded off his Madras Reports with a confident statement on his understanding of town planning:

[240]

Here is the last word of town planning, so far as I know anything of it. The preceding advantages, either separately and in combination, can only be realised in proportion as it is in life-economies, life-efficiencies, life-amenities, that we are striving and learning to keep more and more clearly and constantly in view. For it is people we are planning for: not mere places. To plan places merely, and those from the monetary and the mechanical points of view, has been the essential error, and the permanent source of the material waste and ugliness, and the corresponding social deterioration which have characterised the conventional 'Bye-law Method' ever since its birth, and which are now bringing about its final disappearance.

The promotion of Life: that is what must be constantly before us as our aim in planning. Not only life of trees and life of gardens, but life of workers, life of house-mothers; and above all, the life, health and joy of the children who have so soon to replace them and us. And as this aim becomes clear, the miserable fear of mechanical planners that 'we cannot afford any of these fine things' – is dissipated. As these plans show, we find we can afford all these, and still have something over. We see that this method of town-planning that which views it as life-promoting – ends successfully, even from the standpoint of accountancy, i.e. on the fullest financial balance sheet. . . For individual and family, street and village, for town and city, even for state and empire, what better can we invest in, than in Homes?[11]

After the successful completion of his work in Madras, Geddes' career as town planner and propagandist for Civic Reconstruction really took off. His ceaseless travelling in India and the Middle East over the next decade, and his prolific published output of plans for towns, universities, zoological and botanical gardens in Delhi, Agra, Lucknow and Hyderabad, the huge volume of unpublished letters and the propaganda he engaged in, in Summer Schools, lectures, and published work on civic reconstruction, and regional survey, bear witness to a drive and energy which was quite extraordinary for any one individual and certainly for someone of his age. Everywhere he went, his vitality and enthusiasm generated a response (not always favourable) and much of his influence depended on his personal impact. There were four main stages in his work in these years, and in all of them he experienced varying degrees of success and failure. After Madras came the early invitations, especially those from the governments of the Bombay Presidency and Bengal where Lord Pentland's friends, Lord Willingdon and Lord Carmichael, were governors.

In this early stage Geddes' exhibition work and lectures were as important as his planning advice which was usually little more than on-the-spot advice on specific problems. One town, however, particu-

larly engaged his attention and that was Lucknow. He visited Lucknow
on his arrival from Britain and he produced two reports for the city in
1916 and 1917. His work at Lucknow was the culmination of his early
work in India and was one of his most successful enterprises. During
1915 and 1916 he was greatly helped by H.V. Lanchester who, after his
appointment as the first Town Planning Officer in Madras in 1915, also
opened an architectural office for private work in Lucknow, which he
kept into the 1920s. It was with his help that Geddes was able to
publish three reports on different towns in 1916 and seven in 1917 (for
which the planning work had been carried out at an earlier date,
Lanchester having returned to England in 1916). After Lucknow, which
was within British rule, Geddes had a second phase when he began to
gain planning commissions in the princely states. His work in Baroda
was to be followed by his great effort to justify his whole life's work in
the plan for Indore which he wrote as a personal memorial to his wife
and son.

Then in 1919 there was the new interlude with his adventures in
Palestine. Financially this was the least successful phase of his work,
but it was one he most enjoyed. Jerusalem was the city which Geddes
believed had had the greatest cultural influence on world civilisation,
apart from Athens, where Geddes' old rival from Dunfermline days,
T.H. Mawson, had been given some work. The final phase of Geddes'
career in India extends over the last years, when he was based in
Bombay as the first Professor of Civics and Sociology at the university.
From there he kept up his flow of practical planning work whenever
opportunities arose and he worked on a variety of schemes from the
campus for Osmania University, Hyderabad, to the zoo at Lucknow,
and he wrote a last town-planning report for the Maharajah of Patiala
in 1922. But the most overriding concern of this last phase was his
sustained attempt to launch his civic reconstruction movement on a
world-wide basis. He encouraged, wherever he could, signs of activity
which he described as regional survey work. The term 'regional
survey', however, proved to be an over-arching one covering a wide
range of activities which were often completely unconnected to
Geddes' doctrine of civic reconstruction.[12] He experienced his greatest
hopes and frustrations in these last years.

These various stages overlapped with each other as Geddes,
especially in the peak years 1916–20, always had several projects on
hand at any one time, and was always a propagandist for regional
survey. He was constantly stimulated by the fact that wherever he
went in India he was treated as the 'Guru' and given a completely free
hand in his work. For him this made up for the deficiencies in
resources and trained manpower that were available for actually
carrying out the work. His major aim was always to change people's

perspectives on their city rather than to pay attention to pettifogging detail. But he addressed himself to what he identified as the major planning problem of any colony (or suburb), village, town, or city which he was commissioned to plan. He was thus sustained in the hope that his propaganda and practical planning work would go hand in hand and that, cumulatively, he would generate a new movement in India more suited to the economic and social conditions of the subcontinent than the centrally directed, if modest, efforts by the British Administration to foster industrialisation in India.[13]

While in Madras he had established all the solutions he thought were necessary to Indian urban problems.[14] In Bombay, where he took his Exhibition after Madras, he was invited to visit six towns in the Bombay Presidency: Bandra, Thana, Broach (Bharuch), Surat, Nadiad, and Ahmedabad.[15] His reports on all of them were extremely sketchy but his message was quite clear. At Thana, for instance, the fate of a tank was in balance and his answer was predictably an unequivocal 'No − it must not be destroyed'.[16] The tank was the traditional Indian method for controlling water so that it could be used for maximum social benefit. At Surat the question of alignments of roads had been exercising the municipal officers and Geddes helped out here by giving a few principles on how they should proceed. He threw in some information about traffic circulation at crossroads which he had just picked up from a recently published paper given at the London Surveyors' Institute on a Piccadilly Circus Improvement Scheme.[17] But most of all he wanted to relate the road scheme to plans for the community so that it was not just a matter of achieving a practical result but also an aesthetically pleasing and a socially satisfying one.

In Ahmedabad, though, he met a challenge not immediately amenable to his solutions. Ahmedabad was by far the largest city of the six Bombay towns he visited, the 'Cottonopolis' of India, industrialising rapidly and growing as Manchester had done a century before. Geddes was asked about the city's walls which were hampering outward expansion and traffic flows. Geddes, of course, wanted to keep the walls. However, he had to work hard to make out a case for supporting this. He had already spent a few days looking at the walls in November 1914 on his grand tour of India prior to reaching Madras. He revisited them when asked to report on the problem and walked round them for two days. This was not enough time to produce detailed maps and plans so he had to fall back on exhortation. His line of argument was that Ahmedabad needed a city and town-planning exhibition; for this it needed a civic survey, and the results of this should be housed in a civic museum. Then historic remains such as the city walls would be properly appreciated as a main feature of the city which belonged to the historic traditions of the whole community.

[243]

Thus awakened, people would want to keep them.[18] His advice was not accepted. The old enemy, the demands of industry and convenience won the day and the walls of Ahmedabad fell.[19]

From Bombay Geddes returned to England for his term in Dundee and his Summer Meeting in London in 1915. However, his return passage to England and then back to India again had been paid by the Government of Bengal (in conjunction with Calcutta and Dacca municipalities) who wanted him to bring his Cities Exhibition to Bengal.[20] When he returned in the autumn of 1915 he found planning commissions coming thick and fast. One of the earliest was for Dacca, and Geddes went there this time for a few days and produced a fuller report than he had done hitherto. The poverty and problems of Dacca were appalling and Geddes' optimism that improvements would get carried out was dampened. But his sense of mission carried him on. As he wrote:

> It is largely as an act of expiation that I for one have become a town planner: and as direct observation, and literacy in plan-reading re-appear from their customary submergence by speech and print so increasingly will others, Europeans and Indians alike.[21]

In the introduction to the Dacca Report in the section on 'Geography and Town Plans' Geddes wrote: 'The Town Planning Movement is on this side a revolt of the peasant and the gardener, as on the other of the citizen, and these united by the geographer from their domination by the engineer'. What he sought was a 'change from a *mechanocentric* view and treatment of nature and her processes, to a more and more fully *biocentric* one'.[22]

The single most important theme in Geddes' early Indian reports was the need for all engineers, sanitarians, and planners to develop such a view, starting from an appreciation of the 'natural' environment of the city in its region. To implement this geographical approach, however, was difficult. Very few municipalities had adequate maps of their cities, towns, and villages, and where these existed most were divided up according to electoral wards. The first step for any planning scheme was to plot the topography, geology, and human geography of the area to be planned and all pertinent information should, in Geddes' view, be presented in cartographic form. The planner had to visualise specific problems and the changes proposed in relation to the whole city, and this could best be done on maps. Related to this work was the need for the planner to observe on the ground the conditions which existed and how they could be improved. For Dacca, Geddes goes so far as to suggest a relief model of the Bengal river system should be built, similar to that of the Merseyside relief model displayed at the 1896 British Association Meeting in Liverpool. With maps, with models, with first-hand personal observation, only then was the planner equipped to undertake practical planning work.

After Dacca, Geddes was given the opportunity to plan a model colony for the Eastern Bengal Railway at Kanchrapara, a project which he undertook with the utmost enthusiasm writing up his report as he travelled to the Paris Exposition de la Cité Reconstituée in 1916.[23] It contained all the elements of his doctrine: an emphasis on homes with their own gardens to be built by the people themselves; sewage to be taken to the special gardens and used as fertiliser. Geddes begs the Railway Company to engage a Chinese gardener for a three-year contract to initiate the people in ways of using sewage to produce rich crops.[24] The social and educational needs of the small community are outlined and Geddes wants to encourage traditional social customs and mores in the new community. He suggests encouraging the acquisition of some sacred cows since land is plentiful, as a constant reminder of the sacred which uplifts the community. In these crowded and exciting months Geddes went with Lanchester to plan the small town of Jubbulpore (Jabalpur) for its municipal council.[25] He worked in Lahore, Kapurthala and on the Cawnpore (Kanpur) expansion scheme.[26] But he and Lanchester joined forces to most effect in Lucknow.

The Lucknow Reports: the apogee of Geddes' practical successes

Lucknow, the elegant and well-planned city of the nawabs and kings of Oudh who had developed the city since the eighteenth century, was like a balm to him after the poverty of Dacca or the wastelands of Kanchrapara. It was exactly the kind of city Geddes responded to most warmly. Its magnificent, mostly Muslim, architecture was a source of inspiration. Its economic structure still belonged to the pre-industrial age, its main reputation being for hand-made luxury goods such as gold and silver brocades, chikan and embroidery work, and Indian perfumery. Although it had a population of a quarter of a million in 1881, by the time Geddes was working there, the population was declining, dropping by 9.5 per cent between 1901 and 1921 in the wake of its economic stagnation and decline. The problem was thus not expansion so much as conservation.[27] The initial disaster which brought Geddes to work in the city was due partly to the actions of western engineers who had filled in many tanks and water conduits in the city in the campaign to eliminate malaria. The monsoons of 1915 had been particularly severe and the result of torrential rain produced extensive flooding. It was the perfect context for Geddes' reconstruction ideas.

In the Lucknow Reports Geddes devotes little space to propaganda

[245]

and much more to planning activities. He had Lanchester on hand to produce drawings and a sympathetic response from British officials to encourage his work. His aim, which he succeeded in, was to persuade the town council to set up a town-planning office. He put up his own exhibition in the council chamber and added the projects for Lucknow to it as they were completed. He himself spent most of his time on an applied exercise in planning for people. He spent his entire time in the city exploring it in the company of municipal administrators, visiting trouble spots with his Indian assistants (two of whom he put up for Associate Membership of London's Town Planning Institute), meeting all who wanted to see him, especially the controlling bodies of the various temples. The latter meetings he particularly enjoyed as the temple authorities usually owned land, including gardens, which Geddes longed to renovate. He wrote of this: 'No co-operation has been more gratifying than that with various temple authorities as regards their buildings, sacred trees, and flower gardens and such improvements are also appreciated by city authorities as by the neighbourhoods'.[28]

The major success of Lucknow, however, was the transformation he wrought on the methods of working of the Lucknow Improvement Trust. He managed to gain the ear of Mr Botting, then Municipal Secretary (later to become Chief City Executive Officer), and between them they worked out a process whereby Improvement Trust activities did not dishouse the people and raise land values. As Geddes wrote:

Most conspicuous and most successful in this respect, so far as my knowledge goes, is the Improvement Trust of Lucknow, since this has grown with experience, of able town councillors, guided by our admirably devoted Deputy Commissioner, and a City Secretary no less so. For of its adminstration it is the principle that, though some dispossession, for demolitions, is of course inevitable, it must be accompanied by actual re-housing, and not postponed by mere hopes and promises of this. Its Executive takes a legitimate pride in dealing individually with every case, and in finding a satisfactory new location before taking away the old one. No doubt it may thus, at times and at places, proceed more slowly; but such progress is sure: and it is moreover freed from the drag of popular dissatisfaction and reluctance; it is increasingly winning general confidence and support instead.[29]

The fact that Lucknow's population was declining probably contributed to the viability of this scheme.

Geddes' delight in the response he received from people in Lucknow encouraged him to put forward his usual solution to the problems of waste disposal in the city with more than usual confidence. India must follow the example of China and Ancient Rome:

> The Romans went so far as to create a special God of Manure and Manuring; and thus they incalculably aided the agriculturalists . . . The universal revival and renewal of historic and sacred traditions is now beginning . . . Will it be possible for these to go on without some co-operative and fertile contact between their historic faculties of ancient learning and their modern faculty of medicine, with its Schools of Bacteriology and Public Health?. . . The associations of 'cleanliness and godliness' will again be clearly formulated.[30]

One of Geddes' friends, the human geographer, Professor H.J. Fleure, wrote to him on receiving the Lucknow Report: 'My dear Geddes, your Lucknow Report is a source of much joy to me. Fancy getting such a fine exposition of civic philosophy, with religious considerations as well into the section on latrines'.[31]

In the Lucknow Reports there are no new ideas. The reports themselves do not even add up to a comprehensive plan for the city. Geddes was always too busy to carry out any systematic survey work. But he certainly got across the message that the planner must work with the people. Following his earlier experience in Dublin, he gained permission to seek out the Indian leader of the workers in the city, Ganga Prasad Varma, whom he described as the 'Jim Larkin' of Lucknow. Varma was encouraged to serve on the municipal council and to undertake practical schemes for rehousing the people. Geddes also appealed to the people of Lucknow to become alive to their gardens, since so many in the city had fallen into neglect and disrepair. He suggested an Arbor day on the American model, a tree-planting holiday involving all citizens. Such a scheme was already meeting with some success in New Delhi. Geddes wanted not just tree-planting in Lucknow, but a revival of the art of Moghul gardening.[32]

Many of Geddes' ideas were taken up and implemented and Lucknow established a Geddesian tradition of town planning for at least a decade after Geddes' visits. In the 1920s, a new Chief Engineer to the Lucknow Improvement Trust was appointed, Mr Linton Bogle, graduate of architecture and town planning from the University of Liverpool and student of Patrick Abercrombie. He wrote a book entitled *Town Planning In India* which was published in 1929 and was obviously, though anonymously, influenced by Geddes. He wrote out a credo for the would-be planner which could have been culled from Geddes' Lucknow Reports:[33]

What town planning means:

1. DEFINITE PLAN of orderly development for the Town into which each improvement will fit as it is wanted.	NOT the immediate execution of the whole plan.
2. CARE AND PRESERVATION of human life and energy, particularly child life.	NOT indifference to congestion and insanitation.
3. PROVISION of good building sites.	NOT leaving narrow and awkward shaped plots.
4. ENCOURAGEMENT OF TRADE and increased facilities for business.	NOT interruption of trade.
5. PRESERVATION OF HISTORICAL BUILDINGS and buildings of religious veneration with all their traditions.	NOT destruction of old landmarks.
6. THE DEVELOPMENT of an INDIAN CITY worthy of civic pride.	NOT an imitation of European cities but the utilization of what is best in them.
7. HEALTH, PLEASANT SURROUNDINGS AND RECREATION for all inhabitants	NOT merely expensive roads and parks available only for the rich.
8. CONTROL over the FUTURE GROWTH of the town with adequate provision for future requirements.	NOT haphazard laying out of buildings and roads with resultant COSTLY improvement schemes.
9. ECONOMY.	NOT WASTE.

Geddesian planning in the princely states: Baroda and Indore

Lucknow was a city under British administration. In 1916 Geddes had his first opportunity to work in the capital city of a princely state, Baroda. He found the prospect exciting as he was seeking an alternative to western patterns of industrialisation and he believed he might find this in the princely states.[34] The Maharajas, regardless of their power *vis-à-vis* the British, certainly wielded considerable economic, social, and political influence over the progress of their capital cities.[35] Here Geddes saw an opportunity for inspiring an Indian alternative to British patterns of modernisation if only he could convince the princely rulers of his socio-biological approach. In Baroda the prospect was particularly favourable. The Maharaja had already gained a reputation for his leadership and concern for economic and social change. The state had an excellent education system where first attempts to achieve compulsory elementary education had been made in 1893. A museum had been founded in 1894 and a library department in 1910. The College of Higher Education founded by the Maharaja was to become the University of Baroda after 1918. The Maharaja Gaekwar had provided his city with excellent public buildings, parks, and tanks.[36] The money to pay for it had come from the growing industrial development. A Department of Commerce and Industry had been founded in 1905, the Bank of Baroda in 1906, and an Industrial Advisory Committee appointed by the Maharaja in 1914. Economic and sociological surveys abounded (one, in 1914, a survey of the 800 palace servants) and industrial surveys of state towns. The result was that the Maharaja became known as the reforming Maharaja, and he used his special position to put in hand aspects of social reform which only an Indian ruler could. He was a champion of the revival of Hinduism and was made first Chancellor of the Hindu University of Benares. Geddes had become involved in this scheme hoping for a chance to plan a new Geddesian style university but he was not successful in gaining this commission. At the opening ceremony of the university, the Maharaja Gaekwar spoke of a 'vigorous and practical determination to deal with the difficulties of the present'. What was needed was more research into social problems, better education for priests, better treatment of lower castes, and a struggle for political freedom though 'there can be no rights, no privileges, no genuine freedom without corresponding duties, obligations and self-restraint'.[37] In Baroda he encouraged a flourishing cultural life, his own personal hobby being architecture.

Thus in Baroda Geddes found a model for development, Indianised and already successful. With the Maharaja to give a lead (and in his Baroda Report Geddes also suggests the Dewan should offer prizes for the best kept homes, and so on) he had a pattern to fit his civic reconstruction. Geddes was invited to come to Baroda for two weekends to give lectures and to give advice on Baroda's most pressing planning problem which was congestion in its inner city areas. Geddes went with Lanchester and he was delighted by the large enthusiastic audiences he received for his lectures. He urged them to start a 'Know your city' movement in Baroda similar to those to be found in America. He mentioned the example of the 'Boston 1915' movement of 1911–12 which had just been replaced in 1916 with a new set of objectives by the 'Boston 1920' movement.[38]

The little city of Baroda itself was very attractive. It had benefited considerably from the architectural activities of the Maharaja. It was quite small with a population in the 1921 census of 94,710. It was in fact losing population at this time, mostly because of demolitions in the city centre districts. Geddes rushed in to give his advice and unfortunately on this occasion his hopeful ideas about 'life promotion' turned out to be counter-productive. He tried to apply some techniques of conservative surgery which led to some adverse results. The most congested area of the city was the fort. Geddes ignored the fact that the fort was becoming the commercial centre of the city. It had a leaf pattern of street development with one main artery for traffic and side roads off this which were culs-de-sac. Geddes suggested the opening up of these alleyways and the making of small open spaces to be planted with trees. In fact what occurred was the opening up of the alleys to traffic and traffic congestion was added to the problems of density and the overcrowding of population of that area.[39]

The Baroda Report was typical of Geddes' earlier Indian reports. Largely concerned with attitudes rather than plans, it is didactic, vague on specific problems, and brief on the nature of the transformations the city would face as it developed. It contained his usual recommendations about refuse collection, the whitewashing of house fronts, and the need to direct the energies of the young towards practical activities, in this case the building of a veranda at the boys' school by the boys themselves. He had no time to elaborate on the deeper philosophical concepts behind his doctrine of civic reconstruction. All he could aim at was to 'emotionalise' the response to town planning which he believed was the key towards creating those good traditions which would ensure Baroda's path to her full potential. As he wrote in his report on another capital of a princely state, Balrampur:

> The succession of men and their works in their city should thus be like that of the rings of a tree stem, which are ever being outgrown yet continue of service – not only mechanically keeping their place and burying their past in maintaining the whole tree, but also carrying sap, and like to its fullest and its youngest life, of leaf and flower and fruit, of buds anew'.[40]

The brevity of Geddes' visits to Baroda, the slimness of his report, the emotional and vague nature of his comments, resulted in a missed opportunity. Geddes' potential influence on a reforming Maharaja never materialised.

He was, however, gaining the reputation of not only giving advice on modern town planning, which was a rare expertise, but also making suggestions which were most economical. In comparison with other British engineers and sanitarians, Geddes' techniques of 'conservative surgery', cleansing and renovating, were all very cheap.[41] His demands were only for labour and most of that on a voluntary basis. He was also gaining a reputation for his support for reforms in Indian education and it was a combination of the two which was to bring him his commission in Indore. H.W. Lyons was Professor of Economics at the Indore Christian College and he became a friend and correspondent with Geddes.[42] He brought his name forward to the Maharaja of Indore at a time when the city was suffering one of the worst outbreaks of endemic plague.[43] There had been a growing demand for some action to be taken and the Maharaja was persuaded to engage Geddes with the inducement that he would not produce any costly plans. Indore, despite the fact that it was situated on the confluence of two rivers, also suffered from severe water shortages in the summer months and the Maharaja had already spent crores of rupees on schemes for reservoirs which had been planned by western engineers. These had been badly located, some even above the water-line, and, for all his investment, the Maharaja had not had a single drop of water when it was needed.

The circumstances of Indore, however, were not particularly favourable for an experiment in Geddesian civic reconstruction. The Maharaja Holkar, Tukoji Rao III, was no reformer like the Maharaja Gaekwar. The city was roughly the same size as Baroda, being 93,091 at the 1921 census. But it had few buildings of architectural merit, and its plan lacked the style of Lucknow or Baroda. It consisted of three main areas: the old congested quarter of Juni Indore, which had been pillaged and razed to the ground in the eighteenth century; the modern city and suburbs; and a developing industrial town. What it lacked in elegance it made up for in space as it covered an area of five square miles. With 18,789 occupied houses this gave an average of 344 homes per square mile. Its local economy was still based on its function as a market and it was one of the largest trading centres of central India. Its chief commodities were opium and grain. With the grain came the rats, and with the rats the fleas, and thus the plague — Indore was one of the worst affected cities in India. The new industrial suburb was beginning to be developed as a centre for the cotton industry and with the growth of new mills there were demands for workers' housing. The workers themselves were being recruited from amongst the peasants of the surrounding countryside, and in this instance Geddes' belief in the desirability of maintaining the links between the workers and the countryside through low density housing well supplied with gardens, was not unrealistic. The population of Indore as a whole, however, was mixed in terms of religion and caste.

The religious groupings were 75 per cent Hindu and 21 per cent Muslim, the final 4 per cent made up from small groups of Jains, Christians, Animists, Sikhs, and Parsees. In terms of caste the largest were the Hindu Brahmins (19 per cent of Hindus) the Banias (7 per cent), the Rajputs (6 per cent), the Marathas (5 per cent), and the Dhangars (4 per cent), which was the caste of the Holkar family, the ruling dynasty.[44] Geddes spent his first few days in Indore walking the length and breadth of the city and setting up his workshop on the veranda of the Holkar guest house. As usual his first activity was to get a number of maps of the whole city and to plot on them, as accurately as possible, the main features of the city.

What he found was that the city could boast of only one main street which went across the Khan River into a great square in front of the old palace of the Maharaja. Yet each side of this street were densely populated areas of intricate complexity in their social and religious groupings. The city had 455 Hindu temples, 14 Jain temples, and 131 Mosques. Outside the old city, on the western side of the railway track, lay the new cotton mills, the new town hall called King Edward's Hall, and the state officers' club. The socially superior new suburb was Tukoganj, which had the official residences of state officers. Apart from that, the main buildings of importance in the town were the new hospital, the state offices, the guest house, the English *madrasah*, the gaol, the barracks for imperial service and state troops, and the cenotaphs of deceased Maharajas. The Maharaja himself had four residences outside the city. A palace, Lalbagh, built in the style of, and almost on as grand a scale as, Versailles, and three other residences for parties and hunting lodges.

This was Geddes' raw material. This was the project he threw himself into with the passion of his grief after the death of his wife and son. He hoped to make his report not just a commentary on Indore and its problems and potential, but a major statement on his whole civic reconstruction doctrine. Chapters of the report were sent to England for separate publication in the *Sociological Review*.[45] What Geddes set himself in Indore was to give a practical demonstration of his doctrine. For this the Indian context suited him very well. Indore may not have been beautiful, but its scale and problems made it a possible context for launching civic reconstruction doctrines. In the first volume of his report he began with a preliminary survey which was a *tour de force* displaying his considerable skills of 'reading' a city from direct observation and maps. In the report he traces the city's origins as an old religious centre, then a military centre, and now a state capital, from observation of existing buildings and layouts of streets and squares. He believed

(1) that [Indore's] many good qualities of planning, and its features

generally, date from its origin and development, that is, are of its life and growth and essence; and (2) that its present obvious dirt, congestion, disrepair, deterioration, and even too frequent dilapidation – though these now may, and commonly do, impress us so conspicuously and so painfully, as even to obscure all fundamental merits till a patient survey recovers them – are each and all indeed very largely of our own modern time'.[46]

The first step to the future was to uncover and restore the physical remains of the past.

Geddes, however, did not carry out a survey of Indore. He pleaded lack of time and recommended his readers to study Lanchester's survey of Madras and to look at the work undertaken by the RIBA since 1914 on a survey of Greater London.[47] He was given two Indian assistants by the municipality to help him with his work. But they were fully occupied preparing the cartographic material for the report and working on a project to design solid, yet cheap, homes for the poor. This was the specific problem the Mahatma Gandhi had set Geddes after their correspondence on the Hindi Conference held in Indore in 1918. Geddes himself was mostly occupied seeing and talking to different people. As he wrote to his daughter:

> This is a most interesting job – so varied and complete on all sides of life. This afternoon I have had the geologist of the water scheme; then next the representative of a great Jain cotton millionaire wanting forty acres for mills, bleaching, dyeing, etc. and then a deputation of followers of a new Saint, *The Sadhu* Maharajah, who wants me to plan their earthly paradise in the very front of my new cottonopolis – and round their existing *Duni*, or hermitage of the Sacred Fire, with two Dharmsalas or lodging houses, one for five hundred tramping yogis as they pass from one holy city to another, and the other for family pilgrims.[48]

The first objective in the report, however, and the first practical objective that he set himself in his work, was to deal with the problem of public health. The plague had been the reason for his commission and it had to come first. What he did was to set the municipal sweepers and other workers on to a great programme of cleaning, paving and repairing of the streets and old property in the oldest, most plague-ridden quarters. He wanted the three grain markets removed, amalgamated as one market and put under strict control as a dangerous trade. He advised the acquisition of a number of cats to act as rat-catchers. For the rest he relied on his conservative surgery techniques of renovation with minimal destruction of existing buildings. In the report he writes at length about 'public health in the industrial age', digressing far away from the immediate problems of Indore. His message was that the 'good life' could be lived in cities if only people cultivated what was healthy and life-giving. Better food, better housing, and better recreation for all, were possible, if the sources of wealth and well-being were properly understood. The renovation of houses, the addition of gardens and open spaces, and a responsive people, willing to develop ways of living, the 'plain living and high thinking' of the old religious orders, were the sum of true wealth for any city.[49]

He had arrived in time in Indore to make use of the Diwali celebration as a means of emotionalising this message. He set about making the Diwali Festival of early November 1917 one of the greatest celebrations ever seen in Indore. As he wrote:

> Soon after the beginning of the present study of Indore, the conviction became irresistible that for the arousal of the people from their too neurasthenic submission to plague, their fatalistic acceptance of it, there are needed methods altogether beyond the present cold and conventional ones of sanitarians. To vitalise these, there is required the revival of the best traditional methods of popular appeal — artistic, symbolic, mythic, and thus religious; hence in all these ways emotional, and thus practical. For it is now the commonplace of scientific psychology — though still unrealised by conventional education, and hence disused by the educated and governing classes — that to carry any idea into action, there is needed the corresponding arousal and uplift of emotion, without which no thought, however true, can rise into effective deed. Hence the recent Diwali procession, which despite all previous fears, and subsequent criticisms, has been found so far to justify itself; and I trust its principle yet more.[50]

An extract from the circular advising the Diwali procession is included as Appendix 6 in Volume 2 of the Report, and Dr P. Boardman, Geddes' biographer, has described the success of the occasion.[51] He claims that more than 6,000 cartloads of dirt and refuse were removed from the city by willing workers, anxious to ensure that the procession would pass their homes. The procession route went only through the best cleansed portions of the city. Geddes himself was Maharaja for the day riding on the White Elephant. At the end of the day, after sunset, the effigies of the Plague-Rat with his deadly fleas, and the Dirt Giant, were taken outside the city and burnt on a huge bonfire and there was a firework display. Geddes had learnt from his *Masques of Learning* in Edinburgh and London how to mount dramatic events with mass public participation. He managed to gain the temporary but enthusiastic support of volunteers from amongst the 600 delegates who had come to the city to attend the annual Hindi Language Congress. This he felt was 'full of promise for city betterment'.[52]

However, he well understood that this kind of middle-class support could not be sustained and he directed his attention mainly to the municipal sweepers, all drawn from the untouchable castes whose livelihood was keeping the city clean. He had won their support by seeking their co-operation prior to and during the Diwali procession, and by insisting that the leader of their community should be introduced to him personally. His long-term objective was to elevate the sweepers socially by changing their occupation from the 'loathsome and despised occupation of the sweeper . . . into that of the Chinese gardener'. The sweepers would become responsible for trenching night-soil into specially designed gardens and planting the necessary crops to keep the soil sweet and healthy. With this change would come higher levels of cleanliness, more efficient activity, and a general raising of the quality of urban life.

> The social problem of 'elevating the depressed classes', of which we nowadays hear so much, and see so little, is thus readily soluble for the sweepers, at any rate; and this upon the lines of their own occupation, and not simply that of the verbalistic 'education' of the West. For when we leave our houses in the insanitary street, all served by sweepers, and that too imperfectly we can readily settle into a sanitary suburb served by gardeners'.[53]

As usual Geddes was advocating his method of biological sewage disposal. To get such a scheme implemented in a city the size of Indore was no simple matter. Geddes asked the Principal of Poona Agricultural College, Harold Mann, for his opinion about the possibilities. Mann was cautiously favourable to Geddes' ideas, though qualified his support by the recommendations that there would need to be constant strict administration of the scheme and continuous gardening activity on the land used for it. Mann felt that a whole city might be difficult to manage on these lines though small-scale experiments, one of which he had conducted himself, had been very successful.[54] Geddes was not prepared to be cautious. He wanted to highlight

> the difference between the engineering and the rural attitude and policy in this matter – that between 'all to the sewer' and 'all to the soil' – is in fact parallel to that which was effected in the past generation for agriculture and physiology by the rise of the rural and organically-minded Pasteur, against the long-dominant authority of Baron Liebig, with his too strictly inorganic view and treatment of these subjects. And a generation of thinking peasants is coming: we biologists are but its scouts and pioneers'.[55]

Town planning, Geddes stated again and again, had to begin and end with the people. The prime target to attack was the condition of neurasthenia, or depression, amongst the people which was caused by hostile urban and industrial conditions. Geddes was fearful that even his new industrial estate might create the wrong conditions and he could not help putting in a plea for the manufacture of silk alongside the developing cotton industry. He listed textile industries in a hierarchic order according to their biological impact on the workers. At the bottom of his list was jute and above it, cotton, then linen, then wool and at the top, and the most favoured, was silk. Silk gained this accolade because it was produced by a biological, as opposed to a mechanical process, and it was rich in what Geddes described as 'civilisation values'. These were cleanliness for the culture of the silk worms, a higher status for women responsible for their care, the need for mulberry trees for food, and above all, the constant daily tasks of nurture and development, an antidote to the 'neurasthenic' or depressing effects of labour related to machinery.[56]

Geddes was particularly anxious to press home the message that India must by-pass the effects of 'paleotechnic' industry on her people. There was no need for India to follow the west if only western solutions to Indian economic and urban problems could be rejected. As he wrote in his report:

> While the Western (and engineer-educated) sanitarian as complacently applies his English and Victorian manufacturing town experience to Indian sanitary problems, as do his educational or missionary fellow-students for their corresponding instructional and denominational traditions, this complacent assurance is impossible to the planner who is anything of a geographer and anthropologist. For he sees the people of different climates and environments as adapted through past ages to these. Thus he comes to their ways, their habits, their customs, their institutions, their laws, their morals, their manners, with the ordinary naturalistic attitude of observant and interpretive interests, and not that of superiority. He thus seeks first to learn, to understand, to appreciate, before he attempts to criticise, much less to teach and transform.[57]

Ironically, he was to owe his greatest success in Indore to the work of Western engineers. Geddes had turned his attention to the problem of the water shortage and discovered that the British Public Works Department had sent engineers to lay down a pipe-borne system drawing on local sources of water from outside the city. The problem was that when the scheme had been completed, the engineers left before the new system had been connected to the city's old one with the result that the connection was never made.[58] In time, the location of the new PWD system was forgotten and the city's water shortages continued. Geddes found some maps of the PWD scheme in the municipal office and having done some detective work with the municipal officers, was able to locate where standpipes should be sunk to activate and utilise the system. He was particularly delighted when the PWD system was rediscovered and found to work. But his success was clouded by his appreciation of the fact that he had put the water-carriers out of a job. Since most of the water-carriers, despite their lowly occupation, belonged to a high caste of Brahmins, their plight was accentuated since many occupations were forbidden to them on religious grounds. Geddes suggests that they could become milk suppliers. Aware of the jibe that Indian milk suppliers always watered their milk, he jokingly suggests that at least the added water would be pure.[59]

His success with the water scheme won him the lasting support and approval of the Maharaja and the municipal council. Remaining records of the state administration of the early 1920s suggest that Geddes' impact was considerable well into the 1920s.[60] Each year, instead of Diwali, the birthday week of the Maharaja became the occasion of great municipal cleansing of the city. Trenching grounds were organised, night soil collected in nine depots, thirty sweepers appointed to trench it. Problems for the municipality soon arose, however, because a standing order of the state Home Minister forbade any action in Indore contrary to Geddes' plan. It was a typically inappropriate administrative response to a document which in no way resembled a master plan. Endless delays were caused, and by 1924 an Indore Improvement Trust Act removed at least extension schemes from municipal authority. The problem was that Geddes had left no civic reconstructionist behind him to interpret the spirit rather than the letter of his work. He had also left no detailed master plan.

His main contention was that city development must be a careful evolutionary process as he believed that as yet too little was understood about the intricacies and interrelatedness of city life. Disturbing and remoulding one area could have many unforeseen consequences. There was a need for a greater respect for existing conditions which were the result of an extremely complicated evolutionary pattern from the past. Modern town planners and municipalities were too often unaware of this.

Modern administration, even in its advanced Western forms, had acquired its present division of labour, with its aim of efficiency in each department, irrespective of effects on others, long before the evolutionists of nature and society had at all reached comprehension of these — as not machines, of which the parts are separately constructed, and may be so far separately kept in order or altered —

but as a 'web of life' intricately interdependent throughout, and this in many ways – tens, hundreds probably thousands of ways – of which we only as yet discern the most obvious.[61]

The role of the planner was not only to work towards an understanding of this 'web of life', it was also to prevent unbridled forces wreaking havoc in the city. Prime amongst these were demands for road improvements.

For once Geddes approached this matter with a certain amount of circumspection. His colleague, H.V. Lanchester, had prepared a road scheme for Indore whilst he had been attached to the New Delhi Planning Commission in 1912. While suggesting that Lanchester's report was 'by far the ablest and best' amongst the schemes and plans for Indore presented to him, yet he advised caution.

> The fact is that we town planners have usually not time to master the social, commercial, residential and working class conditions of the quarters which we have to re-plan and we thus tend to be over-influenced towards the better through communications we readily see and with insufficient means of foreseeing the consequences to the quarters we thus cut through.[62]

In fact, Geddes goes broadly against the inevitable trend of more and more traffic and suggests that with the amalgamation of the corn markets to one site, and the removal of industrial activity to the new town, the volume of through traffic would actually decrease.

> For every reason then I essentially leave this old Bazar city as it stands, and without cutting any large new thoroughfares. Yet this is no mere policy of dull conservation, of letting things alone as they are, there is a further alternative:– that of antisepsis and conservative surgery – in plainer terms, cleaning up and clearing up . . . the sanitation of every Mohalla has been carefully gone into and Open Spaces and Gardens are planned wherever possible without costly clearances. By our small removals, straightenings, openings, and re-plannings in detail, a network of clean and decent lanes, of small streets, and open places, and even gardens, is thus formed which is often pleasant, and I venture to say sometimes beautiful.[63]

The one major express way that Geddes does allow himself, the New Express Boulevard, which brings the business quarters of the city into better communication with the railway and the suburbs to the west, south, and east of the city, Geddes supports not only for its convenience but also as an enhancement of the environment. He places it alongside the river, by turns crossing and following the river loops to provide a scenic route. His delight at meeting modern demands with a beautiful solution encourages a flight of hyperbole:

This route would not have been different from the present direct and utilitarian one, had it been laid out purely by the landscape lover . . . had it been designed by the religious man and philosopher, as educators. For Religion and History, in their temples and monuments, Art and Nature in their complemental appeals, are all here upon our way. . . our new road is thus a microcosm of the city, even of the world, sits open to all its attitudes of mind, from modern hurry to contemplative peace.[64]

Apart from his emotional propaganda, Geddes, under the stimulus of Gandhi, did address himself to the crucial social problem of housing. The success of his plans for the new industrial estate in Indore rested on his ability to provide housing for the workers at prices they could afford. Geddes' response was sparked off by his hatred of the 'improved dwellings' for the workers, such as the great tenements of one-roomed 'chawls' built for the cotton workers in Bombay.[65] He was prepared to argue fiercely against those who measure improvements in terms of materials, construction, and the provision of facilities. Concrete or 'pukka' materials, a sound roof and tap-water were not enough. Writing as

> advocate for the condition and well-being of the people, and for their health therefore above all, I have to remind all concerned (1) that the essential need of a house and family is *room* and (2) that the essential improvement of a house for its family is *more room*.[66]

With the help of one of his assistants, Mr A.C. Sinha, a graduate of the Engineering School of the University of Manchester, he worked out a plan for a low cost housing unit, not using 'pukka' materials.

The Sinha-Geddes improved dwelling has an earth floor and mud walls, but both of these are ingeniously strengthened. The floor has a damp-proof course, covered by large sheets of stout matting soaked in tar and then three inches of earth on top of that. The house is set on a plinth which

> can be strengthened at its outer sides with piers of brick; on which may be built a square pillar of brick, with alternate half-bricks left projecting along each of the two adjacent future walls. When the door-posts are also fixed, we can then build up the four walls with mud.[67]

Timbers for the roof and veranda can be found amongst the rubbish from demolished houses and covered with cheap tiles. Geddes estimated that a good one-roomed dwelling on this pattern with cook-room and veranda could be built in pairs for about 470 rupees per house. Two-roomed houses on the same pattern would cost about 750 rupees.[68]

Even these prices, however, were far beyond the resources of the majority of the population and Geddes was well aware of this problem. His main suggestion was to encourage the poor, particularly the sweepers and carters, whom he describes as 'sturdy fellows, handy, willing and often intelligent', to build their own homes: 'What better outlet can a man find for these virtues, or for increasing them, even acquiring them, than in the construction of his own home?'.[69] For capital he suggested housing organisations, founded by the state, by the municipality, and by private philanthropy; co-operative societies based on the English example of Co-operative Tenants Limited; or co-operative banks. He cited, once again, the example of Sir Horace Plunkett's Irish Land Organisation Society, to show what could be done even amongst the very poor. The town planner working in co-operation with these bodies could ensure the orderly layout of plots and adequate provisions of gardens, the essential features for the future health and well-being of newly-developing areas.

The only guarantee against future deterioration and decay, though, Geddes had always maintained, was the culture of the people. Since the Indore Report was to be a full account of his civic reconstruction doctrine, he had no compunction about devoting a substantial section of the second volume of the report to proposals for a new university for Central India at Indore. In seventy-three pages of dense and convoluted prose, he rehearses once again all the arguments he had ever put forward for the reform of higher education since his early experiments in Edinburgh in the 1880s, liberally illustrating his account with examples of initiatives, however modest, where he had had some direct influence. The core of his argument was that modern science and technology had to be made to serve the people, and that this could best be done by study and research at institutions of higher education. Yet such institutions could not simply be established either by public authorities, or by the will of some millionaire. They had to be part of the social and cultural environment, and could only be created by a demand from the people. When such an institution was created Geddes then, as usual, proceeded to suggest that only his pattern of university reform would satisfy the needs of the people, because the new institutions had to meet the economic and social challenges of the twentieth century. The organisation of studies must be synthetic rather than specialist since specialists were often blind to the implications of their work in other contexts. For Geddes, university reform and civic reconstruction were vitally linked because they depended on establishing a new cultural perspective and training people to be sensitive to their environment. The difficulties in the way of achieving an improvement in the condition of society were not economic, they were psychological.

Figure 8.1   Patrick Geddes in Indore, 1919

The circumstances of war had given a great stimulus to the practical application of new ideas. The challenge was to continue that stimulus in peace for the ends of social betterment. It could only be done through a new educational system which released the creativity of individuals and gave them an insight into problems and cultures beyond their local and regional experience. The first step towards this was to abolish the examination system and instead to encourage all students to undertake research, and to travel widely to broaden their personal experience. Their teachers would then provide 'estimates' of the value of their work instead of meaningless examination grades.[70] Anxious not to alienate his readers, Geddes stresses that the reforms he advocates did not mean an abolition of the existing schools and colleges. All that needed changing was their spirit and perception of educational objectives. The new university, though, could not depend on state support, nor would it survive simply on the work of scholars. It had to seek its life's blood in active discussion between city and state about what kind of institution was wanted, it had to be responsive to the potential for change. As usual, Geddes stressed how economical his proposals were, commenting on how the Scottish universities had thrived even when they had been poor and how, at the New University in Brussels in the 1890s, university teachers had served for years without pay.

Finally he suggests that the new university should start modestly with perhaps a small library, an institute, and a garden. Then it would grow with the cultural evolution of the people and would always remain 'of the people'. In this way, cities like Indore might be 'materially small but culturally great'.[71] The message was a new version of his vision for Dunfermline in 1904; the potential of provincial cities to be centres of culture regardless of their size and location, and to provide an environment for life-enhancement through the proper application of modern ideas. Towards the end of this passionate survey of universities past, present, and possible, Geddes permits himself the idea that such a university as the one he proposes for Indore might need a leader. Seemingly contradicting all he has said about the demands of the people, this leader was needed to ensure that a Geddesian style university was created, and it would seem a strong possibility that he was suggesting he would not be averse to being offered the job himself! Obviously there was no one better qualified to do it.

This basic contradiction between Geddes' civic reconstruction message and his own role in helping it come about, made it impossible for anyone to work with him over a long period. All the difficulties his followers at the Outlook Tower had found with him in the 1890s were magnified in India, where there was less to restrain his ebullience and desire to have his own way. The Indore Report is a greatly flawed work because of the character and style of the author. Yet it is full of quotable phrases and Geddes' passionate commitment to put people first in planning comes across with great emotional force. Geddes did not perceive, though, that his work on Indore did not add up to a town plan unless the reader accepted unconditionally his advocacy of the doctrine of civic reconstruction. He was particularly upset by a letter from a Mr Sett in Calcutta who wrote to him after reading the report, to point out that Geddes had not dealt 'with the traffic, engineering and other usual points'. Geddes replied:

> You must permit me to answer, and pardon my saying that you must have missed my treatment of them . . . for besides reforming the Drainage and the Water — engineering tasks both ill-planned previously, and wastefully, I definitely tackle (and I maintain, successfully settle) the *Traffic* question; and this on ample scale for that city, and its future enlargement, though saving some seventeen

lacs upon the sweeping cuts of my predecessor, who was then working in the old style, which Indian cities are still so disastrously following, i.e. the example of Napoleon III in Paris and his imitator in Berlin, both before the modern town planning movement altogether. . . *We now start with the idea that cities are fundamentally to be preserved and lived in; and not freely destroyed, to be driven through, and speculated upon.*[72]

A noble sentiment, which helps to sustain interest in the Indore Report notwithstanding its many idiosyncracies and strange prose.

Jerusalem and Palestine

Geddes not only sent parts of the Indore Report to be published separately in the *Sociological Review*, and his section on 'the History of Universities' to be published in pamphlet form, he also sent copies of the whole report to friends and acquaintances. High up on the list of the latter was Dr M.D. Eder, a member of the Zionist Commission who had been in Palestine between 1918 and 1919 working on reconstruction after the war, and deeply involved in the most eye-catching of the Zionist Commission's new projects, the founding of a Hebrew University at Jerusalem. Eder found a copy of the Indore Report waiting for him in England in May 1919 on his return from Palestine and, as Geddes hoped, he wrote to him 'that there are certain similarities between our needs in Jerusalem and the great Indore city, and we are also considering the formation of a University in Jerusalem'.[73] Geddes had by this time, returned to Dundee for his last term in his old post before taking up the Chair in Bombay which he was scheduled to do at the end of September 1919. He was intensely excited by Eder's response. The prospect of working in Jerusalem seemed to him a summation of all his dreams.

He had actually written to Amelia Defries in 1913 when he was working in Dublin:

But the best example, the classic instance of city renewal (beyond even those of Ancient Rome and Ancient Athens) is that of the rebuilding of Jerusalem; and my particular civic interests owe more to my boyish familiarity with the building of Solomon's temple, and with the books of Ezra and Nehemiah, than to anything else in literature. Jews probably know more or less how the Old Testament has dominated Scottish education and religion for centuries; these were above all the stories which fascinated me as a youngster; and though I lapsed from the church of my fathers well—nigh forty years ago, I still feel these as the great example for the Town Planning Exhibition! .. The improving and renewal of cities might, and should once more, find an initiative, an example, even a world-impulse, at Jerusalem.'[74]

He suggested the carrying out of a survey of Jerusalem, Past, Present, and Possible.

With Eder's letter in his hand, Geddes was dashing off letters to the Zionist Commission giving voice to his enthusiasm, stating his terms (£10 a day or £300 a month with office accommodation) but offering to pay his own passage, subsistence expenses, and the cost of any personal assistance – an arrangement far more favourable to the Zionists than the ones he had exacted for his work in Indian cities.[75] Geddes was fortunate in that his approach to his work, especially his influence on the psychological factors in planning, struck a chord of response from Eder. Eder was a trained psychologist, an early disciple of Sigmund Freud and the first practitioner of psychoanalysis in Great Britain. He had been active in the International Council of Israel Zangwill's Jewish Territorial Organisation, and it was as a representative of IJTO that he had a place on the Zionist Commission set up by the British Foreign Office early in 1918. His first visit to Palestine later that year was to make him a confirmed Zionist, and he believed that Geddes had a great deal to offer the Zionist movement in Palestine. He was to be not only instrumental in securing Geddes his initial invitation, he also remained his staunchest friend and ally in the troubled years ahead.[76]

By mid-1919, when Geddes was corresponding with Eder, the heady days of euphoria for the Zionists after the 'liberation' of Jerusalem and Palestine from the Turk, and the setting up of the Zionist Commission, were virtually over. The function of the Zionist Commission had been merely to maintain a line of communication between the British authorities and the Yishuv (the Jewish population of Palestine), to co-ordinate relief work, and to organise reconstruction after the devastation of the war. The British put an embargo on Jewish immigration into Palestine and on Jewish acquisition of land until the country had been passed over to civilian rule. The most hopeful project for the Zionists had thus been the prospect of establishing a Hebrew University in Jerusalem which was to become the talisman of their passionate commitment. Geddes received news that his application to the Zionist Commission had been successful in August while visiting the Irish Land Organisation Headquarters in Ireland, and he accepted with alacrity.

After considerable difficulty he also managed to arrange to travel at least part of the way with Dr Chaim Weizmann, the leader of the

British Zionists, who joined the same ship as Geddes in Marseilles.[77] Weizmann was at the height of his powers in this period. His service for the British war effort had brought him not only official recognition but had also, with Lloyd George's support, helped him to bring the aspirations of the Zionist cause into a more favourable light as far as the British Government was concerned.[78] He had become the charismatic, idiosyncratic, and enthusiastic leader of world Zionism, whose prominence in the fight for the cause owed not a little to British control of Palestine. Geddes and Weizmann had vigorous discussions on the voyage *en route* for Jerusalem. Geddes shared with Weizmann the idealism, emotional commitment, and belief in the sciences as a guide for the future which made them ideal companions and argumentative sparring partners. They were united in their belief that the reconstruction and development of Jewish settlements in Palestine should be based on empirical practical schemes for agriculture and industry, rather than waiting for political dictate. Their arguments were thus not on principle or approach, but on method and scale.

Weizmann had submitted a note on the university project to the Zionist Commission, and to the Advisory Committee in the summer of 1919.[79] In it he suggests that the two key practical problems for the university were the material ones, buildings, and so on, and attracting talented personnel. He believed that the second would create the most difficulties. The avowed aim of the university was to be a centre of Hebrew culture, and thus all work and teaching were to be conducted in Hebrew. No advanced studies were conducted in the language at this time. Furthermore, in the uncertain political conditions, few Jewish scholars would risk giving up established posts in western countries for the lower salaries, poor working conditions and few students they would have in Jerusalem. Weizmann suggested postponing the development of the university for a decade or so, and offering contracts to young unestablished Jewish scholars who would be given a number of years to master Hebrew and who would, by the time they came to work at the university, have good facilities, as by that time more funds would have been raised. Weizmann recommended that in the interim a small number of research institutes should be set up to give testimony to the strength of Zionist intentions for the future.

Geddes launched himself on this scheme displaying no such caution. In the report he wrote after his seven weeks' stay in Palestine, he valiantly tried to incorporate the Weizmann outline of research institutes, but his major argument was that such an exciting project as the renaissance of Jewish life and culture needed to have a fully-worked-out ideal and plan right from the start.[80] At the outset of his work he had asked the Zionist Commission what he considered was the key question:

[265]

Figure 8.2   A view in Palestine, c. 1920

Is this New University to aim simply at being the latest – the 244th or thereby – upon the list of the World's Universities (in the 'Minerva Jahrbuch der Universitaten' etc.)? Or is it to aim at being also the first of the new and post-war order?[81]

If the Zionist Commission wanted the latter, the answer was a Geddesian style institution devoted to a synthesis of knowledge, unity between disciplines, a close cultural and economic relationship between university, city, and region, and the adaption of novel methods of teaching. Geddes did not rehearse the justification for these views in his own rendering of the history of university development as he had done in the Indore Report. But he does refer to his publications on the subject, and he also suggests that the Zionist Commission should set up a Bureau of Information which would be staffed by someone whose sole job would be monitoring the developments in all subjects at all universities in the world, to ensure that Jerusalem was up to date in the production and dissemination of knowledge. On one very basic issue he parted company with the Zionists. He was against making Hebrew the compulsory language. In

Figure 8.3   A street scene in Palestine, c. 1920

'The Americans are very American, the Germans very German, the French very French
and the English very English, and so on . . . and so in Tel Aviv, etc., we have nice little
houses of the London and other suburban types *before* the Garden Village period of
England, and with no oriental character at all; very expensive accordingly . . . Look at
any good photograph of an Arab hillside village. See the plain walls, but wall above wall;
the flat roofs, but roof above roof. See how these contrast and compare with one another;
see the bright walls and brighter dome roofs in the sunshine, and how as it were they
chime together, with the dark walls and masses in shadow giving deeper notes . . . Here
is architecture in its very essence.'
Source: P. Geddes (1921) 'Palestine in Renewal', *Contemporary Review*, 120: 457–7.

his view the university, as representative of international Jewry,
should be truly international in this respect too.

When he moved from theory to practical planning, Geddes was
delighted to find that the proposed site for the university was on
Mount Scopus. The Zionist Commission had been able to lease a
house, Grey Hill House and its estate, as the embryo of its university
and campus. But what inspired Geddes was the hill site and the
ancient historical associations of Mount Scopus, overlooking on the
one side the city of Jerusalem, on the other the desert stretching down
to the Dead Sea. There were, however, practical problems with this
romantic location. The summit had only a restricted area of level

ground and the early development of the campus had to be contained within the boundaries of Grey Hill House. Geddes wanted his university both to inspire the city of Jerusalem and be part of it. He sought to do this through design, function and organisation. Starting as always with a topographical survey, he immediately engaged the services of the engineer to the Zionist Commission to produce accurate contour maps of Mount Scopus. He had brought his architect son-in-law, Frank Mears, with him from Edinburgh, and he endlessly discussed with him how the design might reflect the great ideals of this new institution. Mears responded by producing the best drawings of his life and he joined Geddes over the next decade in all the vicissitudes of the university's inception and development, fighting to get at least some of their plans realised.[82]

The major problem which confronted them at this early stage was how much of the proposed university should have detailed designs. Geddes decided it must be the entire campus, not because this was realisable in the near future, but because it was important to establish the spirit of the institution right from the start. His argument was that his design did not preclude a much more modest start with the existing buildings of Grey Hill House. He was also anxious that the grand design should not be seen as a blueprint for the future. Future needs might require modification both great and small. But Geddes wanted this university to be a mobilisation of intellectual and educational forces upon the fullest scale. It was to be not just a university, but a tool for civic and regional reconstruction. It was to provide the skilled manpower to revitalise city and region. Everything about the plan was intended to convey this message, both symbolically in design, and in practical arrangements made for different studies. Some buildings such as the Medical Institute, were located, not on the Mount Scopus site, but in the city close to the people and the existing hospitals.

On the Mount Scopus site itself, Geddes' imagination was allowed full rein. He placed the academic buildings on the summit of the hill, and villages for the staff, dormitory villages for students, and industrial villages for workers and craftsmen, on the lower slopes. The academic buildings resembled in form an echo of the old city of Jerusalem, secure within its walls. They were grouped round a central feature, a great hall. The sciences were together in one wing to aid the process of fruitful academic interaction, and Geddes wanted facilities for subjects which were still in their infancy as well as the more established ones. His plans included chemistry, physics, mathematics, mechanics, physiology, experimental psychology, anatomy, bacteriology, botany, and zoology. The humanities wing included not only a library for the study of the creative arts, history and languages, but also educational museums designed to promote the practical exploration of cultural evolution within different societies. At the apex of the plan were the departments of philosophy and sociology, disciplines which in Geddes' view drew upon and co-ordinated all the rest. The key objective of the plan was to bring about a synthesis of modern knowledge which would cross-fertilise the developments in separate disciplines.

The symbol of these ambitious aspirations was to be an architectural feature, a great dome raised over the roof of the university's central hall. For Geddes, the great dome was the most exciting and significant architectural feature of the whole plan. The dome was designed to be the biggest in the world at that time; a feat made possible by the judicious use of the new material, ferro-concrete, which had only been used extensively as a building material during the First World War. The use of this material meant that, despite its size, the dome would not need pillars to support it inside the building. Geddes was particularly delighted by the fact that Mears proposed hanging lamps at the points where, in a conventional design, pillars might have been located, giving the effect of a 'floating' dome. The symbolism of the dome as a sign of unity, as a feature used in the indigenous architecture of the region by Arab as well as Jew, as a reflection too of the Dome on the Rock, Jerusalem's most famous architectural feature, was carried even further by the design of the hall which it was to span. The latter was to be in the shape of a star, the Star of David: the renaissance of Judaism under the Eastern dome. Geddes got quite carried away with the cultural implications of his design, and wrote numerous long and breathless letters to friends and acquaintances, especially if, like the widow of the American soap manufacturing millionaire, Mrs Joseph Fels, they might be enlisted to raise the money to pay for it.[83]

Geddes felt justified in this extravagance, not only because he loved symbolism, but also because he believed the great hall would serve many utilitarian purposes. Its central position meant it would act as an informal meeting place for the members of the university in their leisure moments during the day, rather as the Oxford or Cambridge quadrangle provided informal opportunities for meeting. It would be cool in summer and warm in winter, perfectly suited to the climatic conditions. For more formal occasions it could be used for academic processions and for concerts of music. While its presence nurtured the intellect and spirit above ground, Geddes planned an underground area devoted to physical recreation. There was to be a gymnasium and swimming bath which would not only contribute to the physical well-being of the academic community, it would also raise up the dome even higher off the ground and make it more visually dominant. The great dome was to be echoed by a number of smaller domes on other buildings, the whole aiming to provide a spectacular skyline to Mount Scopus.

[269]

GENERAL VIEW FROM SOUTH, PROPOSED HEBREW UNIVERSITY IN JERUSALEM.
SUBMITTED TO THE ZIONIST ORGANISATION
By Professor Patrick Geddes and Mr F.C. Mears, Architect.

Figure 8.4

THE UNIVERSITY OF JERUSALEM.

Rosenbloom Memorial Building

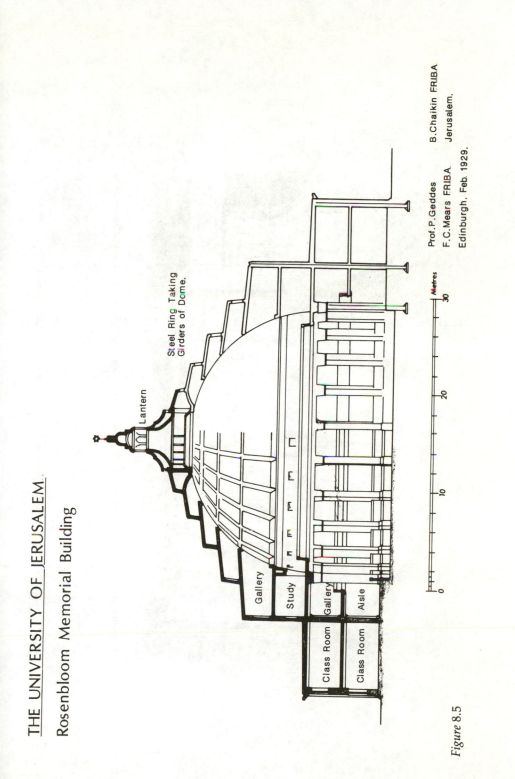

Steel Ring Taking
Girders of Dome.

Lantern

Gallery

Study

Gallery

Aisle

Class Room

Class Room

Metres

0   10   20   30

Prof. P. Geddes   B. Chaikin FRIBA
F. C. Mears FRIBA   Jerusalem.
Edinburgh. Feb. 1929.

Figure 8.5

Greyhill House behind pillar   Entrance to Einstein Institute   Meteorological Tower   South Gate of University

Figure 8.6   Interior quadrangle, University of Jerusalem

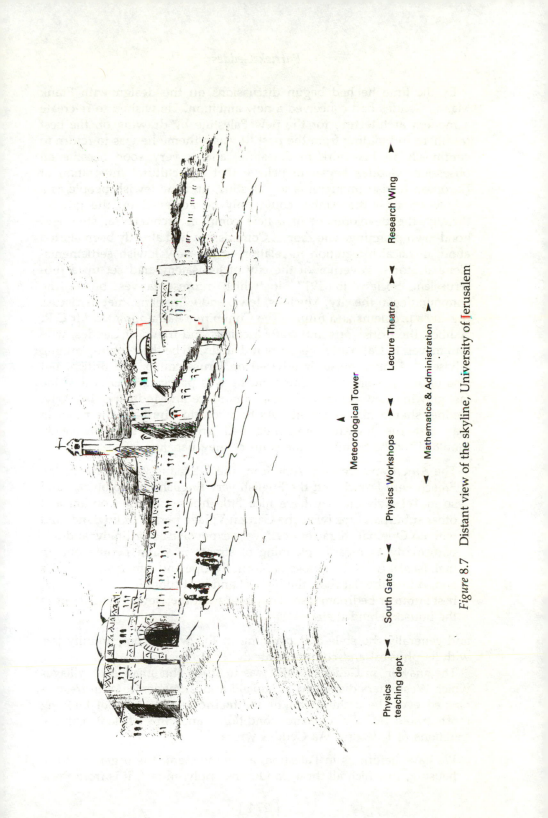

Figure 8.7 Distant view of the skyline, University of Jerusalem

Physics
teaching dept.

South Gate

Physics Workshops

Meteorological Tower

Lecture Theatre

Mathematics & Administration

Research Wing

By the time he had begun discussions on the design with Frank Mears, Geddes had conceived a new ambition. He wished to re-create a modern architecture for the new Palestine by drawing on the best traditions in building from the past. It was a theme he was to return to continually in his work in Palestine, and it very soon became an obsession. Geddes began to believe that the cultural integration of European Jewish immigrants to Palestine, and the Jewish people as a whole amongst the Arabs, could only be achieved on the ground through the development of a new common architectural style and good town planning. The Zionist Commission had already been alerted about physical integration in relation to the new Jewish settlements. General Storrs, governor of the city of Jerusalem, had set up a pro-Jerusalem Society in 1917, in which representatives of all the communities in the city, Muslims, Jews, and Christians, met to discuss the city's problems and future. The Committee had appointed Mr C.R. Ashbee, the British arts and crafts architect and friend of Geddes, who had moved to Cairo after his work in Dublin, to be their town-planning adviser.[84] Ashbee not only advised on Jerusalem for the British, but also on the Jewish settlements, and he was particularly critical of the fast-growing new centre of Jewish commerce and industry, Tel Aviv. Geddes shared his sentiments. As he wrote to Mrs Fels, the problem was that the Jewish immigrants had arrived in Palestine ready moulded by the culture of the country they had just left, so that:

The Americans are very American, the Germans very German, the French very French and the English very English and so on . . . and so in Tel Aviv etc. we have nice little houses of the London and other suburban type *before* the Garden Village period of England, and with no Oriental character at all; very expensive accordingly; and this particularly as regards planning of streets . . . and people put up high *(snow)* roofs, like those of North Europe and America, with tiles and so lose the flat roof, the highest and best room in the house, the best Summer bedroom also. They also spoil the view of neighbour in the house behind theirs,

and generally the style and planning of the whole commmunity jars with its physical and cultural context.

The answer, in Geddes' view, was to study the small Arab villages which Westerners dismissed as squalid and unhygienic. With freshly-opened eyes the architect might see methods and styles of building most suited to the climatic conditions and the ancient cultural traditions of Palestine. As Geddes wrote:

We have before us in Palestine, as everywhere, the urgent need of housing, to which all these conditions apply; in fact, it is from these simple tasks of good house-building, at once traditional, modern, and progressive, that greater things will in due time arise, with skill adequate to execute them. Look at any good photograph of an Arab hillside village. See the plain walls, but wall above wall; the flat roofs, but roof above roof. See how these contrast and compare with one another; see the bright walls and brighter dome-roofs in the sunshine, and how as it were they chime together, with the dark walls and masses in shadow giving deeper notes. These simple houses and small domes make up the essential picture, ranging from sunrise joy to sunset glory; and thus they justify the bigger dome that here and there gives them value. Here is architecture in its very essence.[85]

Geddes' taciturn son-in-law, Frank Mears, found inspiration trying to achieve this ideal in an alien land, and his designs for the Hebrew university are masterpieces in a reconstituted 'traditional' style. His plans show buildings of simple, solid masses, sparingly pierced by windows, and capped by domes, grouped round the central feature of the great dome. Below the university, on the lower slopes, the university villages were built in the Arab style, nestling naturally against the hillside along the contour lines. Geddes engaged a Jewish sculptor to make a model of the university in stone and sent a copy of Mears's drawings to architectural journals in the west. The British Zionists, however, had already discovered that their early optimism about the university project, which had seemed within their grasp at the ceremony of the laying of the foundation stones in July 1918,[86] was now slipping ever further away. At that time Dr Weizmann had just returned from his meeting with Emir Feisal I in Transjordan, who seemed in sympathy with Zionist reconstruction work. Now, in 1919, the prospect of the British Mandate was on the horizon, and the Zionist Commission, dominated by the British Jews, was disbanded in favour of a World Zionist Organisation, which was more in the hands of the Americans.

Two leading American Zionists concerned with the university project, Judge Brandeis and Dr Magnes, did not see the project, or the future of Zionism in Palestine, in the same light as Weizmann. Brandeis disagreed with Weizmann over the methods of achieving a future Jewish state in Palestine; Magnes, a fanatical rabbi destined to be the first President of the University, was a fervent advocate of the revival of Hebrew and quite unsympathetic to the Geddesian synoptic vision. Weizmann took Mears's drawings of the university to the United States in the winter of 1919 on his abortive mission to heal the rift between the American and the British views of the future. But many drawings and plans of 1919, and the stone model, were lost

[275]

when the Zionist Commission offices in Jerusalem were ransacked by rioters in the early months of 1920. Geddes had left them there when he had, very belatedly, gone on to Bombay in November 1919.[87]

But even though his stay in Palestine had been so short, Geddes had not devoted himself entirely to the university project. He had managed, with Lanchester's help, to get an official request from Colonel Storrs to comment on the town plan for Jerusalem that Storrs had commissioned from Mr McClean, the municipal engineer in Alexandria, soon after he had taken over control of the city. McClean's plan had been exhibited at the Royal Academy in London where Lanchester saw it and wrote a scathing attack on it in *The Observer* of July 1919, suggesting that:

> We cannot but accuse (Mr. McClean) of being, to say the least, somewhat disrespectful towards his art as a town planner in submitting a scheme for the laying out of the selected area so lacking in even superficial study of the site and conditions . . . The plan as it stands being endorsed 'approved by the Municipality of Jerusalem, July 20th 1918' it is obvious that but a short time was given both to its preparation and its consideration by those in authority.[88]

Storrs was particularly sensitive to this criticism. To a most unusual degree, he felt a concern for the future of Jerusalem which he had found in a totally dilapidated and run-down condition at the end of the war, and which he wished to renovate and conserve.

Ironically, the edicts which he issued to control the nature and extent of building in the city in April 1918, before he consulted any architect or planner, were probably the single most lasting influence on the physical environment of Jerusalem in the twentieth century.[89] Geddes had no time to do more than get to know the city in his short stay, which he did on endless perambulations around it at all times of the day, and sometimes, too, at night.[90] He did not attempt to carry out any systematic survey work. But he quickly decided on three objectives. To strengthen and emphasise the methods put forward by McClean for preserving the Old City, to encourage the archaeological excavations of the Old City of David as an essential activity for the regeneration of civic spirit; and to insist that new suburbs were laid out with more concern for contours of the land and the most economical (and more beautiful) grouping of houses.[91] With his Indian experiences fresh in his mind, Geddes was completely undaunted by the squalor and overcrowding in the Old City. He wished to relieve some of the congestion by moving population out, particularly the Jewish community to new suburbs. The Jewish quarter had suffered some of the worst overcrowding as the community stayed together for protection.

Geddes endorsed McClean's plans to put parks around the Old City to separate it and keep it distinct from the new and he added, for practical measure, that he hoped the soil from the archaeological excavations would be taken down the valley to improve the gardens below which suffered from thin, poor soil. But his main passionate plea was for a museum or museums of Palestine to build up a picture of the evolution of the city and region in a manner similar to the one he used in his Cities and Town Planning Exhibition and his *Masques of Learning*. The revitalisation and development of Palestine would depend on the perceptions of those living in Palestine of their past, present, and future, and thus these museums would be not just repositories of historical relics but the means to recreate the individuality of Jew, Muslim, and Christian, whose fortunes had been so mixed under Turkish rule. Geddes' Jerusalem Report differed very little in details or approach from his Baroda Report; his plans for the Hebrew University differed little in outline from the plans he had drawn up (as it turned out abortively) for the Hindu University at Benares, except that he personally found Jerusalem the most inspiring of cities. The different cultural and religious contexts did not inhibit him. He saw himself as being above the wranglings of sect and creed and thus able to direct the emotional resources generated by such allegiances into channels which could produce results. As he wrote:

> Just as the hygienist, whether at home in Europe, in India, or in Palestine, is in practice but little concerned with either the religion or the politics of the public which he is called to serve, so it is also with the town-planner. Essentially occupied as our hygienist and planner with general and community interests, a substantial impartiality is thus necessary and customary in their professions; indeed, without it their plans would be vitiated in all men's eyes and not simply their own. Yet such general impartiality does not preclude active enquiry, nor an all-round critical attitude; it, indeed, compels these. Moslem, Christian, and Jewish ways of living and working, in country and in town, have all to be observed; and each are found to present advantages and disadvantages — often unexpected ones. The task is not to plan or plead in favour of any race, any social or religious grouping; it is to search out the qualities and defects of each type, in its housing, agriculture, etc.; thence to combine such local lessons, and warnings, with those of more general experience, and thus to plan for the benefit of each and all concerned.[92]

It is illuminating to compare and contrast Geddes and his whirlwind methods of working with C.R. Ashbee, the architect retained by the pro-Jerusalem Committee to advise them on ways of beautifying the city. Ashbee went on to become civic adviser to the civil administration of the British Mandate under Herbert Samuel. He lived and worked in

Jerusalem, making hundreds of sketches, and personally supervising the conservation and repair work being carried out in the Old City. One of his greatest achievements was to find, in the true arts and craft manner, old craftsmen capable of making the right kind of tiles to repair the damaged Dome on the Rock. He persuaded them to come to Jerusalem, make a kiln, and produce the tiles. It was a painstaking, slow, but ultimately very rewarding action which rejuvenated not only the old buildings, but a craft industry as well. Geddes, on the other hand, was more interested in talking to people from all communities, and trying to interest them in civic reconstruction. In many respects, in Jerusalem he was less successful in this than he had ever been. When he eagerly returned in April 1920 with his Cities and Town Planning Exhibition, he found no interest whatsoever in his lectures or exhibits, in the uneasy peace after the rioting.[93]

Ashbee offered him help and support, but he knew how little Geddes cared about the realities of the political circumstances. Geddes was so convinced that his kind of work would solve social and political problems that he would make no compromise with the pressure of the moment. Ashbee felt that Geddes was probably right in the long term except that, after all, evolution can never be planned, especially in the uncertain circumstances of Palestine. However, if Geddes had accepted that premiss, his work would have been totally undermined. As it was, he made common cause with the Zionists who were working for the future, and did his best to help with their plans. On his second visit, in 1920, he left behind the problems of Jerusalem and the Jewish University and visited Jewish settlements throughout Palestine.[94] He spent most time in Haifa advising on the development of the port as well as on the educational institutions of this little town. But he enjoyed himself most at Tiberias. Here he found a chance to offer suggestions for the future by drawing on the most ancient of traditions from the past. The hot spring of Tiberias had been 'the most ancient, famous and enduring of the world's health resorts'. Why not renovate them as a tourist attraction of the twentieth century? Tourists would then discover not only the baths but the 'whole Lake District' of this region, which would thus benefit from a boost to its local economy.[95] It was a solution which delighted him, with all the elements of civic reconstruction plus concern for the material well-being of the people.

After this second visit, however, Geddes did not return to Palestine until 1925. The problems of the university project lingered on. Frank Mears visited Jerusalem regularly, hoping to get some return for all his work. But problems intensified, although three research institutes – one devoted to medical studies, another to chemistry, mathematics, and the biological sciences, and the third devoted to the study of Jewish literature and culture – were established. On the strength of this it was decided to hold a grand opening ceremony in 1925, and Lord Balfour was asked to be the guest of honour. To mark the occasion the governing board of the university, which had now taken over control from the Zionist Organisation, launched a competition for a new library building. Geddes had been invited back for the opening and he spent the time promoting Mears's drawings. Mears's plan, though, was not acceptable, because he was not Jewish.[96]

Mears therefore took a Jewish architect, Benjamin Chaikin, into partnership and, perhaps with a certain bad conscience about the way in which poor Mears had been treated over the past years, the university governing board awarded the prize and commission to the Mears/Chaikin plan.[97] It was built: the only part of the plans ever to be realised, a solitary example of Geddes' new style of architecture for Palestine, using traditional form in modern ways.[98] Geddes, with a semblance of his old energy, which he had lost during a serious bout of illness in 1924 (in his seventieth year),[99] used the opportunity of this visit to Palestine to revisit the Jewish settlements where he had worked in 1920, and to comment on the growth of Tel Aviv as it was recorded in municipal plans in 1914, 1921, 1924, and 1925. This was to be his last such commission, and he addressed himself fully to the problem of this bustling, little settlement, only established in 1909, but growing very rapidly. He wrote, though, this time for his Jewish friends and not for the world at large as he had done in Indore. The style is thus more intimate, and Geddes' way of incorporating his concern for the cultural evolution of the city, alongside its physical growth, is much more closely integrated in the text. After all the disappointments of Jerusalem and the Hebrew University, Geddes felt that the commercial and financial centre of Tel Aviv was perhaps better material for him to work with.[100]

He was also beginning to come to terms with the post-war world. As he wrote to Eder in May 1925

> I am really planning Tel Aviv just now. I am impressed by the *life* of Tel-Aviv — the real *live* Jewish city, free from the mutual inhibitions which are so tragic everywhere in Jerusalem (and everywhere else!). Here too is the mass of the youth of Israel needing education. Goodness knows I am not going back on Universities — as the main interest of my life — but here are demands no Universities yet meet . . . a new sociology and civics, a needed ethics and etho-politics and psycho–biology and so on, which no university is yet reaching.[101]

His major physical recommendations involved zoning of industrial and other areas, keeping the main thoroughfares running north/south and leaving the east/west roads as minor 'home-roads' bordered by gardens to enhance the quality of the residential environment. Civic buildings should be grouped together as should all the Geddesian cultural institutes. Gardens and agricultural stations should be close to the schools to ensure the awakening and encouragement of the children in these activities, which he believed were of equal importance to town based activities of finance and industry in promoting the country's future.

The whole question of the location of social and educational institutions he refers to again and again.

> It has too often been the case, in the history of cities, that their Culture Institutes have been postponed until adequate sites for them are no longer obtainable. Modern cities (British and American especially) are thus discovering their needs when too late adequately to supply them save at great expense, and then in too scattered locations.

His point is that it was vital to ensure the

> *proximity* of these institutes, so as to prevent their mutual forgetfulness, which in time hardens to exclusiveness, and thus to failure of usefulness all round: and just when duly intelligent and understanding and sympathetic co-operation are most required. This condition of proximity, and for mutual interaction, is fundamentally necessary.[102]

Of course, Geddes wanted to reserve the best site available in the whole city for this purpose, regardless of land values and cost.

He was also concerned about density of dwellings and population in the city. He wanted low density housing with gardens in the usual Garden Village style and his recent work in India gave him added arguments in its support. He wrote:

> Here is the medical answer. Imagine yourself a working man's wife, with her full marketing basket on one arm, her baby on the other, and another coming within: so now, tell me how many stairs would you like to climb up? You never thought of that before. Again, though the child mortality of the next and even great port — Colombo — is still too high, it is under a third of Bombay's, and the lowest of all great cities of the tropical East: and why? Because in the main still much of one and two storey houses, and largely with gardens — because its founders were the Dutch, who brought their gardening interests, skill and taste with them to Ceylon, and have diffused them throughout the population.[103]

Geddes wrote the Tel Aviv and Jaffa Report in Edinburgh after he had left Palestine for the last time. He had been offered a site to build himself a house on Mount Carmel, one of the settlements for which he had produced a detailed plan, as a gift and an inducement to keep him coming to Palestine and taking part in the development of Jewish settlements. But he declined, as he wished to make his future home in Europe and in some ways he felt he had done what he could for Palestine. As a non-Zionist, he had committed himself to the Zionist cause as a working model of civic reconstruction, and he had inspired the Zionists he met in these years with his vision and enthusiasm. He had defended their aspirations against widespread international hostility in articles published in leading journals such as the *Contemporary Review*. As he wrote to Eder from Montpellier in 1924:

> You must be amused by the way in which one of my many affirmations of Zionism turns the flank of some of your unfriends in France here — when I explain you (Israel) as turning from (or rather getting beyond) your recent and present leadership, in science and finances or politics etc. etc. to that of Regionalism — as antidote to the Statist and Imperialist centralisations from which great capitals ruin their provinces . . . the 'Holy Land' the 'Sacred City' are not these, as they can be got into harmonious progress, the *Eutopia* of Man today? And the true after-war Campaign.[104]

But his practical influence had been slight. His impact on higher education, cultural institutes, and the Hebrew University of Jerusalem were minimal. He, in his own words, 'stirred-up' Samuel's Town Planning Committee of 1920. Yet, whilst he had sympathisers like Ashbee, his polemical approach antagonised the administrators. He was proud of the fact that on his visit of 1920, which had otherwise been a failure, he had prevented, using arguments based on town-planning principles, the location of two military barracks on prime sites in Jerusalem, on a Damascus road and on Talpioth. As an apologist of town planning, he had introduced new considerations into the discussions of practical problems. But his ideas were generally outweighed by other pressures. The Department of Trade and Industry took his suggestion on the baths of Tiberias seriously, getting the project costed and the water chemically analysed, but the funds for development were not forthcoming. The brunt of the disappointment of Geddes' hopes in Palestine was, in fact, borne by Frank Mears, who felt that his efforts to secure the contract for the Hebrew University had cost him the good chance he might have had to win the commission for the buildings of a new university college in England at Hull.[105] Furthermore, the hope that he would stimulate a new style of Palestinian architecture never materialised. During the 1920s, numbers of Jewish architects were coming to Palestine from Germany, and the most eye-catching of Tel Aviv's new suburbs was built in the style of the Bauhaus and the Modern Movement.

[281]

Geddes' investment of time in Palestine over the period had been necessarily limited. He not only had the responsibility for setting up the new teaching department of Sociology in Bombay; he was also involved in a great effort to promote the doctrine of civic reconstruction world-wide, trying to recapture some of the momentum of the movement he had experienced especially in the years 1911 to 1915.[106] His effort to achieve this throughout the period 1918 to 1926 was the thread which tied all his varied activities together. His Indore Report was his manifesto. His work in Palestine he hoped would be a practical example of civic reconstruction which would attract world-wide attention.

Barra Bazar and the problems of the Geddesian approach

But there were always obstacles to face even in India. Geddes had found himself in 1919 in a practical planning exercise playing out the role of King Canute in his own chosen field of town planning. He had been asked to comment on a plan for a central district of Calcutta, Barra Bazar, drawn up by the Calcutta Improvement Trust. This was the financial and commercial centre of the great city, containing the Mint and with access to the docks, which were beginning to grow rapidly with India's changing economic fortunes. The clash between the interests of the market and the interests of the people was direct and stark. There was little chance that the area would be saved for the residents, even though Geddes could point to evidence of the insensitive removal of people from their homes, with consequent violence and social unrest. He wrote:

> From life-long knowledge of most of the great capitals and industrial centres of the West as well as general knowledge of many others as from Paris to Berlin or Petrograd, from Dublin to Chicago – I have come to know how intimate and intense has been, and now is, the connection between 'City Improvement' of the older demolition type, and their social unrest . . . the recent militant activity of Berlin, and its subsequent revolutionary outbreaks now manifest to the world (1914–19) have both been predicted, upon its town-plan, for a good many years past, in the ordinary demonstrations of the Cities Exhibition. Again, both before and after recent troubles in Ireland, I have had peculiar opportunities of investigating such connection as might be between Irish urban unrest (so much more serious than the older rural form) with the deplorable condition of Dublin and other cities. And that this factor of causation has been a deep and intimate one I have thus come definitely to know.[107]

In Barra Bazar Geddes found himself committed to trying to put his civic reconstruction doctrine, with its commitment to places and people, in an urban context most hostile to such priorities. Apart from the competition for space, the existing buildings were often large streetless slum blocks. The CIT plan for the area had to pay for itself as there were no extra funds for redevelopment to relieve the traffic congestion and to deal with the appalling slums. Predictably it opted for a programme of demolition which would enable roads to be widened, property values to be increased, business needs to be fulfilled, and the poorer people evicted from the area. For Geddes there was no chance of 'conservative surgery' here. He was faced with these huge blocks of streetless slums. But he hung on to his doctrine. The blocks should be demolished but replaced by special buildings four storeys high; the lower two storeys would be for business use, the upper two for residential accommodation. If people were to remain, then the Mint should be removed to another area where there was room for its expansion. Its removal would, of course, leave room for gardens. Additional narrow lanes should be built as 'home-roads' and the main arteries widened to take the traffic. Overcrowding would be regulated by the good old Victorian public health measure of registering the numbers allowed in any one dwelling, marking this with plaques over each door and employing inspectors.

It would appear from the report that Geddes had to some extent appreciated that his socio-biological approach would appear nonsensical to both the CIT, the municipality, and the business men of Barra Bazar. He attached a number of appendices to the report trying to put his message across more strongly: warning about demolitions and revolution, arguing for the continuing need for city and local survey, defending the appropriateness of traditional Indian houses such as courtyard homes, advocating the need to regulate the height of building, and the possibilty of erecting cheap housing and the importance of gardens and trees. Geddes had become a victim of his own propaganda, and he had become totally unresponsive to any other viewpoint than his own. The Barra Bazar Report was completed in March 1919. The confusion it illustrates between doctrine, planning problems and objectives, and a realistic grasp of what was possible, was symptomatic of Geddes' last phase in his attempts to promote social reconstruction.

He was committed to stay and work in India at the University of Bombay, and he continued to get the odd commission for town-planning reports, or commissions for such projects as the Lucknow Zoo, or laying out the campus for the Osmania University at Hyderabad. But as his work in India came to an end, there was no sign that he had initiated a new civics movement there. A number of individuals, one or two cities such as Lucknow and Indore, bore marks of his influence. If his ideas survived it was because they were nurtured by the personal dedication of some individual who had responded to him. The effect of all his practical work was not so much to initiate a new era, as to provide a body of published work which became a major influence on the small number of professional British planners who managed to read them. Yet even for Geddes' admirers, his message of civic reconstruction after social survey could still appear confusing and the Indian and Palestinian reports have been scanned mostly for their practical advice on dealing with specific problems in third world countries.[108]

Notes

1. See A.D. King (1976) *Colonial Urban Development: culture, social power and environment* London: Routledge, and A.K. Dutt (1959) 'Critique of Town Plans of Jamshedpur', *National Geographical Journal of India* 5 (4): 205–211.
2. Their hostility to each other caused embarrassment for Frank Mears who was collaborating with Lutyens on the Theosophical Building in London and who tried to recommend Lutyens to Patrick Geddes when Geddes was on his first visit to India 'I hope you aren't at daggers drawn – and I see your point of view quite clearly. Yet one shudders to think of the results of the employment of perhaps *any* other architects there than Lutyens and Baker . . . Lutyens seems to me the only architect we have who understands "fine" building – he unifies his work in an extraordinary way. From what I know of him I see no need whatever for you to be at cross purposes'. Mears to Patrick Geddes, 20 December 1914, Geddes Papers MS10573, NLS. Geddes's hostility was roused by Lutyens' disregard of Indians and their contemporary circumstances in his activities. He refused to employ Indian artists in Delhi. Geddes tried to get heads of major art schools in India to write to complain to the Indian government about this. (Patrick Geddes to unknown correspondent, 1922, Geddes Papers MS10516, NLS.
3. F.C. Temple (1919) *Report on Town Planning in Jamshedpur* Jamshedpur: Jamshedpur Social Welfare Services.
4. P. Geddes (1915) *Reports on the Towns in the Madras Presidency visited by Professor Geddes 1914–15*, Madras: Government Press, pp.86–8.
5. The first selection of material to illustrate these themes was made by H.V. Lanchester as acknowledged by Miss J. Tyrwhitt in her editor's note in J. Tyrwhitt (ed.) (1947) *Patrick Geddes in India*, London: Lund Humphries, p.6.
6. Geddes, op.cit., p.2.
7. Ibid., p.12.
8. Ibid., p.26.
9. Ibid., p.38.
10. Ibid., p.21.
11. Ibid., p.96.

12. See chapter 9, p.290.
13. H.E. Meller (1979) 'Urbanisation and the Introduction of Modern Town Planning Ideas in India, 1900–1925', in K.N. Chaudhuri and C.J. Dewey (eds) *Economy in Society: essays in Indian economic and social history*, Delhi: Oxford University Press, pp.330–50.
14. Geddes held his Exhibition in Madras for two weeks in January 1915, gave a course of six lectures at the University of Madras 1–6 February 1915 on 'Cities in Evolution', and gave a course on planning techniques for surveyors from Madras State.
15. This was reprinted in 1965 by the Maharashtra State Press, Bombay under the title: P. Geddes (1915) *Reports on Re-Planning of six towns in the Bombay Presidency*.
16. Ibid., p.3.
17. Ibid., p.9.
18. Ibid., pp.23–4.
19. K.L. Gillion (1968) *Ahmedabad: A Study in Indian Urban History*, Berkeley: University of California Press, pp.149–50.
20. Geddes was paid 7 guineas a day expenses with extra fees for the exhibition and lectures. He also had his fare to England and back paid. The burden of his expenditure was shared in a complicated agreement between state governments, municipalities and bodies like the Calcutta Improvement Trust – *Proceedings of the Government of Bengal, Municipalities Department*, January 1917, London: India Office Records Library.
21. P. Geddes (1917) *Report on Town Planning: Dacca*, Calcutta: Bengal Secretariat Book Depot, p.7.
22. Ibid., p.17.
23. P. Geddes (1916) *Report on Model Colony at Kanchrapara*, Eastern Bengal Railway, p.7.
24. Ibid., p.13.
25. P. Geddes and H.V. Lanchester (1917) *A Report to the Municipal Committee: Town Planning in Jubbulpore* Jubbulpore: Hitkarini Press.
26. At Cawnpore (Kanpur) he had to comment on the plan drawn up by the same man whose 'improvement scheme' (involving the removal of a corner of a Muslim temple) had sparked off serious rioting in 1914. Geddes expressed 'fierce satisfaction' in
      tearing up such a document with its (1) vast and elaborate estimates for public waste, (2) its persuasive report and (3) its utterly and astoundingly incompetent plans on which of course the whole inverted pyramid rests . . . I believe there is enough mischief in this proposed City scheme alone, to light a discontent enough to lose the Indian Empire, at any rate to intensify all elements for estrangement beyond any other positive proposal I have struck. (Patrick Geddes to H.J. Fleure, 4 April 1917, Geddes Papers MS10572, NLS).
27. R. Mukerjee (ed.) (1952) *A City in Transition: social problems of Lucknow*, J.K. Institute of Sociology, University of Lucknow.
28. P. Geddes (1922) *Town Planning in Patiala State and City: A Report to H.H. The Maharaja of Patiala*, Lucknow: Perry's Printing Press, p.66.
29. Ibid., pp.20–1.
30. P. Geddes (1916) *Town Planning in Lucknow: a report to the Municipal Council*, Lucknow: Murray's London Printing Press, pp.65–6.
31. H.J. Fleure to Patrick Geddes, 25 February 1917, Geddes Papers

MS10572, NLS.
32. Geddes, *Lucknow Report*, op.cit., pp.67—76.
33. J.M. Linton Bogle (1929) *Town Planning in India*, London: Oxford University Press, p.6.
34. H.C. Malkani (1957) *A Socio-Economic Survey of Baroda City*, Unpublished typescript, University of Baroda.
35. S. Rice (1931) *Sayaji Rao III, Marahaja of Baroda*, London: Oxford University Press; P.W. Sergeant (1928) *The Ruler of Baroda: an account of the life and work of the Marahaja Gaekwar*, London: John Murray.
36. *Gazetteer of the Baroda State*, (1921) vols.I and II compiled by Rao Bahadur Govindbkai, H. Desai, and A.B. Clarke.
37. Ibid., p.185.
38. P. Geddes (1916) *Report on the Development and Expansion of the City of Baroda* Baroda: 'Lakshmi Vilas' Press, p.21.
39. This might not have happened had all Geddes' recommendations been followed, including those for a new suburb and development plans for the village of Anandpura. But the late Professor Achewal, who was Professor of Architecture at the University of Baroda in the early 1970s, suggested to the author that Geddes concentrated too much on preserving the 'old world' charm of the city, parks and palaces and so on, and too little on economic survival in the future (interview with Professor Achewal, 21 November 1972, University of Baroda).
40. P. Geddes (1917) *Town Planning in Balrampur. A Report to the Hon'ble the Marahaja Bahadur*, Lucknow: Murray's London Printing Press. p.80.
41. As Geddes himself wrote to the Dewan of Patiala, 15 September 1922, Geddes Papers MS10516, NLS. 'There are finer architects than I, and bolder planners too: but *none so economical* in India or Europe'.
42. Lyon was engaged in socio-economic reconstruction work in Indore in a small way — see his papers, one on combatting the influenza epidemic and other diseases 1919 — Geddes Papers MS10631, NLS.
43. *Report of the Administration of the Holkar State for 1925*, Indore: Holkar State Papers, 1926, p.27.
44. *Indore State Gazetteer*, 1906, pp.281—5.
45. Section entitled 'Public Health in the Industrial Age' was published in the *Sociological Review* XI, spring 1919.
46. P. Geddes (1918) *Town Planning towards City Development: a report to the Durbar of Indore*, 2 vols, Indore: Holkar State Printing Press, vol.I, p.9.
47. Ibid., p.12.
48. Patrick Geddes to Norah and Arthur, 13 October 1917, Geddes Papers MS10501, NLS.
49. Geddes (1918) *Indore Report*, vol.I, p.2.
50. Ibid., p.120.
51. P. Boardman (1978) *The Worlds of Patrick Geddes: biologist, town-planner, re-educator, peace warrier*, London: Routledge & Kegan Paul, p.295.
52. *Indore Report*, vol.I, p.180.
53. Ibid., p.74.
54. Ibid., vol.II, Appendix III, pp. xiv—xv.
55. Ibid., vol.I, p.81.
56. Ibid., pp.28—31.
57. Ibid., p.73.
58. Ibid., pp.40—1.
59. Ibid., p.165.

60. These were *Reports of the Administration of the Holkar State 1923–26,* located in the library of Lalbagh Palace, Indore.
61. Geddes (1918), *Indore Report* vol.I, p.136.
62. Ibid., p.160.
63. Ibid., p.161.
64. Ibid., vol.II, pp.187–8.
65. See chapter 7, p.209.
66. Geddes (1918) *Indore Report,* vol.I, p.85.
67. Ibid., p.81 and Plan III.
68. Ibid., vol.II, p.113.
69. Ibid., pp.113–15.
70. Ibid., p.65.
71. Ibid., p.73.
72. Patrick Geddes to Mr Sett, 20 March 1919, Geddes Papers MS10516, NLS. His predecessor, to whom he was referring, was H.V. Lanchester.
73. M.D. Eder to Patrick Geddes, 11 May 1919, Central Zionist Archives, MS Z4/1721, Jerusalem.
74. Quoted in A. Defries (1927) *The Interpreter: Geddes, the Man and his Gospel,* London: Routledge, p.260.
75. Patrick Geddes to Secretary, Zionist Bureau, 16 June 1919, Central Zionist Archives, MS Z4/1729, Jerusalem.
76. Correspondence between Patrick Geddes, D. Eder and F. Mears is located at the Zionist Central Archives, File L12/39, Jerusalem.
77. Zionist Central Archives, File Z4/2790, Jerusalem.
78. D. Lloyd George (1938 edn) *War Memoirs,* vol.I London: Odhams Press, pp.348–9.
79. Chaim Weizmann (1919) 'Note on the University Project', typescript, Central Zionist Archives, File L7/108, Jerusalem.
80. P. Geddes (1919) 'Preliminary Report on University Design', typescript, Central Zionist Archives, 20/12/1919, Jerusalem.
81. Ibid., p.5.
82. Mears's drawings, photographs of the model, and detailed plans of the scheme are located in the Papers of Sir Patrick Geddes, University of Strathclyde.
83. Part of this letter is published in P. Mairet (1957) *A Pioneer of Sociology: life and letters of Patrick Geddes,* London: Lund Humphries, pp.186–8. A full copy is in the Central Zionist Archives, MS L15/720G, Jerusalem.
84. A. Crawford (1986) *C.R. Ashbee: architect, designer and romantic socialist,* New Haven and London: Yale University Press, pp.173–94.
85. P. Geddes (1921) 'Palestine in Renewal', *Contemporary Review* 120: 475–7.
86. In his speech on that occasion, Dr Weizmann spoke of the university project as 'the Jewish Dreadnought', *Official History of the Hebrew University of Jerusalem 1925–50,* Jerusalem, 1950, p.67.
87. He later wrote of the bitter experience of leaving maps, plans, model, and typescripts in the Zionist Offices and under the guardianship of the Civic Adviser – Patrick Geddes to Dr. Van Vriesland, 1 July 1925, Central Zionist Archives, File A114/12, Jerusalem.
88. Copy of Jerusalem Plan and Lanchester's attack in *The Observer* in Central Zionist Archives, File Z4/10.202, Jerusalem.
89. R. Storrs (1937) *Orientations,* London: Ivor Nicholson & Watson, p.363.
90. In an interview, the now late Lord Bentinck, a young writer and lawyer in 1919 for the Zionist Organisation, described what excitement Geddes

generated about Jerusalem and how he often took individuals with him on his walks (Interview with author, July 1971).

91. P. Geddes (1919) *Jerusalem Actual and Possible: a preliminary report to the chief administrator of Palestine and the military governor of Jerusalem on town planning and improvements*, November, typescript, copy in Central Zionist Archives, File Z4/10.202, Jerusalem.

92. P. Geddes (1921) 'Palestine in Renewal', op.cit., p.481.

93. This was hardly surprising in this deeply-divided community. In Palestine there were 40–50,000 Jews, only 5,000 of them Zionists, and 600,000 Arabs at this time.

94. C.R. Ashbee (1923) *A Palestine Notebook 1918–1923*, London: Heinemann, p.168–70.

95. P. Geddes (1920) *The Hot Springs of Tiberias*, typescript, Central Zionist Archives, File A107/825, Jerusalem.

96. M.D. Eder to R. Weltsch on difficulties of promoting P. Geddes and Colleagues because they were not Jewish. Central Zionist Archives, File 24/3497, Jerusalem.

97. Patrick Geddes to Dr van Vriesland suggesting B. Chaikin will be made a partner, 1 July 1925, Central Zionist Archives, File A114/12, Jerusalem.

98. It still exists and can be visited on the Mount Scopus site.

99. P. Boardman, op.cit., p.352.

100. P. Geddes (1925) *Town Planning Report – Jaffa and Tel-Aviv*, typescript, copy in author's possession and one given to Tel-Aviv Museum.

101. Patrick Geddes to M.D. Eder, 2 May 1925, Central Zionist Archives, File L12/39, Jerusalem.

102. Geddes to Dr van Vriesland:
     Putting it quite clearly from our distinct British points of view; that of complete distinctness of Training College from University is the traditional English method – and that of their close association is the traditional Scottish one. The English method is that of caste separation . . . In Scotland . . . the ranks of the elementary teachers have always contained a very large proportion of our University men. (15 April 1925, Central Zionist Archives, File Sz/565, Jerusalem).

103. P. Geddes (1925) *Jaffa and Tel-Aviv* op.cit., p.14.

104. Patrick Geddes to M.D. Eder, 17 November 1924, Central Zionist Archives, File L12/39, Jerusalem.

105. Mears to Dr Weizmann, 12 October 1926, Central Zionist Archives, File L12/39, Jerusalem.

106. This will be discussed in the next chapter.

107. P. Geddes (1919) *Barra Bazar Improvement: a report to the corporation of Calcutta*, Calcutta: Corporation Press, p.34.

108. J. Tyrwhitt (ed.) (1947) in co-operation with H.V. Lanchester and Arthur Geddes ,*Patrick Geddes in India*, London: Lund Humphries.

CHAPTER 9

Regional survey
in the international context:
the legacy of Geddes

THE LEGACY OF GEDDES' LIFE WORK IS NO STRAIGHT-
forward matter. The preceding chapters have analysed his activities
and charted his frequent failures. Yet Geddes' contribution was of a
kind which, despite disasters, continued and continues to have
relevance in many ways. The outstanding quality that Mumford and
others have seen in him was due to the unique perspective that he
cultivated on modern society and its development. By concentrating on
a generalist, synthetic approach to current knowledge, Geddes was
able to sharpen his perception of social changes. He gained for himself
an independent viewpoint from which he could assess and criticise the
assumptions and prejudices of others.[1] This did not mean he was free
from his own assumptions and prejudices. But he was able to look
freshly at key areas of social life such as the education of children and
adults and the control and enhancement of the urban environment.
This was his greatest gift and the legacy that he has left to subsequent
generations. Not the least significant element of this legacy was the
freedom he left for others, in his generalist approach, to take from his
ideas only those which were relevant or useful to them at the time. The

actual use of his ideas has thus depended very much on the intentions of those who have drawn upon them.

In his later years Geddes tried to initiate others into his unique perspective by promoting his idea of regional survey. What happened to the regional survey movement both during his lifetime and afterwards illustrates how interpretations of his work are moulded by current concerns. His work for 'regionalism' has been hailed as his greatest achievement and his most damaging failure.[2] The polarity of interpretations would have delighted him. His answer would have been to invite critics and supporters to join him in practical survey work. It was the only way he knew he could put across his views on educational reform, cultural evolution, and the relationship between social processes and spatial form which, together, were the key elements in his evolutionary social theory. It was a combination which was never likely to be totally successful, whatever the value of the component parts. But some initial success in the regional survey movement between 1918 and 1920 had aroused Geddes' natural optimism. At the end of the First World War, Geddes' regional survey mission became inextricably mixed up with the short-lived aspirations of post-war 'Reconstruction'.[3] In the immediate aftermath of the 'war to end all wars' there was widespread euphoria and a desire to return to 'normal life' or to 'reconstruct' the peace-time Europe. There was little understanding of the ideological, administrative, and economic factors which were to place limitations on these objectives. Geddes' social reconstruction doctrine was thus able to flourish as, for a moment, between the end of the war and the early 1920s, the social realities of the world were kept at bay, or at least distorted by the abnormalities of the return to peace.

In the course of the early 1920s the concept of the region and the meaning of reconstruction were reformulated on a more pragmatic basis everywhere in Europe. In Central and Eastern Europe, economic relationships of the pre-war period had been thrown into total chaos and confusion, not only by the war, but also by the creation of the new nation states which broke up the pattern of regional economic specialisation.[4] It was becoming increasingly clear that these difficulties were neither temporary nor a direct result of the war only, and that future economic progress in Europe was going to be much harder to sustain. The idealism, though, surrounding the birth of the League of Nations, was not entirely crushed. The remnants of the old pre-war international organisations for promoting peace rushed to affiliate themselves to the League of Nations. But while Bergson was creating his International Organisation of Intellectual Activities, Rabindranath Tagore floating the idea of a World University, and Geddes thinking the time might be ripe for a world union of provincial cities to bypass the warmongering state capitals,[5] the new post-war world was refusing to assume the shape of the old.

The process of modern industrialisation and urbanisation was being experienced on a world-wide basis, speeded up as the technology of steam and steel was being complemented by the internal combustion engine and science-based industries. Yet, at the same time, regional specialisation in production for world markets was being seriously curtailed by the internal problems of nation states. In political as well as economic terms, the United States of America had taken over from Europe as the leader of the world; Russia was in a state of civil war as counter-revolutionaries fought to overthrow the fledgling revolutionary state. But for the first two years after the war the implications of all this were not really understood. The experience of the war and the problems of peace left everything in a state of uncertainty. The emotionally charged term 'reconstruction' was interpreted very differently in different contexts. The most precise about it were the French. They wanted to 'reconstruct' their devastated regions and make the Germans pay. In Germany reconstruction was a shaky affair: the initiation of a new republic in the ashes of the old empire. But defeat, the burden of war reparations, and an abortive revolution denied the Germans any share of after-war euphoria. In Britain the euphoria was perhaps greatest of all, and the term 'reconstruction' most imprecisely defined. As victors, the British expected not only to put the clock back to 1914 and restore British world supremacy, but also to reward all citizens for their war-time effort by ensuring a better life for all.

Such ambivalence exactly suited Geddes' ideas on social reconstruction. The sense of common purpose which had been experienced in wartime seemed sufficient for a great voluntary effort to ensure higher evolutionary development. 'A better life for all', of course, was not conceived solely in economic terms. In Britain it meant better education, a more enlightened public health programme, better housing, and improved environmental conditions: all activities which had depended very much in the past on voluntary effort.[6] In the aftermath of the war, however, the British Government became determined to take these matters up as public policy. The hastily-set-up Ministry of Reconstruction was supposed to work with the Local Government Board to marshal ideas and mobilise resources to fulfil these objectives. Giving impetus to the proceedings was the government's fear that demobilised troops would not easily find work and might vent their dissatisfactions in political violence. Lloyd George went to the country with the election pledge to build 'Homes for Heroes'. Housing was the symbol of government inspired 'social reconstruction'.[7]

'Regional Survey' and the town-planning profession

This placed the newly-established planning profession in some quandary. During the war the planners had eagerly fed on the idea of reconstruction, helping the Belgians to plan for their future and promoting the Geddesian idea of the survey of city and region as the necessary prerequisite for the future. Yet at the end of the war Britain had no devastated areas requiring urgent reconstruction, and there was no obvious need for city development apart from slum clearance. Yet the government was poised on the edge of a new and significant foray into the housing market, providing municipally-financed housing, supposedly on such a scale that planning of some kind was going to be essential. The planners got their status reaffirmed when once again town planning was tacked on to a housing Act, the 1919 Housing and Town Planning Act, which stated that local authorities were obliged to prepare surveys of their housing needs and to draw up plans for dealing with them.[8] Yet the government refused to act other than in an *ad hoc* manner, dealing individually with each local authority on this matter. This left the planner in the role of political manipulator, the committee man able to co-ordinate local authorities, architects, and local vested interests, and help them to agree on some specific objective. This was hardly the exciting prospect for the future that had kept alive the idealism of the town-planning movement in the difficult years of the war.

Geddes had always preached in his social reconstruction doctrine that the planner was more than an administrator. By the quality of his practical work, and the ways in which he organised space and buildings, the planner was to have a social impact as the liberator of the people. Unwin, Abercrombie, and Pepler, with their experiences of Garden Cities, Garden Suburbs and Model Estates, or 'problem' areas such as Dublin or the South Wales Coalfield, were exponents of this view.[9] There was a hope that in the immediate post-war reconstruction euphoria it might be possible to put across this message. It was a question of propaganda and persuasion. In Belgium, the planners, helped by the need for total reconstruction, were able to work along these lines. In Britain there was a renewed upsurge of voluntary effort, backed by commercial interests, and a second Garden City was begun at Welwyn in 1920. But to make a more immediate effect on the thinking of local authorities, the planners turned to the methods and ideas of Geddes.

Pepler was President of the Town Planning Institute in 1919 and he travelled the country giving lectures on city development along Geddesian lines.[10] But the problem was that it was housing and not regional surveys which most concerned local authorities. As wartime economic controls were quickly dismantled, the local authorities were left without the likelihood of substantial financial help from central government. They were also ill-equipped to administer large housing schemes. In many quarters local authorities were, anyway, less than enthusiastic about establishing a large public sector in the housing market. As problems mounted, the campaign to initiate 'social reconstruction' in Britain began to falter, and its decline left the planners without even a platform for their propaganda. Geddes himself was, of course, absent in Palestine and India. But eventually in 1921 he did come back to Britain and addressed the Town Planning Institute on 'Regional and City Surveys as affording policy and theory for Town Planning and Design'.[11]

Unfortunately, though, by this time the post-war impetus towards reconstruction in Britain was dead. The government was beginning to realise that 'putting the clock back' was perhaps an unattainable goal. Far from creating a 'better life for all' the depression in 1921, which followed the 'restocking' boom after the war, ushered in Britain's longest experience of mass unemployment, which was to last throughout the inter-war period. The total failure of the housing policy in the space of two short years from the 'Homes for Heroes' election pledge to the reality of a mere 1,239 houses completed and a further 14,594 under construction, has been the subject of much study. It was a startling reversal for stated public policy. As Abrams has pointed out:

> Nowhere were war-time pledges more vehement and far-reaching than in respect of housing. Nowhere was suffering and social dislocation more apparent than in the need for houses. And nowhere was the failure to provide relief more clear and calamitous by the end of 1920. The history of the housing programme is the history of a rout.[12]

With this failure, the British planners were left in an impasse of powerlessness. Yet, curiously enough, so were their Belgian counterparts who had wielded such enormous influence after the war. The problems seemed to lie in the growing divergence between the objectives of the planners and the economic, social, and political realities of the time. The Belgian planners had drawn very heavily on modern international town-planning practice, but they had been particularly receptive to current British ideas, which had been widely discussed since the London Conference of 1915 on Reconstruction in Belgium. Prominent amongst the latter were the ideas of Geddes. In many respects, Belgian reconstruction work was a practical attempt to apply Geddesian social reconstruction techniques. Three elements were shared by the Belgian planners: the emphasis on the historical origins of cities and villages; the desire to achieve a harmonious whole which was aesthetically pleasing by careful renovation and reconstruction of old buildings; and a stated objective of giving 'places' back to the 'people', had a particularly Geddesian flavour.[13] Verwilghen and others even carried out surveys, though the relationship between survey and city development was not spelt out. Geddes' emphasis on the importance of 'culture' in this respect would have been impossible to test in the particular circumstances of the post-war world.

A recent commentator on Belgian reconstruction has described it as 'the last convulsion of a closed period'.[14] He suggests that the development of modern planning had taken place in the wake of urbanisation and industrialisation, initially for reasons of public health and latterly, since the war, by the political necessity of securing minimum standards for all, especially in housing. There was a continuum between the pre-war and post-war periods in this respect. Yet for a brief moment, for entirely extraneous reasons mostly related to the war, there was a passion to reconstruct the mediaeval past. It was based on an ideology which presupposed that a state of social harmony existed between all citizens who would agree on this objective of reconstructive planning; and it was morally justified on the grounds that the planners were giving back to the people something of great worth and beauty, a symbol of their national life. The financial constraints that these reconstruction projects encountered, and the prospect of placing the improvement of public places as matters of higher priority than housing, were difficult to handle. After the first flush of restoration, planners began to lose control as economic and social forces put them under pressure. By the mid 1920s, even their aesthetic judgement was under attack by the forerunners of the Modern Movement.[15] This work, however, remains a monument to their attempt to restore to the Belgians the physical heritage of their illustrious past and is also, in part, a tribute to the inspiration of Geddes' perspective on city life.

In Britain Abercrombie and Pepler did their best to use Geddesian ideas, this time to retrieve an ideal of planning practice which in Britain had barely blossomed without any commissions for large-scale projects. What was working in their favour was a growing realisation of the economic and social significance of the region. With the onset of depression in 1921, the incidence of high unemployment was markedly greater in those areas dominated by Britain's old export industries: the coalfield regions, the textile towns, the areas dominated by the iron and steel industries, and shipbuilding. In the early 1920s it was not yet accepted that the economic decline of these areas was inevitable and there was little likelihood of full recovery. But still the social consequences of high unemployment were enough to cause concern. With his Dublin experience freshened by the recent survey he had carried out there, Abercrombie believed that Geddes' approach to the region might give planners a new role in community reconstruction in areas suffering economic decline.[16] The Ministry of Health had been given overall responsibility for housing reform under the 1919 Act and had set up 'Regional Planning Departmental Committees' which, by 1921, covered five areas: the South Wales coalfield, South-east Lancashire (with headquarters at Manchester), the Doncaster region, South Tees-side, and Deeside. Abercrombie was invited to contribute to all of them.

Abercrombie undertook this work using Geddes' biological framework as his reference point. The argument he put across was that the health of a region lay in the interaction of people with their place, and an appropriate balance between town and countryside. Such an approach not only revitalised the people to enable them to overcome their difficulties, it also made sound business sense. The first approved regional scheme in England was the Doncaster Regional Planning Scheme of 1922. The opening up of the local coalfields from the latter years of the nineteenth century, in an area of small towns, countryside, and small factory areas, had created a number of problems which obviously benefited from being treated at a regional level. When Addison, as Minister of Health, gave his approval to the scheme, he reiterated Abercrombie's propaganda that 'town-planning is pre-eminently a business proposition'.[17] What Abercrombie actually achieved in these early plans, however, was hardly a breakthrough in planning methodology. The surveys indicate a desire for Geddesian breadth but the related planning proposals were much narrower. The gap between the vision and the proposals was compounded by a lack of perception of possible future economic trends, a vital prerequisite for achieving the route to 'Eutopia'.

Abercrombie's Geddesian perspective made him oversanguine that he had a sociological understanding of the interaction of place and people, and he bolstered this belief by the wholehearted espousal of Geddesian social reconstruction doctrine. While he was engaged on the work of these regional committees, he also published articles on education and university reform, on the need for civic societies, and he gave his support to the evolution of the League of Nations as an organisation to promote active and peaceful co-operation between citizens in the new post-war world.[18] Geddes' ideas on social reconstruction were important to help Abercrombie sustain and build up the image of the planning profession as a socially important occupation. As head of the Department of Civic Design at Liverpool, and editor of the *Town Planning Review*, he was professionally involved in promoting propaganda on behalf of planning. He did share a natural affinity with Geddes in his approach to cities, though he was no longer an uncritical disciple.[19] To the Geddesian framework of regional survey, he was to add his own shrewd business sense.

He was delighted when the first local authority to commission a regional plan invited him to undertake the task. The city was Sheffield[20] and he wrote to Amelia Defries in 1927 that:

> the hard-headed business man is beginning to recognise that the Geddesian method is the only safe one — Sheffield, the hardest-headed town in this country, has found schemes under the Town-planning Act (the politicians' solution) not enough; they begin just about where you should be ending; you can't plan for the future growth without improving the centre; you should not build houses without studying where the people want to work; you can't understand what the future of Sheffield will be, unless you know something about her past; in a word, you need a Civic Survey. . . It was fitting that Sheffield, the mostly coldly scientific of our technical cities and the one whose historic legacies and difficult site make town-planning obviously an involved problem, should be the first to adopt publicly the Geddes method.[21]

In 1924, at the Town Planning Institute Conference, Abercrombie and Pepler both read papers on 'Planning Industrial Regions' and 'Regional Planning' respectively, yet once again the hoped-for breakthrough in regional planning refused to materialise despite all their efforts. The economic problems of Britain since 1921 had intensified. The British Government pursued a policy of deflation in order to try to restore the Gold Standard with the pound at its pre-war parity, and British exports had failed to find new markets. The coal industry was reaching a state of crisis with coal-miners taking wage cuts to keep the price of coal down, though the world-wide demand for British coal was continuing to fall. In 1925 Britain returned to the Gold Standard. It was a move meant to help create a stable world monetary system which would encourage trade and thus help British exports.[22] But the pound was overvalued and the circumstances of world trade were no more favourable than they had been before and, in the case of coal, the demand continued to fall. In 1925 the miners decided to strike. The prospect that the miners' case might be widely supported by the Trade Union movement (which it was, eventually, in the General Strike of 1926) galvanised the government into appointing a Royal Commission to consider the coal industry as a whole. When the Royal Commission was deliberating in 1925, Abercrombie agreed to take part in a symposium organised by the Sociological Society with Victor Branford and Geddes, various engineers, public health experts, and others on 'the Coal Crisis'. It was an act of propaganda to demonstrate that current problems had a regional base and that 'sociological' town planning could bring relief.

With this support from Abercrombie and the planners, Geddes had come out of retirement in Montpellier to mount his last public attempt to salvage something of his social reconstruction doctrine. Papers from the symposium were published in a volume entitled *The Coal Crisis and the Future: a study of social disorders and their treatment*. With its companion volume entitled *Coal*, it was the last of the 'Making of the Future' series, and it proved to be an odd assortment of papers. The fact that it was actually published at the time of the General Strike meant that it appeared very dated even from the outset, and caused no response whatsoever. Abrams suggests though that this volume was a demonstration that Geddes saw that a mature sociology was an interdisciplinary study which drew on a number of different sciences.[23] Geddes had, in fact, always held that view. *The Coal Crisis*, however, demonstrates that Geddes was now less certain than he had been that his social reconstruction doctrine would appear the obvious answer to complex economic and social problems to all concerned. The volume is divided into two unequal parts, 'Teamwork under the Town Planner' and 'The Condition of Eutopian Repair and Reconstruction'. The latter, by Victor Branford, is a tired reiteration of Le Playist doctrine, occasionally enlivened by a passionate review of the potential of long-term planning in the provision of energy resources and the development of natural resources such as afforestation.[24] Branford's business acumen seems to co-exist in an evermore uneasy relationship with his desire to portray the mystical and spiritual resources of mankind. He had been greatly stimulated by the recent exhibition of 'The Living Religions of the World' held at the Imperial Institute in 1924, when the Sociological Society had been allocated the task of trying to create an amalgam of the fullest possible life for man drawing from all religions. Branford's grasp of the Coal Crisis was patently non-existent.

Abercrombie's essay on the East Kent coalfield on the other hand, is a description of his work on this project, full of practical ideas about suiting plans to the industrial and social requirements of the region. As in the Doncaster plan, it was to some extent a case of bolting the stable door after the horse had gone. Of the four collieries to be established in Kent, two were already in production by 1913. Still Abercrombie was able to make propaganda out of his activities. He persuaded Lord Milner to come out of his retirement to use his organisational skills and his prestige in conjunction with the Archbishop of Canterbury to bring together the local authorities, industrialists, miners, and owners to convince them of the need to plan to avoid producing unplanned chaos.[25] Such leadership was very important for establishing agreement over the objectives of an exercise in planning, such was the general hostility to the whole procedure. But Abercrombie's 'regional' message is now muted. Whilst 'regional survey' is the essential preliminary to plan, the plan is a matter of environmental arrangements for maximum social benefits, with some concern about future social needs, especially educational institutions. There is a suggestion that Canterbury might be a suitable location for a new University of Kent.[26] But Abercrombie had lost his belief that survey work could solve economic and social problems. The fact that he merely recommends his East Kent work as an example which might be followed in other coalfield areas where there were strikes and uneconomic pits shows that he had not grasped the important differences between the Kent coalfield and other older established mining areas. As Geddes was forced to say, after seven long and depressing years: 'Serious after-war discussions are increasingly making clear how intricate and difficult is the general situation throughout Europe, indeed over the whole world'.[27]

Problems about energy sources, technological change, industrial decline, and changing world markets made a purely 'regional' approach to the future totally inadequate. The disillusionment which followed the puncturing of the high hopes for regional survey amongst the planners was considerable. Geddes' ideas – survey, diagnosis, plan – were used as a planning tool, and embedded in planning education. But now there was more concern for political answers to social problems, and planning regions with master plans which could be put on some political agenda and professionally carried out. The ethos of the regional survey which was to try to involve people in decisions about their local environment based on their understanding of their heritage had only been given some form through the voluntary labour of a tiny minority of interested people from the middle classes.[28] The ideal that people could have some kind of control over their future on a regional basis seemed even more ludicrous in the face of the growing understanding of Britain's and the world's economic problems. The Great Depression which began in 1929 marked the nadir of the hopes of those who had believed so enthusiastically in the Geddesian view of regional survey. Government policy on the 'Special Areas' of high unemployment, when it came in 1934, fell far short of direct economic regional planning.

In these difficult years, sensitivity to the historical origins and built form of cities seemed of ever less importance. A growing lack of concern over these matters was bolstered by the recent tendency of post-war architects, excited by the Modern Movement, to denigrate the importance of the architecture of the past. The architecture of the twentieth century had to be new, bold, and appropriate to the modern technological age.[29] Examples of the past were irrelevant and cities needed to reflect modern trends. Perhaps Geddes' greatest failure was his failure to put across the evolutionary idea he himself believed so firmly: that there was a very direct and important relationship between social development and the built environment. His early work in Edinburgh had been born out of his desire to rekindle the greatness of Edinburgh and her university in the late eighteenth century once again, by bringing students back to the very houses where formerly the great scholars had lived. Abercrombie and Pepler, who did most to try and give some professional base to Geddes' regional survey, did not fully share Geddes' passionate socio-biological belief that the roots of one's culture, including the heritage of the built environment, were the vital means of achieving the potential for individual growth. As architects, they were not steeped in the literature of the pioneering modern geographers trying to define the importance of human geography.

Their response to Geddes had been part emotional, part practical, as he seemed to be the only source of assistance in making planning a universal professional activity. They wanted to achieve the transition from the idealism of the Garden City movement to the concept of town planning as a normal function of local and national government, without losing that excitement about the future that that idealism engendered. Geddes' advocacy of his regional survey as a route to Eutopia seemed to provide the answer, especially since, for British planners at least, propaganda was as important as practice in these early years. Not least helpful in this latter respect was Geddes' emphasis on his 'rustic' vision of urban problems and his hostility to the idea of centralised planning. In the immediate aftermath of the First World War the town planners were trying to establish their professional importance at the very time when planning was thought to be inimical to all the best traditions of British life that had flourished in the pre-war era, and especially the freedoms of citizens which the war had been fought to defend.[30]

Pepler, on behalf of the planners, went to the length of defining the 'freedoms' that planning could bring to the community in contrast to unplanned anarchy.[31] He claimed that the 'freedoms' of citizens had been defined by Geddes as 'the right relationship between Folk, Work and Place'. Since there were always competing needs, the role of the planner was 'to secure the best use of land in the interests of the community'. Pepler struck an even richer vein in 'the traditions' of England by suggesting that planning was now vital to the preservation of the great British countryside — all that made England 'a green and pleasant land'. Pepler spoke of 'our almost total failure to cope with new forces which modern science and invention had let loose on our old towns and countryside'. Alerted by Geddes' regional perspective, both Pepler and Abercrombie were in the vanguard of the conservationists attempting to preserve rural England. Pepler had been a leading member of the Town and Country Planning Association since 1919.[32] Abercrombie became a founding member of the Council for the Preservation of Rural England in 1926. It seemed positive action at last as regional survey began to fade.

Lewis Mumford and the legacy of Geddes

In England Geddes remained an inspiration to the faithful few at the Town Planning Institute and Le Play House, the new headquarters of the Sociological Society set up in 1920, but the major impact of his ideas was to come through the work of an ardent young American disciple who came to London in 1920 to work with Branford. Lewis Mumford has left a number of testimonies to the importance Geddes had on the evolution of his ideas. Yet it is significant, perhaps, that he did not meet Geddes personally until 1923. He first came across Geddes' work in 1915 as a young 18-year-old student, when he found the 1904 Dunfermline Report and became passionately interested in city development. He corresponded with Geddes but they did not meet since Geddes was occupied with work in Palestine and India. Instead he was introduced to the social reconstruction doctrine by Victor Branford who had been engaged in a veritable frenzy of organisation, authorship, and publication at the end of the war.

Apart from the 'Making of the Future' series, Branford edited a series of pamphlets for the Sociological Society: 'Papers for the Present' and 'Regional Survey Papers'. Mumford met him while he was working on the lectures and papers which he was to collect under the title *Interpretations and Forecasts: a study of survivals and tendencies in contemporary society*, a volume which was preceded by a shorter study *Whitherward? Hell or Eutopia?* Mumford was to take over temporarily the editorship of the *Sociological Review* in the year that Le Play House was set up, and on his return to the US his first book was to be *The Story of Utopias* published in 1922. In the same year Mumford wrote an essay on 'The City' for a symposium on 'Civilisation in the United States', which he claimed was 'the first historic analysis of its kind to be published in the US'.[33] Geddes' influence on the young Mumford had been filtered by his contact with Branford. His greatest and most abiding interests were to be philosophical ones on the condition of man and his most elaborate creation, civilisation.

The city which concerned Mumford most at this time was New York, which was then in the grip of its Master Regional Plan designed to emphasise New York's metropolitan status as well as to outdo the Burnham Plan for the Chicago region which had so dominated the 1910 first International Exhibition of Town Planning in London. In the

Geddes/Branford doctrine of anti-metropolitan regional renewal,[34] Mumford found support for the hostile stance he was to take to the official New York plan. Mumford's work was to have all the passion and moral commitment of a social reformer which made him particularly receptive to Geddes' moral mission of 'social reconstruction' to save cities and the civilisation of the future. He became the focus for a small group of architects, environmentalists, and social commentators in New York who were reacting against the kind of planning being developed by the New York Regional Plan. The 'best' utilisation of the region's resources begged the question of who would benefit most from large scale environmental planning. A small group, with Lewis Mumford as an active co-ordinator, were to form themselves into the Regional Planning Association of America, held together by a modern interpretation of the hallowed American belief in freedom – the freedom that all Americans, regardless of their socio-economic status, should have a chance to share in the opportunities of the future.

Mumford reached a peak in his organisational activities to promote the alternative view in 1923, when he at last managed to persuade Geddes to come to the United States to give lectures and to meet him and his friends. The history of those few short weeks of Geddes' visit has been written by Mumford himself. He wrote it in 1966 as an expiation partly to assuage the guilt he felt at failing to help Geddes with his social reconstruction work at that time, and subsequently failing to write a biography of him, and partly because he had been stung by the often repeated comment that all his original ideas emanated from Patrick Geddes.[35] Yet Mumford himself in his earlier publications has done more than any other writer to keep alive interest in Patrick Geddes' ideas and to hail him as the greatest unrecognised genius of the twentieth century. Mumford has thus put himself in a position whereby he controls both the source of information and the legend of Patrick Geddes. It is a position he has exploited to the full. At that first meeting in 1923 they were, of course, set on a collision course. Geddes was looking for a young collaborator in the west who would have the time and energy to help him. He sent Mumford a fee so that he could devote himself full time to helping him and greeted him as 'the image of my poor dead lad', Alasdair, who might, had he lived, have been the collaborator of Geddes' dreams. Mumford wanted Geddes to give backing and encouragement to his friends of the Regional Planning Association of America and to publicise his message on whirlwind lecture tours.

The picture he chooses to paint of Geddes many years after the event is of an old man out of touch with reality, cut off by his genius and vibrant personality from normal intercourse with others. He depicts himself as a rather naïve young student scholar, desperately trying to absorb the master's message whilst he was emotionally fighting against any personal commitment. Certainly what he wrote in the 1920s bears the mark of being dazzled by the man and his mission. In the eulogy he wrote on Geddes in 1925, in *The Survey* (New York) which Amelia Defries uses as a preface to her book, *The Interpreter: Geddes, the man and his Gospel* (1927), Geddes showed no feet of clay.[36] At that time Geddes was still Mumford's major source of inspiration for his own philosophical stance, which he defined as searching for a new way to meet the challenge of living in the modern, 'machine' age. Geddes' main use to Mumford in the early 1920s, though, was his appeal to the members of the Regional Planning Association of America. Geddes' insistence on the importance of culture in the life of cities, the need to stimulate cultural life in provincial centres (to offset the damaging, centralising powers of the metropolis) and the part played by town and country in promoting the values of civilisation, were just what the RPAA wanted to hear.

Another RPAA member, the forester and environmentalist, Benton Mackaye, was particularly receptive to Geddes' advocacy of practical action and found Geddes' description of his own work as 'geotechnics' a revelation. It helped him to find the confidence to express his ideas in his path-breaking book *The New Exploration* of 1928. He wrote about geotechnics and regional planning, and the practical need for the rehabilitation of US regions. He drew on the ecological ideas of the American, George Perkins Marsh, and the English Utopian planner, Ebenezer Howard, but he makes no reference to Geddes. Only much later in life, like Mumford, he pays tribute to the stimulation he received in 1923 at the brief weekend meeting he had with Geddes.[37] This suggests that what the young members of the RPAA were most in need of at the time was recognition of their work. This Geddes was able to provide with enthusiasm.

The problem for all would-be followers of Geddes, even Lewis Mumford, was Geddes' insistence on his own particular theory of knowledge and the use of his 'thinking machines' and 'notations'. He found Geddes' mode of thought, his notations, an insuperable barrier.[38] Instead he confined himself to acknowledging the debt he owed to Geddes and his Cities and Town Planning Exhibition in helping him to formulate his own approach to analysing cities, past, present, and future. He used with particular effect in his books, *The City in History* and *The Culture of Cities*, Geddes' periodicity in charting the evolution of cities. He also used Geddesian terminology especially for urban growth after the Industrial Revolution period: the 'paleo-technic' and 'neotechnic' ages and the spread of 'conurbations'. The emotive language that Geddes employed about the cyclical nature of urban growth 'from acropolis to necropolis', drawing on his love of symbolism from ancient Greek culture, was also grist to Mumford's mill.[39]

But Mumford was unable to accept Geddes' doctrine of social reconstruction and his theory of life, as presented in the Geddes and Thomson volume *Life: outlines of biology* of 1931. Even in his mature assessment of Geddes in 1966, he can only point to the magnetism of Geddes' personality as the main source of his inspiration, and that he found Geddes and Branford the most 'alive' men he had ever met. In the concluding sequence of his own philosophical treatise *The Condition of Man* (1944), he cast Geddes as the personification of all that man could hope to achieve.[40] Yet Mumford established himself in the critical years of the Second World War and its aftermath as the main interpreter of Geddes. His advocacy cast Geddes as the 'guru', but as 'guru' without a formulated doctrine except for a non-socialist commitment to 'the people', for generations of architects and planners. Mumford wrote in 1966 'I fear that there is a popular tendency to reduce his rank to that of a mere father of modern town planning; a depreciation that I have never shared'.[41] But he, in his work, was most directly responsible for producing this result.

Urban sociology and the legacy of Geddes

Geddes was able to find common ground with the 'rebels' of the RPAA, but he was not to do so with the fledgling American urban sociologists in Chicago. Geddes had lectured there with success on his first visit to the United States in 1898 and a member of the School, Charles Zueblin, had been a regular visitor to Edinburgh and the Outlook Tower.[42] The natural science approach of the first professor, Albion W. Smith, to sociology had helped to make an ecological urban sociology a possibility. But in the 1920s the Chicago sociologists were leaving an evolutionary and historical approach behind and concentrating instead on the 'psychological' factors involved in understanding cities. Park, Burgess, and McKenzie were engaged in a struggle to match ecology with the organism through a study of environment and psychological factors.[43] Both Geddes and Mumford, while welcoming their pioneering work *The City* (1925), were rather dismissive about it. Mumford suggests that 'the advance of the Chicago School of urban sociologists had been somewhat tangential' since much of the research was concentrated on the pathology of the city: on, for example, juvenile delinquency or 'the mind of the Hobo'.[44] Geddes was more enthusiastic about the willingness of American sociologists to study psychological attitudes but suggests that the Chicago School had barely begun the 'adequately scientific treatment of urban communities'.[45] But what he meant by the latter was the adoption of a Geddesian method of 'scientific' study, with exhibitions, regional surveys and conferences co-ordinating the input from different disciplines. He never gave up the idea that his graphic notations provided the means for establishing urban studies.

The idiosyncratic and rather confusing way in which Geddes tried to demonstrate the use of his notations to others, however, did not win him any support. The Chicago School studiously avoid any mention of him and his work. But the most vitriolic attack on Geddes was to come from British sociologists, especially those attempting to nourish a tradition of urban sociological studies in Britain. Ruth Glass, reviewing the development of urban sociology in Britain in 1955, wrote that 'Patrick Geddes, Victor Branford, and their partners from the planning field like Raymond Unwin, left hardly any traces'.[46] At no point does she take Geddes' notations and social philosophy seriously. He was part of the 'amateur' world which undermined the early development of sociology in Britain. Only from the 1930s were publications such as the *Sociological Review* no longer dominated by the amateurs, and with

> this new professionalisation, the anxiety about towns, the interest in them, even the references to them becomes scarce. For town planners however, Geddes' 'sociology', as interpreted by his most prominent and persuasive disciple – Lewis Mumford – remained *the* sociology. To this day, planners in this country have taken little notice of other traditions and developments in the social sciences.[47]

Ruth Glass does not place all the blame on Geddes and his school for retarding the development of viable urban sociology in Britain. His partners in crime were the anti-urbanists. She holds forth at length on the extraordinary phenomenon of the British, the most urbanised nation in the world, resolutely refusing to accept this fact and nurturing a romantic ideal of the countryside as the essence of the true Britain. Indirectly she was actually proving Geddes' point that the study of place requires an emotional commitment and that the future is dominated by the cultural values of past and present. She was completely dismissive however, of the Geddes/Branford School. She writes:

> Surveys of towns and regions as well as investigations of particular social aspects relevant to town planning, have therefore by and large not been sufficiently systematic to have made a cumulative contribution to the knowledge of urban environment and society.[48]

She suggests that deficiencies of data were compounded by deficiencies of vision – 'The few, isolated attempts to view the urban social scene in wide perspective belong to the Victorian and Edwardian eras'.

The Regional Survey movement of the 1920s

To explain why the wide view on cities and social life was lost in the inter-war period is an interesting question to which there is no straightforward answer. But one factor which provides some kind of yardstick of what was happening was the Regional Survey movement of the 1920s, in which Geddes' played a major role. The Regional Survey Association was never more than a thinly-supported organisation, concentrated particularly in the Home Counties with one or two active groups elsewhere run by people who had had direct contact with Geddes.[49] Yet, for all this patently limited activity, those who partook in it really believed that they were part of a world-wide movement. This was not only because regional survey was so vaguely defined that it could cover any locally-based activity. There was a discernible movement on a world-wide basis after the First World War when the importance of place, city, and region did have a new significance. In 1925 a Professor Schmidt of Leningrad gave a paper in London to mark the Bicentenary of the Russian Academy of Sciences with the title 'On the Development of the Regional Survey in Russia'.[50] He described a vast, largely voluntary movement with more than 1,000 societies dedicated to exploring their local regions, and the foundation of 300 or so museums, 21 biological stations and 16 nature reserves. One example was the little town of Murom in Vladimir Province which in the fateful year, 1917, began a regional survey and started a biological station on the River Oka, to study questions of hydrology, hydrobiology, meteorology, and archaeology. Schmidt's thesis was that nationalism and love of country was extending the scientific study of Russia to the farthest corners of the country, educating the people and stimulating the progress of the national economy.

The USSR was perhaps a special case where a torrent of ideas and activities had been unleashed by political change. The impetus behind the movement in Britain, though, had much to do with the work of a group of men seeking to promote a professional approach to their subject and encourage public interest in it. These were the British academic geographers and two of the leading practitioners, H.J. Fleure and C.B. Fawcett, were to publish important monographs in the 'Making of the Future' series at the end of the war. For both these scholars the region was a particularly important concept, and Geddes and his organisation gave them a platform for their work. H.J. Fleure perhaps had become most closely associated with Geddes. His personal origins as a Channel Islander and his work at the University College of Wales at Aberystwyth, in the heady years of the creation of Welsh national identity and culture, made him particularly receptive to Geddes' stance and ideas. His passionate commitment to human geography and introducing the work of European geographers to Britain, especially that of Vidal de la Blache, only reinforced this. Since 1907 he had held Summer Schools for teachers of geography at Aberystwyth to which he had invited Geddes, modelling his work on the Oxford Summer School which, of course, in its turn was modelled on the Edinburgh Summer School. Geddes had responded by inviting him to contribute to the fateful Summer School on civics that he had held in Dublin in August 1914, and the two men corresponded at length during the war.[51]

Fleure was invited in 1919 to address the Sociological Society on the subject of regional surveys. His studies in anthropology took him one step beyond the impasse experienced by Herbertson, who had tried so valiantly to define the region. For Fleure, regional survey was the means to study man in his environment, and he frequently quoted de la Blache's dictum 'a ne pas morceler ce que la Nature rassemble'.[52] Fleure was very active in promoting regional survey techniques, especially in the early 1920s. It was not just a matter of geographical education; he also believed, after the terrible experience of the First World War, that 'universal mutual knowledge between peoples of different environments ought to be an element in education for peace'. He continued to be prominent in the British Regional Survey movement until 1930, when he moved from Aberystwyth on being invited to become the first occupant of a newly-created Chair of Geography at Manchester. There he built up his department, continued his work at the Geographical Association, and also published much. His perspective was still that of the anthropological geographer, but now other Schools of Geography were emerging and he no longer promoted regional survey as the only method of study. He was still publishing work on European cities, the races of mankind, and the geographical background to modern problems. But his time was devoted increasingly to the definition and development of geography within the school system and determining 'the new outlook' in geography.

Fawcett's relationship with Geddes was not as close as that of Fleure. Fawcett's use of the regional concept in his monograph, *The Provinces of England: a Study of Some Geographical Aspects of Devolution*, was also somewhat different.[53] Whilst Fleure's regionalism gave him the basis for the study of the different European peoples and their culture, Fawcett's was directed towards a fresh understanding of the economic, social and political base of different regions in England. Fawcett wanted to suggest the need for the devolution of central powers of government and the way in which regional autonomy could be strengthened. At the centre of the argument were questions about land utilisation. These were to be explored, not on a regional basis but on a national scale a decade later by L. Dudley Stamp, who initiated the great Land Utilisation Survey of Great Britain as a pioneer attempt to discover the facts of the matter. But before this, a number of land surveys were carried out by an army of volunteers, including the pupils of elementary schools, and this activity proved a great stimulant to an extension of geographical studies in schools. It was symptomatic of the exhilaration which had also helped to unleash the new 'regional' perspective.

Geddes and Branford had very little to do with most of this activity but Le Play House set itself up as a Regional Survey centre to advise and encourage this kind of work.[54] Perhaps the best definition of the regional survey movement comes from C.C. Fagg, leader of the regional survey section of the South-East Union of Scientific Societies, who resorted to an analogy from nature, likening the movement to a river in limestone country, which is only the visible flow of a much vaster if somewhat slower movement of underground water in the same direction. With his Anglo-centric vision, he places Geddes as the 'mainspring' of the river though, of course the movement had many tributaries.[55] The 'golden age' of this generalist approach to regional studies was the 1920s. A decade later, the regional perspective was no longer new, and regionalism had begun to take on the specific burden of the 'depressed regions'. The academic use of the concept was narrowed and refined and deliberate attempts were made, especially amongst the social scientists, to disassociate their work from the Geddes/Branford school. The social survey of Merseyside, begun in 1929, was built on the Booth/Rowntree, rather than the Geddes/Branford, model.[56] Geddes' hope that his kind of survey work would produce the answers to economic and social ills, instead of merely charting them, had been effectively dead since the mid 1920s and the *Coal Crisis*.

In its heyday, however, the movement had been an interdisciplinary one, able to attract adherents from a number of different backgrounds. Its leadership had always been confined to a small elite drawn from the Sociological Society, the Geographical Association and the Town Planning Institute. The hon. presidents of the regional survey section of the South-East Union of Scientific Societies reads like a roll-call of the most trusty warriors of the cause: V.V. Branford (1923), Sir E. Ogilvie (1924), G.L. Pepler (1925), A. Farquharson (1926), P. Geddes (1927), C.C. Fagg (1928/29), H.J. Fleure (1930).[57] The major powerhouse of the movement, however, remained Le Play House under the direction of Victor Branford, Alexander Farquharson, and Margaret Tatton. The latter two, together with F.J. Adkins, were responsible for developing one of the most popular activities of Le Play House, the Le Play House Educational Tours. These were described as study tours in which the participants were expected to undertake field-work using regional survey techniques. The organisation grew from the old-established International Visits Association which Geddes had helped to establish at the Sociological Society from the earliest days. The post-war (regional) perspective that was now so widely shared gave a great boost to the tours which flourished under the guiding hand of Miss Tatton.

They attracted a loyal band of supporters who went annually on these overseas visits.[58] One of the most notable of these was Sir E. John Russell, the soil chemist who had become director of the Rothamsted Agricultural Research Station in 1912. The 'regional' perspective was important to him in his work, and he was also willing to promote the idea as a sound one for educational purposes, and became chairman of the Geographical Association on regional surveys until this work was taken over by Dudley Stamp's Land Utilisation Survey. What he loved most, however, were the overseas educational tours. He acted as a leader on countless occasions, taking his last trip in 1959 at the advanced age of 87. Le Play Educational Tours conducted at least seventy-one overseas trips in its thirty years' existence, and probably more than ten trips within Britain. Geddes' early ideas on the wandering scholar had become transmogrified by their social context. There was no attempt made to assess critically the results of the surveys undertaken on these trips. The aim was to educate the participants by training their powers of observation and by bringing them, through direct personal experience of foreign lands, to a wider understanding of world citizenship. One of Geddes' pedagogic aphorisms 'Let's go and see for ourselves' was considered ample justification.

The influence of Le Play House as a centre of the regional survey movement in Britain was limited.[59] In the early 1920s, the initiative had been with the planners, but by 1924, when their bid for regional planning surveys appeared to be faltering,[60] a symposium was organised by the Sociological Society on 'Regional Survey: the next step'. But there was little fresh response. In the late 1920s the small band dedicated to the Le Play/Geddes Regional Survey method of sociological analysis became ever more dissatisfied with the response of the Sociological Society. In 1930 dissatisfaction flared up into schism under the influence of personal conflict between leading members, Alexander Farquharson and Miss Tatton. The Geddes/Branford supporters formed themselves into the Le Play Society, a separate organisation from the Sociological Society. The death of Branford in 1931 made the break a final one. Like all evangelical sects, the Le Play Society dedicated itself to a closer adherence of the Le Play/Geddes method of regional survey. But its inception as a separate organisation really meant the end of the 'generalist' phase of the survey movement. Leading geographers and anthropologists such as Mackinder, Fawcett, Fleure, Stamp, Dickenson, Pelham, Hilda Ormsby, A. Davies, and Estyn Evans were still willing to lend support, and the educational tours were kept going under the chairmanship of K.C. Edwards, then a young lecturer in geography at the University College of Nottingham.[61]

But there was now a growing hostility to Geddes' approach to survey work which became pronounced in the course of the 1930s. Geddes had finally been beaten by the mathematical statisticians despite his lifelong effort to support his belief that the natural sciences, with their understanding of space and organisms and the interaction between the two, provided a much better basis for the study of communities.[62] Unfortunately the regional survey work had never approximated to Geddes' ideal, partly because of the practical difficulties involved, and partly because few understood the concepts which Geddes had tried to refine with his notations. His publications in the *Sociological Review* of these years such as 'Social Evolution: how to advance it' (1929) and (with Victor Branford) 'Rural and Urban Thought: a contribution to the theory of progress and decay' or in 1930 'Ways of Transition: towards constructive peace' are no more than a reiteration of the ideas from the 'Making of the Future' series, this time without the optimism engendered by the prospect of peace after the war. But there was no response. The increasing professionalisation amongst social scientists, the overwhelming problems of the Great Depression, and the outmoded appearance of major British cities with their Victorian centres and crowded slums, destroyed the last vestiges of nineteenth century neo-Romanticism about the city.

Geddes' last major work: 'Life: outlines of general biology'

In the relative obscurity of his retirement years there was one old friend, J.Arthur Thomson, who continued to try to make him put his ideas down on paper. Geddes had a dozen ideas for possible books but, with patience, Thomson at last got him to write a final statement of his biological perspective on life, which Thomson appended to his own mammoth effort at writing a complete biological textbook. It was published in 1931 under the title *Life: outlines of general biology* in two large volumes containing 1,500 pages. Textual evidence suggests Geddes was responsible mainly for the last 300 pages, under the chapter headings: 'Biology amongst the Sciences', 'The Biology of Man', 'Biology in its Wider Aspects', and 'Towards a Theory of Life'. The emphasis of his contribution, however, is not on the development of scientific theory so much as on the implications of biological theory for the social sciences. The general tone of his contribution is defiant, confirming once again his views on life, society, and the importance of natural history, though he recognises that his ideas are less likely to get any response than ever before. Mumford and others, however, have pointed to this work by Thomson and Geddes as one of the most important and most neglected contributions made by natural scientists in the twentieth century and suggest that here one may find at last a full statement of Geddes' philosophy.[63]

While the latter claim may have some foundation, the former cannot be sustained. Deep knowledge of botany and microbiology was no longer a sufficiently rigorous intellectual basis for understanding the nature of the universe and physicists particularly thought that biologists had little to contribute to new developments. But the Thomson/Geddes book has moments of more than antiquarian interest. Geddes' contributions are, as usual, more difficult to appreciate as he continues to structure his contributions in his totally idiosyncratic way. He tells the reader the location in which he is writing particular sections and mentions particular meetings and conferences which he has just attended and which prompt him to develop certain ideas. The result is a wide-ranging discussion but many instances of repetition of the same material. For all the energy and defiance, it is an old man writing now, far less sure of himself than his younger self tackling the same issues in his pamphlets of the 1880s. In fact, while many points are made sharply and with insight, there is evidence of an attempt to disguise muddled thinking by constant amplification.[64] Yet the pages are well worth reading. Geddes does his best to engage the interest and emotions of his readers with a level of application often lacking in the publications of his middle years. On occasions he manages passages of considerable quality, welcome indications of the kind of magic he could sometimes draw upon to enthrall those who listened to him.

One such example of this is the description he gives of the deep sea-bed and its natural history. It is worth quoting this short passage to illustrate how he could transmit an emotional response even to the most extreme examples of an emotionless environment. He describes 'life in the great abysses of the ocean' in the following terms:

the floor of the Deep Sea shows vast undulating plains like sand-dunes, but covered with shiny mud. No scenery, no sound, no vegetation, not even rottenness. But many animals have colonised these inhospitable depths, some anchored, others slowly swimming, as if half asleep, and others walking delicately with stilt-like legs on the treacherous ooze. Sluggish existences there, devouring one another in a grim sequence of reincarnations, the last link in the chain depending on the ceaseless snow-shower of moribund minutiae sinking through the miles of water from the surface overhead. There is enormous pressure, many tons on every square inch of the body, but the tissues are so interpenetrated by water that the pressure is not felt. The current of life flows slowly and centenarians flourish. There is no light, save the fitful gleams of luminescence from fixed animals that sparkle like Christmas trees and from free-swimmers gliding slowly past like illuminated miniature gondolas. Otherwise utter darkness. Also intense cold, near the freezing point, due to the downsinking of icy water from the Polar Regions. What an eerie world, covering a hundred million square miles, more than half of the Earth's whole surface, a world of eternal night and eternal winter, soundless stagnant and monotonous, a plantless world with a stern struggle for existence.[65]

Numerous students have paid tribute to Geddes' descriptive powers which gave them a sense of really seeing things for the first time. In these last chapters of *Life: outlines of general biology*, however, what Geddes was trying to establish was not just a way of seeing the marvels of natural history, but his unique sense of a new conceptual surface which would bring together the essential factors for the study of man and his environment. By constantly referring back to his basic triad: Place, Work, and Folk, Geddes believed that he had achieved such a conceptual surface which was capable of producing new insights to the age old problems of avoiding an over-simplified and deterministic analysis of human behaviour and social life. The gist of his argument was that time, space, and organism could be studied from many different viewpoints, and each would produce new ideas and understanding. He refers to the new developments in relativity theory and genetics, but his interest is focused at all times on the implication of modern knowledge in the practical context of human society. Geddes writes time and time again of the fact that there is 'a web of life' which was becoming ever more complex with advances of culture, and ever less likely to be understood by the fragmented perception produced by specialist knowledge.[66] The relationships between space and substance, between subjectivity and objectivity, had been practically and conceptually severed. With the scientific capabilities of the twentieth century of bringing the greatest benefits to mankind that humanity had ever known, we did not as yet understand the consequences of our actions.

Starting from a survey of the origins of man, he shows the similarities between man and animals except for one major difference: 'Man has the power, if he would only exercise it more, of guiding his conduct in reference to ideals'. In the past there have been pinnacles of human achievement which have never been superseded. Yet over time there has been a gradual increase in the 'average of culture' because there has been 'an increase in social susceptibility or impressionability'. This has come about because of evolution and natural selection, and Geddes believed it was the main method of progress. As far as mankind's future was concerned 'impressionability to the ideal, which the social heritage always expresses, is the saving grace'.[67] Such ideas were at the root of Geddes' passionate advocacy of the goal of Eutopia, the Utopia which was realisable here and now. He saw his work as a means of liberating people and maximising individual potential. From his stance within the natural sciences, Geddes takes on all those who were dedicated to different paths in the hope of achieving human progress. He reserves his most acid comment for the advocates of state socialism which he warns his readers to spurn. Drawing from natural history the examples of ants and bees as some of the most socialised insects, he warns that social instincts can lead to a huge sacrifice of lives which are organically incomplete. He describes the honey bee as 'the short-lived martyr to extreme state socialism' who has to work to support, not only the queen, but also the useless drones.[68]

Geddes' alternative requires an understanding of the 'web of life' or the complicated interaction between individual and environment which can come only in three specific ways: first, by an educational revolution which concentrates on the education of the child through the senses, especially through trained observation; second, the improvement of the physical quality of people through the medical application of new biological knowledge; and finally, a new understanding of the human influence on ecology, both natural and man-made. In pursuit of these goals he produces his 'tour de force', his final chapter 'Towards a Theory of Life'. But there is no great clarity in the analysis he puts forward. He advocates, as usual, parallel surveys, biological and social, the regional survey, and the social survey.[69] The valley section with its Le Playist interpretation of occupations re-emerges as ever, coupled with Geddes' own brand of evolutionary history.[70] While he has undoubtedly established his special perspective, he is in the long run beaten by the level of generalisation at which he works. He produces cycles of syntheses and critiques rather than workable theories.[71] His analysis of the structure of human behaviour owes more to his own personal experiences and prejudices than it does to any stated theoretical approach. He was well and truly trapped by an intellectual impasse of his own making.

Geddes' vision and originality might have provided the basis for a new study of human ecology had he been willing to speculate within verifiable limits. But he was not willing and, in the course of developing further his new conceptual surface, his 'theory of life', he

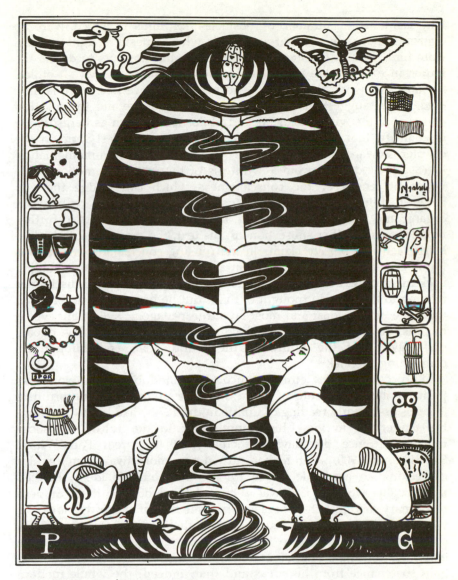

Figure 9.1   Arbor Saeculorum: the 'tree of life'

The symbolic tree 'has its roots amid the fires of life, and is perpetually renewed from
them. But the spirals of smoke which curl among its branches blind the thinkers and
workers of each successive age to the thought and work of their precursors. Two
sphinxes guard the tree and gaze upward in eternal questioning, their lion bodies
recalling man's origin in the animal world, their human faces the ascent of man. The
branches symbolize the past and passing developments of society, while the bud at the
tree-top suggests the hope of the opening future. Issuing from the smoke-wreaths at the
top, you can also see the phoenix of man's ever-renewed body and the butterfly or
Psyche of the deathless soul of humanity.'

Source: P. Geddes (1906) 'A First Visit to the Outlook Tower', pamphlet, Edinburgh: Patrick Geddes
          and colleagues, p. 26.

had fashioned a number of obstacles for himself which were to overcome not only him but also those who earnestly sought to learn from him. Two of the most daunting were his attempts to define the importance of the life-spirit and his determination to seek an ethical basis for his work. As far as the former was concerned there is little evidence to suggest that Geddes had really got beyond the mystical *élan vital* of the 'vital' biologists and philosophers such as Bergson. He is adamant that his life-graphs, set out in their fullest form in the chapter 'Towards a Theory of Life', are based on a non-mathematical logic, a logic derived from a synthesis of thought and emotion.[72] He writes at length of the consciousness which can bridge the gap between things and thoughts, the sciences and the arts, or between organic and psychic interrelations. But the more he resorts to further notations and further subsections in his text, the more elusive the life-spirit seems to be. He claims in his rousing conclusions that he has found a way out of the environmental determinism which was holding back the progress of human geographers and ultimately all disciplines relating to the life and thought of man. But he has not got any further than one of his oldest and most favourite pedagogic syllogisms which he has taken from Schiller, that while philosophers may debate endlessly about the future of the world 'love and hunger are completing the task'.[73]

The problem of the ethical stance Geddes adopts in his work was perhaps the most acute of them all. One of the most unfortunate legacies of his early reading of Comte in the 1870s, had been to confirm his tendencies to view his commitment to the future as 'social religion'.[74] Since the days particularly of the great Paris World Exposition of 1900, when he had turned ever more away from the idea of seeking an academic milieu for working out his ideas, he had pursued his work as if it was a crusade and he, the great modern missionary amongst the people. Whatever the quality of his ideas he was never likely to succeed in either the short or long term, in convincing many others to follow him through faith alone. He was extremely fortunate that geographers and town planners were able to gain sustenance from his ideas and that, indeed, the whole modern concept of town planning gave him a practical arena in which to work. The impact he had on many who came into contact with him in the 1920s is admirably captured in the book written about him by Amelia Defries, an American journalist, with the title *The Interpreter: Geddes, the Man and his Gospel* (1927). It is a work of devout hagiography which conveys a strong emotional response to Geddes himself. His ideas are produced as 'the gospel' without any critical analysis of their worth. He was delighted by it and it brought him some comfort in his declining years as he faced the overwhelming evidence of his own failures[75]

In fact the Thomson/Geddes volumes with their bold title *Life: outlines of general biology* marked the end of an era. The age of the grand synthesis of knowledge of life built on relatively simple definitions of Darwinian evolutionary theory was virtually over. A few months before the Thomson/Geddes volumes came out there had been another of similar type, *The Science of Life* by H.G. Wells, G.P. Wells, and Julian Huxley, which suggests that the genre was not quite dead. But these projects appear to have been inspired by men who belonged to a former scientific generation. Geddes had shared many interests with H.G. Wells over the years which had developed from their common starting point as students of T.H. Huxley. Wells had written on Utopia, now subject to the concept of evolution, and had been equally mesmerised by the challenges of modern knowledge and technology.[76] But Wells had made a far greater impact than Geddes with his ideas. What they shared was a common obsession with the social implications of modern knowledge rather than the knowledge itself. When they wrote these grand biologically-based works both were, in effect, amateurs. Geddes openly admitted to Thomson that he had not even kept up with current work in the natural sciences, and he relied on him to keep him informed.[77] *Life: outlines of general biology* remains relatively unknown and unread.

Regional Survey at Montpellier: the Collège des Ecossais

Geddes himself had no great hopes of his writings having a wide impact. Instead he devoted his remaining years to a project at Montpellier in the South of France where his doctors advised him to live for health reasons after his breakdown in 1924. The area around Montpellier is one of the most varied and interesting natural regions in Europe and Geddes wanted to dedicate his last energies to encouraging regional studies there. The home base of Professor Flahault, whose work had inspired his own perception of the region in the 1890s, gave him a refuge from a harsh world which had mostly not understood him. He wanted to turn his retirement home at Montpellier into a complete, living symbol of his life's work.

This bizarre scheme, which occupied him for longer and longer intervals from 1923 onwards to his death in 1932, bore all the hallmarks of a typical Geddesian folly. It cost far more than he could afford, though war damages compensation for the loss of the Town and City Planning Exhibition, and a second marriage to a rich and rather vague wife helped in this direction.[78] The work was sustained by a small group of family and friends, especially Geddes' youngest son, Arthur, and T.R. Marr, an old devotee of Geddes' from the Outlook Tower days in the 1890s. It was completely impractical since the site for the building had no water supply and the surrounding land, which Geddes wanted to make into a garden, was wild and infertile. Above all, the grand conception of the plan which was to be for a group of international halls of residence for overseas students studying at the University of Montpellier, who would band together to promote world citizenship, bore no relation whatsoever to the reality. Lewis Mumford was angry that Geddes appeared to be wasting his time instead of putting his ideas on paper.[79] But Geddes was unrepentant and in some respects he was fairly astute. While no one now reads *Life: outlines of general biology*, the Collège des Ecossais at Montpellier (after a very chequered history) has served to stimulate a local group of planners and architects, who have formed themselves into a Patrick Geddes Association.[80]

Much of the emotional commitment to Geddes and his ideas has been transmitted through the links built up at Montpellier. The symbolism of the Collège des Ecossais has continued to intrigue subsequent generations, though as indications of Geddes' idiosyncratic genius rather than as a message for the future. The physical legacy of Geddes' work in these years consists of two extraordinary buildings: one, the Collège des Ecossais, an ill-planned, battlemented building with a small tower from which it is possible to view the distant town of Montpellier and its environs; and the other, the Collège des Indiens, a rather dull structure with a flat modern-style façade, which was the gift of the only other national group which raised some money for this project. This was the stuff on which Geddes built his dreams whilst he put his passion and commitment into creating his last garden out of the wilderness, his final creation to illustrate his evolutionary perspective through the symbolic use and juxtaposition of specific plants.[81]

The whole project itself, colleges, garden, and the acquisition of a nearby chateau[82] (that Geddes made in 1926 against the advice of his friends), were supposed to be symbolic of Geddes' life work. His son, Arthur, brought hurriedly from India to work on the construction and decoration of the Collège des Ecossais, was given most instruction about the decoration of the building rather than its overall plan. On the battlements, his fancy unreined, Geddes had decorative friezes depicting the owls of Pallas Athene, Goddess of Wisdom. On one, the two owls were apart, the Arts and the Sciences; on the next, the two were in flight, symbolising synthesis and synergy. There was an olive branch representing constructive peace and Geddes' motto 'vivendo discimus' – by living we learn. A final frieze depicted in Geddes' words 'the Owl of Pure Science – interpretative and penetrative' – symbolised by a divining rod which could seek the Water of Life. He used birds often, mostly three doves depicting synthesis,[83] synergy, and sympathy, and a Phoenix of Reconstruction who was to come to life again with the renewal of the regional spirit, the informed interaction between environment and society. The halls of residence did not attract many students because of their distance from Montpellier and their physical discomforts, quite apart from the need for all residents to be sympathetic to and interested in Geddes and his ideas. Students were also expected to live alongside a number of elderly people: artists, writers, and acquaintances of Geddes, to whom he had offered, in moments of compassion, free board and lodging.[84]

While Geddes was spending his time and energy creating his base in France, the Outlook Tower in Edinburgh had once again fallen on difficult times. Chairmanship of the Outlook Tower Committee had been taken over by an old friend from the days of the Edinburgh Summer Schools in the 1890s, Mrs Craigie Cunningham, who was a passionate devotee of Geddes and all he stood for.[85] But all her efforts and enthusiasm were not enough to prevent the accounts running into a deficit. Geddes himself was no longer any help as he put his money into his French ventures instead of sending support to Edinburgh. By operating outside the university, and even outside the growing system of extra-mural adult education, Geddes ensured that his institutions could survive only with the dedicated help of family and friends. As time passed it was inevitable that these people would find the burden too great, though one or two, particularly his son Arthur, spent the rest of their lives working to keep alive the legacy of Geddes in Scotland and in France.

Geddes' legacy to the making of the future

The legacy of Geddes, however, was not to depend on these relics of his practical projects. Geddes' fame has continued to grow steadily since his death in 1932, and in the last decade or two new interest has arisen in his life and work, not only in Britain, Europe, and India, but also in America and Japan. His name and what he stands for have had potent emotional appeal in a number of contexts. Generally his appeal has been on two levels: first, as an emotional inspiration to environmental groups engaged in some way in resisting particularly well-established anti-democratic trends in modern society, who need all available means to legitimise their position; second, as a record of one man's odyssey through many of the problems faced by those concerned with reconciling 'place' with 'people', either in practical terms, such as planners, educationalists, and environmentalists, or academic ones, especially developments in geographical and ecological studies. For much of the half century since his death the appeal of the first has obliterated that of the the second.[86] Those closely concerned on a professional level with the disciplines most nearly related to his work, biology, sociology, geography, and town planning, have shown greatest antipathy to him.[87] By virtue of the rational nature of all disciplines, the emotion of Geddes' mission is the greatest anathema. On the other hand, in those contexts, such as twentieth century Scottish nationalism, which feed on deeply-felt emotions, Geddes' name has been honoured as a pioneer. The Scottish Nationalist poet, Hugh McDiarmid (Christopher Grieve) was a frequent visitor to the Outlook Tower in the 1920s.[88]

Yet as academic specialists fiercely reject him, his supporters have continued to emphasise the importance of his stand as a generalist with interdisciplinary interests. Geddes' major intellectual problem was his determination to build a grand general social theory based on the natural sciences which would supersede all others. This overriding, over-ambitious desire, born in a spirit of nineteenth century positivism which made it possible to believe that, with the application of modern knowledge, evolution and progress could be made to go hand in hand, was an impossible goal. It placed Geddes beyond the range of other attempts in the late nineteenth century to make some headway in trying to understand the city on a scientific basis. Max Weber's work on the city, for instance, remains stimulating and fruitful, while Geddes' social theory is forgotten. It is easy to point out that Geddes was careless about defining his terms and even using his own concepts precisely. For him it was enough to be able to point to physical illustrations of what he was trying to show in, for example, Old Edinburgh.

His interpretations of what he saw were highly personal, which impeded the problem of establishing reference points for charting the interaction of social processes and spatial form. In conventional academic terms there is no doubt that Geddes was a failure. But then he himself deliberately set out to defy convention. He treated academic subjects in a cavalier fashion. His survey technique was based, he declared, on history and geography. Yet he was no historian and only a talented but amateur geographer. He picked up his ideas on history as 'a drama in time' from popularly held views on history which were current in the 1870s and 1880s.[89] His positivism made him receptive to an historicist approach to history. For him, what was important was not the quality of historical scholarship, but the use history had for understanding social evolution. Where history has been a part of the education of planners and architects, it is this kind of Geddesian 'history' which has been most often used since it has a direct relevance in training the eye 'to read' the complex structure of the urban environment.

THE LEARNING OF THE SCHOOLS

Figure 9.2

'Copy and cram, jaw and pi-jaw: a poor turn-out from Gallery and Museums, from
Lecture room and Union, and from Chapel! Is not this the reason – and no wonder –
that the ablest of our graduates so often become the most cynical? . . . Briefly stated, our
essential educational policy, that of seeking to develop sense, experience and feeling, by
means of the heritage of Art, Science, Literature, Politics and Religion . . . has been, and
still is in the main, wrongly and falsely handled.'

Source: Professor P. Geddes (1924) 'Education in Return to Life', pamphlet, Bombay: Bombay
Vaibhav Press, pp. 11 and 16.

This, for Geddes, was the object of the exercise. He made no pretence otherwise. His own historical excursions into the past of Edinburgh which was a crucial element in his survey of the Edinburgh region, were undertaken in the spirit of full-blown neo-Romantic fantasy in which Celtic origins and the cult of the individual figured prominently.[90] The unacademic nature of all this was irrelevant to Geddes. The purpose of his kind of history was to make him sensitive to the physical remains of the past, and to make him an excellent adviser on conservation in the future use of historical buildings and sites. His coining of the terms 'paleotechnic' and 'neotechnic' was in this spirit. They made the observer of nineteenth century cities responsive to the physical implications of change since the Industrial Revolution period. They were tools for analysing the environment, not an explanation of social change. The latter he believed, in any case, was determined by cultural factors.

Geddes was obsessed by the idea of cultural evolution.[91] This more than anything else was to make his path through the social sciences, through town-planning ideology, through the regionalist movement, one which was littered with wrecks and confusion. Most confusion centred on his sociology of cities. The transformation of the British nation from a rural to an urban dwelling people was, arguably, the greatest change in the cultural evolution of the people that had ever occurred. Geddes was the only British sociologist who took cognisance of this fact and tried to study it in a practical way. He made the study of cities his province. But his unconventional and unacademic stance left him isolated. Few were prepared to see his notations and 'thinking machines' as a route to understanding the complex interaction of factors which make up urban life. There was little hope that this kind of methodology and field-work would become an acceptable part of academic sociology. The result of Geddes' intervention here was to keep 'urban sociology' from becoming an integral part of the social sciences in Britain for a couple of decades.

Some of Geddes' followers like T.R. Marr recognised that Geddes' 'civics' was not an academic subject. What Marr hoped was that it would prove a practical basis for philanthropists and social reformers working in the city. Marr, with others of Geddes' friends who belonged to the Edinburgh Social Union and the Outlook Tower Association, was conscientiously devoted to practical endeavour to improve the social life of the poor. What Geddes' doctrine of civic reconstruction offered was a moral basis for their work with which they could counter the challenge of socialism.[92] While the adherents of Barnett's University Settlement of Toynbee Hall were moving cautiously towards the idea of limited state intervention to help solve social problems, Geddes' position remained totally hostile to any such idea.

[320]

Cultural evolution depended on the interaction of individual and community, city and region. Direct action by the state would be an unwarrantable intrusion Geddes believed, which could easily inflict far more damage than it could offer relief.

The problem was, as always, Geddes was not interested directly in social reform. His stance left individuals like Marr stranded between their desire for practical action and their understanding of what to do. But if Geddes left the social reformers behind, he became, and was to remain, a cult figure amongst another group, architects, town planners, and environmentalists united by their dedication to improve the condition of the people. Geddes' legacy of inspiration amongst this elite, a tiny but highly influential minority in their professions, has been constantly invoked. He was wanted for two reasons: his emotional commitment to cities, civilisation, and the future of the people, and his advocacy of regionalism as the basic unit for environmental engineering. The first generation of leading planners in Britain: Abercrombie, Pepler, and Lanchester, were dependent on Geddes' vision as the justification of, not only their work, but the very methods they used to carry it out. In the early days of establishing planning practice on a professional basis they needed to believe that they knew what was best for both city and people. Subsequent generations in different contexts found a similar need. Geddes' objective of cultural evolution was broad enough to cover some very free interpretations of what he stood for.[93]

His most influential publication after his death was *Cities in Evolution*. This remains a unique work. Its combination of factual detail with prophetic insight can strike the reader as electrifying and revelatory. For a moment there seems to be a prospect of a philosophy of life in which the grandest abstractions are related to the ordinary affairs of the world with a compelling intensity. Yet while chapters seem to be packed with information about the nature of modern civilisation and what must be done to ensure favourable evolutionary trends for the future, there is no coherent structure to the book. It is an amalgam of all Geddes' personal idiosyncrasies.[94] What the book offers is the strength of his personal perceptions. Geddes had found ways of expressing how 'to see' and 'to feel' modern cities and urban life. What he was after, as a cultural evolutionist, was the 'soul'. As he himself puts it:

> He is no true town-planner, but at best a too simple engineer, who sees only the similarity of cities, their common network of roads and communications. He who would be even a sound engineer, doing work to endure, let alone an artist in his work must know his city indeed, and have entered into its soul — as Scott and Stevenson knew and loved their Edinburgh.[95]

Much of the legacy of Geddes rests on this kind of inspiration with its inbuilt justification for the responsive planner or architect to trust his own judgement. Geddes' regionalism contained similar strands of prophetic inspiration. For the *cognoscenti*, his Dunfermline Report of 1904 was, in Howard's phrase, like 'a beam in the darkness'.[96] It was not widely read, nor were Geddes' Indian reports, largely because so few copies of them were published. This left those that had read them (in the case of the Indian reports, the tiny number who went to the library of the Calcutta Improvement Trust to look at the complete set) as a specially privileged small group who could keep Geddes' name alive by reference and quotation. Geddes was always eminently quotable. He specialised in producing catch-phrases to sum up his views which are instantly memorable. For incipient regionalists, even his basic Place, Work, Folk was a reassuring starting point. Geddes' concept of regionalism, with the emotional commitment to cultural evolution, its emphasis on the uniqueness of place and people, its promise of hope for a better life for those away from centres of political and cultural domination, lifted the whole prospect of regional planning from a mere manipulation of resources to a crusade for mankind.

By the time of Geddes' death, experiments in regional planning unconnected with Geddes had begun to acquire this additional lustre. The most outstanding of these was the work of the Tennessee Valley Authority whose activities in the 1930s became a symbol of all that regional planning could achieve.[97] Geddes actually had some indirect influences here, not only through Mumford and Benton Mackaye, but through the influence he had had prior to the setting up of the Tennessee Valley Authority in encouraging Howard Odum to develop his regionalist institute in the south. Odum's Institute for Research in Social Science at the University of North Carolina was to establish a distinctively southern school of sociology which owed not a little to Geddes' ideas. Odum himself worked empirically on questions of race, agricultural change, and migration. But his regionalist perspective was based on Geddes' and Branford's 'third alternative': between unbridled capitalism and socialist state intervention. He hoped to train workers to undertake planned rural-urban regional development maximising the use of resources for the greatest benefit of the whole community.[98] This, together with the eye-catching attempt to restore the fertility of a region made barren by man's greed, was what had made the work of the TVA something more than an exercise in regional planning.

The case for recognising the significance of the legacy of Geddes' ideas as propaganda can be made much more strongly with reference to the development of regional planning studies in Britain. In 1935, as the work of the TVA was impressing the world, a small group was set up in London under the auspices of the Architectural Association calling itself the School of Planning and Research for Regional Development. It was the brainchild of E.A.A. Rowse, who more or less single-handedly, and without financial backing, sustained its activities throughout the 1930s.[99] Rowse was an advocate of the Geddes/Le Play method and he passed on his passion and commitment to the cause to the small numbers of students with whom he came into contact.[100] In the aftermath of the Second World War, these students provided the nucleus of professional expertise which was urgently called upon for reconstruction and development throughout the world. Geddes' ideas thus became very widely diffused. At the Architectural Association itself during the war, it was recognised that the post-war world would require more trained planners. An attempt to meet this need was made by the School of Planning which set up a correspondence course for which 1,600 students enrolled. At the end of their courses they were all encouraged to come to London where their studies were supervised by Miss Jacqueline Tyrwhitt, another ardent Geddesian. She emphasised four aspects of planning which she believed were essential: the need for the activity to be multidisciplinary, the use of the region as a planning unit, the necessity of a holistic approach, and the importance of economic and social factors. All four were derived from her Geddesian perspective.

The problems of Third World countries, particularly the problems of modernisation and the effective use of resources, became key issues in the 1950s, of special concern to planners and architects. In 1953 a Conference on Tropical Architecture was held in London from which was to grow the School of Development Studies at University College London. Housing the poor of the Third World as they poured into the cities called for interdisciplinary co-operation between teams of experts always working within stringent economic restraints. Some of those recruited to this work had been trained at the Architectural Association and Geddes' ideas became known in wider circles. But his name tended to be invoked for the inspiration he offered rather than his practical techniques. The value of his legacy was what people believed he stood for, especially the cultural and economic autonomy of the region. This seemed a desirable objective as the problems inherent in all kinds of planning, but particularly regional planning, became obvious. Geddes' reputation remained and remains a talisman and a rallying cry.

Planning for Third World countries was a highly specialised activity to some extent outside the mainstream of what was happening in Europe and America. In the euphoria after the end of the Second World War there was an even greater commitment than there had been in 1918 to reshape the future for the benefit of 'the common man'. Here, despite the efforts of Miss Tyrwhitt and others, what this meant in ideology and building form was the adoption of the Modern Movement. Germany and the Bauhaus, Le Corbusier and the technical city, devoid of any reference to history or emotion, became the hallmarks of the new era. The international style of high-rise building, now possible in all places and climates with the help of modern technology, became the norm. In Britain there was a determination to eliminate at last the lingering problem of urban slums as well as to rebuild the bombed cities. There was wholesale demolition of city centres which brought in its wake the destruction of the 'social heritage' and the disruption of the 'web of life' on a scale which Geddes had feared ever since planning legislation had first reached the statute book in 1909.[101]

Even while planning students were still trained in Geddesian techniques, survey, diagnosis, plan, and so on, the most essential element of Geddes' message, the critical relationship between social processes and spatial form, was ignored in relation to the existing environment. In the new redeveloped areas, where this critical relationship was taken into account, it was comfort, space, convenience, trees, and gardens which constituted the elements of a 'good environment' with no reference at all to history.[102] Planners left behind the biological and evolutionary approach which Geddes had tried to stress. He predicted dire results if his message was ignored. By the 1960s the social failures of modern planning practice had reached crisis proportions. New 'planned' environments were becoming ghettos, vandalism was rife, residents were experiencing stress and tension. Above all, the cycle of growth, blossom, decay, deterioration, had not been broken after all. Some of the new estates built to ensure slum clearance from the inner city areas were themselves becoming slums.[103] Geddes and his unfashionable concerns for history and geography became fashionable again.

His name was evoked as a source of inspiration as a pioneer who produced answers to these problems almost a century ago. Indeed Sir Robert Grieve, who did most to ensure one of the outstanding successes of regional planning after the Second World War, the Clyde Valley Plan, which was developed under the auspices of the Strathclyde Regional Council, has paid tribute to the inspiration of Geddes. He suggested, perhaps over-modestly, that Geddes, and his closest planning disciples such as Abercrombie, had not had the chance that he was given to work out their ideas in practice.[104] They were only able to suggest how it might be done. The Strathclyde region, extended well beyond the area originally designated for the Clyde Valley Plan, provided a context in which town and country, metropolitan area and remote rural areas, could be treated together in an administrative context, under the Scottish Office, which gave some flexibility and coherence to the whole activity. In Scotland Geddes' reputation has continued to gain lustre over the years. Edinburgh, the city which had found it so difficult to respond to Geddes' initiatives during his lifetime, began to give recognition to his work, especially since the Lawnmarket and Royal Mile, which he contributed so much towards conserving, has now become a major tourist attraction.

The fact of the matter is that Geddes did address himself to the kinds of problems for which each generation will have to seek new solutions.

As Sir William Holford said in 1954 at the centenary celebration marking Geddes' birth, which he addressed as president of the Town Planning Institute: 'I cannot escape his influence. The Greek epigram on Plato is applicable to him: "Wherever I go in my mind I meet Geddes coming back"'.[105] His work repays study as he had, apart from his own highly-trained powers of observation, the widest experience of urban life in Europe, India, and America at a formative period of world urbanisation in the twentieth century. He also had a political stance which owed much more to the now extinct nineteenth century form of anarchism espoused by Kropotkin, which gave him a sense of social progress in terms of the immediate environment and possible improvements, rather than vaguer concepts of individual rights, justice, and equality. From the earliest pamphlets he wrote in the 1880s, he had been offering 'the third alternative', neither conservatism nor state socialism, (which he describes as 'Lib-Lab Fabianism'), while insisting that the objective was to enable every individual, regardless of wealth, status and class, to achieve his or her personal potential. It is a position which has kept Geddes fashionable amongst contemporary anarchists, such as Colin Ward, and supporters of co-operative and self-help community ventures.

As evolutionary biologist, and outstanding gardener, his approach to planning was organic and related to people before place, except that as an evolutionist he was very aware of how place influenced people and thus trends for the future. This concern with people made him first and foremost an educationalist. His tenuous connection with the progressive school movement in Britain and France through his contacts with Cecil Reddie of Abbotsholme and Edmund Demolins and the Ecole des Roches, and with adult education through the rather specialised and small-scale Edinburgh Summer Schools, do not place him as a central figure in the mainstream of modern educational developments. But his pedagogic syllogisms about the nature of education and learning reached far beyond his personal influence, and his educational ideas repay investigation and analysis.[106] During his lifetime many men and women found new opportunities opening up for them through their contact with Geddes.

Yet, in his work at the Outlook Tower, he was reaching out well beyond the individual. He had a strong vision of what he called 'cultural evolution', the interconnections between academic freedom, national or regional identity, cosmopolitanism and citizenship.[107] His work had fed his belief that such a combination was the only way to regenerate society and reach higher levels of social evolution. The function of a Geddesian style university would have been the dual one of applying modern knowledge to everyday life, and sustaining a vigorous regional culture, the cultural role taking pre-eminence in its importance. The questions he raised about the nature of university education were stimulating even if some of his particular answers were eccentric. His central proposition, that students of the social sciences need to explore the relationship between historical evolution and its geographical location in the built environment is still challenging.[108] Very little is yet known about the connections between social processes and spatial form.[109] Geddes' essays in interpretation, especially his Indian Reports and his work in Palestine, for all their idiosyncracies, contain insights on this subject. The search for a regional identity, the built environment which encapsulates its form and history, and the conscious cultivation of cultural diversity, are issues which remain perennially pertinent. Geddes' genius helped him to identify these issues and alerted him to both the dangers and the potential for human development that lay in the future.

Notes

1. R. Daston, M. Heidelberger, and L. Kruger (eds) (1986) *The Probabilistic Revolution* vol.I *Ideas in History*, London: Bradford Books.
2. Architect-planner Arthur Glikson publicly acknowledged his debt to Geddes and his regional vision in 'The planner Geddes?' *News Sheet of the International Federation for Housing and Town Planning* The Hague, May 1955, quoted in P. Boardman (1978)*The Worlds of Patrick Geddes: biologist, town-planner, re-educator, peace warrior*, London: Routledge and Kegan Paul, pp.439–40. The hostile view is given in M. Hebbert (1980) 'Patrick Geddes Reconsidered', *Town and Country Planning* pp.15–17.
3. For a discussion on housing and reconstruction see M. Swenarton (1981) *Homes fit for Heroes*, London: Heinemann.
4. W. Ashworth (1987) *Short History of the International Economy*, London: Longman 4th edn, pp.228–30.
5. Geddes' thesis was that the concentration of power in the metropolitan capital cities was a decisive factor in encouraging governments to wage war. If provincial cities formed friendly and cultural links, this would be the greatest investment for ensuring peace in the future. Regional renewal and regional co-operation were the 'third alternative' to war or revolution – P. Geddes and V.V. Branford (1917–19) 'The Drift to Revolution', *Papers for the Present*, pamphlet issued for the Cities Committee of the Sociological Society by Headley, London, pp. 44–6.

6. F. Prochaska (1988) *The Voluntary Impulse: philanthropy in modern Britain*, London: Faber.
7. P. Abrams (1963) 'The Failure of Social Reform 1918–1920', *Past and Present* 24: 43.
8. M. Hawtree (1983) 'The Emergence of the Planning Profession', in A. Sutcliffe (ed.) *British Town Planning: The Formative Years*, Leicester: Leicester University Press, pp.64–104.
9. G. Cherry (ed.) (1981) *Pioneers in British Planning*, London: The Architectural Press – M. Miller 'Raymond Unwin' pp.72–102; G. Dix, 'Patrick Abercrombie' pp.103–30, and G. Cherry, 'George Pepler' pp.131–49.
10. H. McCrae (1975) *George Pepler: knight of the planners*, research monograph, University of Strathclyde, Department of Urban and Regional Planning, p.7.
11. Ibid., p.7.
12. Abrams, op.cit., p.44.
13. P. Uyttenhove (1985) 'Les Efforts internationaux pour une Belgique moderne' in *Resurgam. La reconstruction en Belgique après 1914*, catalogue, Bruxelles: Crédit Communal de Belgique, p.64.
14. M. Smets (1983) 'Research on the Belgian Reconstruction after World War 1', *Planning History Bulletin* 5 (2): 24–8.
15. E.A. Rose (1988) 'Life imitating art – Le Corbusier centenary', review article, *Planning Perspectives* 3 (2): 217–23.
16. Published as P. Abercrombie, S. Kelly, and A. Kelly (1922) *Dublin of the Future: the new town plan*, London: University Press of Liverpool/Hodder & Stoughton.
17. S.V. Ward (1984) 'Local Authorities and Industrial Promotion 1900–1939: rediscovering a lost tradition', in S.V. Ward (ed.) *Planning and Economic Change: an historical perspective*, Oxford: Oxford Polytechnic, Department of Town Planning, working paper 78, pp.25–49.
18. A list of Abercrombie's publications is given by A. Manno (1980) *Patrick Abercrombie: a chronological bibliography, with annotations and biographical details*, paper 19, Planning Research Unit, School of Town Planning, Leeds Polytechnic.
19. The relationship between Abercrombie and Geddes, at its high point in 1913 at the Ghent International Exhibition, had cooled considerably in the aftermath of the Dublin competition. Abercrombie's entry, although awarded the prize, was criticised by Geddes as lacking the perspective of regional survey. Geddes mentions the rift to Norah in a letter, 14 September 1922, Geddes Papers MS10502, NLS: 'I told Abercrombie privately his survey was less adequate than it should have been and thus than Ashbee's – but the latter was unable to finish, or – –? (A (naturally) did not like this!) Best let him alone: no letters from me certainly'.
20. For an account of why Sheffield was a pioneer see R. Marshall (1985) 'Town Planning Initiatives in Sheffield 1909–1919', *Planning History Bulletin* 7 (1):41–48.
21. A. Defries (1927) *The Interpreter: Geddes, the Man and his Gospel*, London: Routledge, pp.324–5.
22. S. Pollard (ed.) (1970) *The Gold Standard and Employment Policies Between the Wars, 1919–39*, London: Methuen.
23. P. Abrams (1968) *The Origins of British Sociology 1834–1914*, Chicago:

University of Chicago Press, p.119.

24. Victor Branford (1926) 'The Conditions of Eutopian Repair and Reconstruction', pp.108–9 in P. Abercrombie, V.V. Branford *et al*. *The Coal Crisis and the Future: a Study of social disorders and their treatment*, the 'Making of the Future series', London: Le Play House Press.

25. P. Abercrombie (1926) 'The Planning of a New Coal Field', in Abercrombie, Branford *et al.*, op.cit., p.40.

26. Ibid., p.39.

27. P. Geddes (1926) 'Introduction: a national transition' in Abercrombie, Branford *et al* op.cit., p.1.

28. See p.149.

29. R. Fishman (1977) *Urban Utopias in the Twentieth Century: Ebenezer Howard, Frank Lloyd Wright and Le Corbusier*, New York: Basic Books, pp.188–204.

30. Controls over the economy were dismantled as quickly as possible after the war – R. Lowe (1978) 'The erosion of state intervention in Britain 1917–24', *Economic History Review* 31: 270–86.

31. H. McCrae, op.cit., p.15.

32. Pepler's biographer suggests that 'he was among those who felt deeply that to divorce a man from the land and from Nature was fundamentally wrong and in the early years following the First World War suggested some form of development which enabled people to participate part-time in agricultural work', H. McCrae, op.cit., p.3.

33. Lewis Mumford (1945) *City Development*, London: Secker & Warburg, p.7.

34. See P. Geddes and V.V. Branford (1917) *The Coming Polity*, the 'Making of the Future Series', London: Williams & Norgate, Chapter VIII, 'State and City'

35. Lewis Mumford (1966) 'Men and Ideas: The disciple's rebellion; a memoir of Patrick Geddes' *Encounter* 27: 11–21.

36. Lewis Mumford (1927) 'Who is Patrick Geddes', in A. Defries, op.cit., pp.1–7.

37. B. Mackaye (1950) 'Growth of a New Science', *The Survey* LXXXVI (10): 432–42. He suggests here that Geddes supplied him with the word 'geotechnics' which he came to define as 'the applied science of making the earth more habitable' which he saw as the purpose of his life's work with the TVA and later.

38. He described the notations as a 'heap of husks' that Geddes offered him when he sought 'the living kernel of his wisdom' – L. Mumford (1966) 'Men and Ideas', op.cit., p.14.

39. Lewis Mumford (1961) *The City in History*, London: Secker & Warburg, chapter 8.

40. Lewis Mumford (1944) *The Condition of Man*, London: Secker & Warburg, pp.382–90.

41. Mumford (1966) op.cit., p.20.

42. Zueblin (1899) wrote the article on the Outlook Tower 'The World's First Sociological Laboratory' *American Journal of Sociology* IV (5): 577–92.

43. E.W. Burgess and D.J. Brogue (eds) (1964) *Contributions to Urban Sociology*, Chicago: University of Chicago Press.

44. Lewis Mumford (1945) *City Development*, op.cit., p.7. But see R.E. Park (1925) 'The City: suggestions for the investigation of human behaviour in the urban environment', in R.E. Park, E.W. Burgess, and R.D. McKenzie (eds) *The City*, Chicago: University of Chicago Press,

pp.1—46.

45. See Geddes' review of Park *et al.* (eds), op.cit., in *Sociological Review* XVIII (1925): 167—8.

46. Ruth Glass (1955) 'Urban Sociology in Great Britain: a trend report', *Current Sociology* IV (4): 12.

47. Ibid., p.12.

48. Ibid., p.13.

49. A loosely Federated Regional Survey Association was formed in 1913. The most active centres were run by teachers who had attended Edinburgh Summer Meetings such as those at Kings Langley and Saffron Walden. Haslemere in Surrey had some enthusiastic volunteers. The WEA and Training College at Bingley, Yorkshire co-operated on a survey. There was also a Civic Survey group in Leicester. Regional Survey Conferences were run by the Geographical Association, proceedings published in *The Geographical Teacher* VIII (1915—1916): 89—102, 164—172; and by the Cities Committee of the Sociological Society, proceedings published in *Sociological Review* 11 (1919) and 12 (1920) under the title 'The Third Alternative'.

50. *Sociological Review* 18 (1926): 106—9.

51. Many letters have survived and are in the Geddes Papers MS10572, NLS.

52. Alice Garnett (1970) 'Herbert John Fleure 1877—1969' *Biographical Memoirs of Fellows of the Royal Society* 16: 259.

53. T.W. Freeman locates Fawcett's contribution to the 'regional approach' in a broader non-Geddesian movement which incorporated Odum and Moore in the US, Vidal de la Blache in France, and Swedish geographers at Lund University — T.W. Freeman (1961) *A Hundred Years of Geography*, London: Duckworth, pp.122—3.

54. The records of the Institute of Sociology (Le Play House) are kept at the University of Keele. Ancillary information is in the archive: the Papers of Sir Patrick Geddes, part 3 section 2, University of Strathclyde.

55. C.C. Fagg (1928) 'The History of the Regional Survey Movement', *South-Eastern Naturalist and Antiquary* 32: 71—94.

56. D.C.Jones (1934) *The Social Survey of Merseyside*, Liverpool: University Press of Liverpool, 3 vols.

57. Their presidential addresses were published in the *South-Eastern Naturalist and Antiquary* under the following titles: V.V. Branford (1923), 27, 'The Sciences and the Humanities', pp.60—70; E. Ogilvie (1924), 28, 'The Educational Value of the Regional Survey, pp.33—42; G.L. Pepler (1925), 29, 'The Regional Survey as a Preliminary to Town Planning', pp.81—9; P. Geddes (1927), 31, 'The Movement Towards Synthetic Studies and its Educational and Social Bearings', pp.83—94; H.J. Fleure (1930), 33, 'Regional Surveys and Welfare', pp.73—82.

58. S.H. Beaver (1962) 'The Le Play Society and Field Work', *Geography* XLVII: 225—40.

59. A periodical was launched by a group connected with Le Play House in 1924 with the title *Observation*. It was an attempt in their own words: 'to develop and popularise the scientific attitude of mind and the use of scientific methods in fields easily accessible to everyone'. Its appeal was very limited.

60. Regional survey committees (composed principally of local authority representatives) had been set up in twelve places for town planning purposes: at Doncaster, Manchester, Deeside, Rotherham, S. Teeside,

Mansfield, N. Tyneside, S. Tyneside, the Wirral, W. Middlesex, N.E. Surrey and Thames Valley — P. Abercrombie (1923–4) 'Regional Planning', *Town Planning Review* 10: 109–18.

61. S.H. Beaver, op.cit., p.238.
62. The work of Sir R.A. Fisher, which was decisive in ensuring this outcome included his three books *Statistical Methods for Research Workers*, *The Genetical Theory of Natural Selection* , *The Design of Experiments* — see D.A. MacKenzie (1981) *Statistics in Britain 1865–1930: the social construction of scientific knowledge*, Edinburgh: Edinburgh University Press, pp.183–213.
63. Lewis Mumford (1966) 'Men and Ideas', op.cit., p.19.
64. For example he constantly refers to the links between medicine and biology as if this proves all his theories, see vol.II of *Life: Outlines of General Biology* pp.1207, 1209, 1231, 1240, 1438, etc.
65. Ibid., pp.1194–5.
66. Ibid., p.1199 for definition of what he means by 'the web of life'.
67. Ibid., pp.1167–8.
68. Ibid., p.1197.
69. Ibid., p.1384.
70. Ibid., pp.1395–8.
71. Ibid., pp.1400–17.
72. The developments of his notations in *Life: Outlines of General Biology* are reproduced on pp.50–1.
73. Ibid., p.1123.
74. See chapter 2, p.21.
75. Patrick Geddes to David Eder, 22 June 1927, Central Zionist Archives Z1/3497 'Pardon also a personal point: but get from your library Miss Defries' book about me, just out 'The Interpreter Geddes: the Man and his Gospel' (an alarming title but she says due to Tagore)'.
76. H.G. Wells (1905) *A Modern Utopia*, London: Nelson.
77. Patrick Geddes to J. Arthur Thomson 8 February 1925, Geddes Papers MS10555, NLS.
78. For details see P. Boardman, op.cit., pp.391–2.
79. Lewis Mumford (1966) 'Men and Ideas', op.cit., p.18.
80. Organised by Professor A. Schimmerling who is also an editor of the architectural journal with a sociological perspective *Le Carré Bleu* (Paris).
81. A number of interviews, including one with Professor Marres of Montpellier and the late Mrs. Janine Geddes provided evidence that Geddes succeeded in creating a very beautiful garden despite the inhospitable hillside. Mrs. Geddes did not enjoy acting out mythological masques on her visits when a young girl.
82. Chateau D'Assas, P. Boardman, op.cit., pp. 377–9.
83. This was the symbol he had used since the earliest days which is seen most often on the publications of P. Geddes and Colleagues from 1895.
84. Some students very much enjoyed contact with Geddes, nevertheless — Interview with Dr. Pheroze Bharucha, 8 November 1972, Bombay. Bharucha was a student there in the late 1920s. He emphasised Geddes' vitality, energy and good humour and the charming companion he made on 7-mile after-dinner walks to Alsace!
85. Some of their long-standing correspondence is preserved in the Geddes Papers MS10569, NLS.
86. Due in part to the efforts made by his youngest son Arthur to sustain the

full emotional content and commitment of his father's social reconstruction doctrine and the tendency for Geddes' biographers to indulge in hagiography, for example, A. Defries, op.cit.; P. Boardman, op.cit. (1944 and 1978); and P. Mairet (1957) *A Pioneer of Sociology: life and letters of Patrick Geddes*, London: Lund Humphries.

87. With some notable exceptions, P. Abercrombie, G. Pepler, A. Glikson, and R.Grieve in town-planning; K.C. Edwards, A. Learmouth, and B.T. Robson in geography. See also George Gordon (ed.) (1986) *Regional Cities in the UK 1890—1980*, London: Harper & Row, pp.239—42.

88. For a discussion of this see M. Cuthbert (1987) 'The Concept of the Outlook Tower in the Work of Patrick Geddes', unpublished MA thesis, University of St Andrews, chapter 5.

89. Geddes' historical mentor was his friend, Professor J.S. Stuart-Glennie, of the University of Edinburgh. See his article on 'The Desirability of Treating History as a Science of Origins' *Transactions of the Royal Historical Society* (1890): 229—40. See also chapter 4, p.101.

90. The best illustration of this is the four editions of *The Evergreen*, winter, spring, summer, autumn, published by P. Geddes, Colleagues & Co., 1896.

91. The idea has been taken up from time to time by historical geographers during Geddes' lifetime and more recently. A recent discussion is L. Newson (1976) 'Cultural evolution: a basic concept for human and historical geography', *Journal of Historical Geography* 2 (3): 239—55.

92. The 'Third Alternative' to war and revolution.

93. W. Lesser (1974) 'Patrick Geddes: the practical visionary' *Town Planning Review* 45: 311—27.

94. Abercrombie wrote a long and eulogistic review of the work in *Town Planning Review* 6 (1915—16): 137—42. Lewis Mumford (1950) wrote 'Mumford on Geddes' *The Architectural Review* (August): 81—6, to greet the second edition, abridged and edited by J. Tyrwhitt, with the help of Arthur Geddes and G. Pepler brought out in 1949. A third edition, a full reprint of the original, was published in 1968 by Ernest Benn, with an introduction by Percy Johnson-Marshall.

95. *Cities in Evolution*, London: Williams & Norgate (1915) p.364.

96. E. Howard to Patrick Geddes, 2 August 1904, Geddes Papers MS10536, NLS.

97. D. Schaffer (1986) 'Ideal and Reality in 1930's Regional Planning: the case of the Tennessee Valley Authority', *Planning Perspectives* 1 (1): 27—44.

98. H.W.Odum (1951) *American Sociology: the story of sociology in the U.S. through 1950*, London and New York: Longmans, Green, pp.355—6.

99. E.A.A.Rowse was the principal of the Architectural Association School and founder of the school of Planning and Research for National Development in 1935. R. Unwin was the chairman, and G.Pepler was a member of the executive committee of the school.

100. Planning of a city is not merely the laying out of so many buildings and streets. It is the ordering of the lives of thousands of human beings. So great a work can only be undertaken by those who have received the fullest training possible. . . . The cost? We have in the face of death always found money to foot the bill. Today we face death as surely as if one held a knife at our throat — whether it be peace or war, it is nevertheless death! (E.A.A.Rowse (1938—9) 'The Planning of a City'

Journal of the Town Planning Institute 25: 167–71).
101. For example, the rebuilding of Plymouth and Bristol city centres after the war, and the later redevelopment of Birmingham city centre to accommodate the motor car.
102. Maxwell Fry (1941) 'The New Britain must be planned' *Picture Post*, 10 (1): 16–20.
103. P.Hall (ed.) (1981) *The Inner City in Context: the final report of the Social Science Research Council Inner Cities Working Party*, London: Heinemann.
104. Urlan Wannop (1987) 'The Planning Case for Regions and its Vindication in Strathclyde', unpublished paper, Conference on British Regionalism 1900–2000, Regional Studies Association and Planning History Group, September. See also Urlan Wannop (1986) 'Regional Fulfilment: planning into administration in the Clyde Valley 1944–84', *Planning Perspectives* 1: 207–29.
105. Quoted in Boardman, op.cit., p.448.
106. Geddes' ideas, as interpreted by L. Mumford, are influencing current developments in environmental education in Japan – Sadahiko Fujioka (1981) 'Environmental Education in Japan', *Hitotsubashi Journal of Social Studies* 13 (1): 14–16. See also Toshihito Andoh (1988) 'The Genesis of Environmental Education in Britain: Patrick Geddes and Edinburgh, 1880–1914', unpublished PhD thesis, Hitotsubashi University, Tokyo.
107. 'Why is progress thus so slow in the social levels, so that cities are still so unwholesome and under-housed, country villages so often proportionately yet worse off, as compared with what they should and so easily might be? The wheels of politics seldom fit and move ours, nor as yet conversely. What intermediate agency is needed, beyond the occasional 'crank', more useful than he gets credit for? One answer, at least, is surely – CITIZENSHIP', *Life: outlines of general biology*, vol.II, p.1233.
108. The relevance of this idea is discussed in P.D. Goist (1974) 'Patrick Geddes and the City', *Journal of the American Institute of Planners* 40: 31–7.
109. Some work is being done on this in the field of critical human geography: D.Gregory and J.Urry (eds) (1985) *Social Relations and Spatial Structures* London: Macmillan; R.D.Sack (1980) *Conceptions of Space in Social Thought: a geographic perspective*, London: Macmillan.

Select Bibliography

A. Notes on archival material

B. Works by Geddes:
        a) Natural sciences
        b) Social sciences
        c) Works relating to India and Palestine

C. Works on Geddes:
        a) Books
        b) Books containing substantial comment on Geddes
        c) Articles

D. Theses on Geddes
E. Select bibliography of secondary scources

A. Archival material

The main Archival collections of Geddes material are:

1. The Geddes Papers, National Library of Scotland, (catalogued but without descriptions).

2. The Geddes Papers, University of Strathclyde, (catalogued with descriptions).

3. The Branford Papers (containing the papers of LePlay House and the early Sociological Society but without any catalogue), University of Keele.

4. The Geddes collection, assembled by Julian Houston and Marshall Stalley, Rutgers University Library, USA.

5. Geddes Papers, catalogued without description, Central Zionist Archives, Jerusalem.

There is much Geddes archival material scattered in other places, especially municipalities and local record offices where he worked in the UK, Ireland, India, and Israel.

Select bibliography

B. Works by Geddes

Arranged in chronological order, with some grouping between the
natural/social sciences, major periodicals, and the Indian and Palestinian Town
Planning Reports.

a) Natural sciences

1879 'Chlorophylle animale et la physiologie des planaires vertes' *Archives de
Zoologie expérimentale et générale*, Paris.

1875–89 *Encyclopaedia Britannica*, 9th edition, continued in 10th and 11th
editions, articles on Reproduction, Sex, Variation and Selection,
Morphology.

1888–92 with J.A.Thomson, *Chambers Encyclopaedia* articles on Biology,
Botany, Environment, Evolution, and minor articles such as
Coelenterates, Flower, Fruit, etc.

1883–84 'Restatement of Cell Theory', *Proceedings of the Royal Society,
Edinburgh*.

1885–86 'Theory of Growth, Reproduction, Sex and Heredity', *Proceedings of
the Royal Society, Edinburgh*.

1886 'A Type of Botanic Garden', *Transactions of the Botanical Society of
Edinburgh* XVI.

1886 with J.A.Thomson, 'History and Theory of Spermatogenesis', *Proceedings
of the Royal Society, Edinburgh*.

1888 'Letter of application for the Chair of Botany, University of Edinburgh',
privately printed.

1888 'The Rise and Aims of Modern Botany', Dundee: University College.

1888–9 'A Restatement of the Theory of Organic Evolution', *Proceedings of the
Royal Society, Edinburgh*.

1889 with J.A.Thomson, *The Evolution of Sex*, London: Walter Scott. First
volume in the Contemporary Science series, (ed.) Havelock Ellis.

1890 'On the Origin of Thorny Plants', *Annual Report, British Association for the
Advancement of Science*

1893 *Chapters in Modern Botany*, London: John Murray. University Extension
Manuals series (ed.) Professor Knight.

1911 With J.A. Thomson, *Evolution*, London: Williams & Norgate, Home
University series.

1914 With J.A. Thomson, *Sex*, London: Williams & Norgate, Home University
series.

1924 With J.A. Thomson, *Biology*, London: Williams & Norgate, Home
University series.

1931 With J.A. Thomson, *Life: Outlines of General Biology*, London: Williams &
Norgate, 2 volumes.

b) Social sciences

1881 'Economics and Statistics, viewed from the standpoint of the preliminary
sciences', *Nature* 29.

Select bibliography

1881 'The Classification of Statistics and Its Results', *Proceedings of the Royal Society, Edinburgh*, (reprinted, Edinburgh: A. & C. Black, 1881).

1884 'An Analysis of the Principles of Economics', *Proceedings of the Royal Society, Edinburgh*, (reprinted, London: Williams & Norgate, 1885).

1884 *John Ruskin: Economist*, pamphlet, Edinburgh: Brown.

1885 'On the Application of Biology to Economics', *British Association for the Advancement of Science Annual Report*.

1885 'Report of Patrick Geddes, as the representative of the Edinburgh Social Union, to the Industrial Remuneration Conference', *Industrial Remuneration Conference, Report of the Proceedings and Papers*, London: Cassell.

1886 'On the Condition of Progress of the Capitalist and the Labourer', *The Claims of Labour*, Edinburgh: Co-operative Printing Co.

1886 *'Viri Illustres Academiae'*, (ed.) Edinburgh University Tercentenary, Edinburgh: Pentland.

1888 *Co-operation versus Socialism*, pamphlet, Manchester: Co-operative Printing Society Ltd.

1888 *Everyman His Own Art Critic: an introduction to the study of pictures*, pamphlet, Glasgow Exhibition, Edinburgh: William Brown.

1888 'University Extension', *The Scottish Review*, Edinburgh.

1890 'A Theory of the Consumption of Wealth', *Annual Report, British Association for the Advancement of Science*.

1896 'Life and Its Science', and 'The Scots Renascence' vol.I, Spring; 'Flower of the Grass', vol.II, Summer; 'The Sociology of Autumn', vol.III, Autumn; 'The Megalithic Builders', vol.IV, Winter; *The Evergreen*, Edinburgh: Patrick Geddes, Colleagues and Co.

1902 'Note on a Draft Plan for an Institute of Geography', 'Edinburgh and Its Region, Geographic and Historical', 'Nature Study and Geographical Education', all in the *Scottish Geographical Magazine* XVIII.

1903 'A Naturalists' Society and Its Work': an address to the introductory meeting of the Dunfermline Naturalists' Society', Part I and Part II *Scottish Geographical Magazine*, XIX.

1904 'On Universities in Europe and in India: and a needed type of research institute, geographical and social, five letters to an Indian friend,' reprinted from *The Pioneer* August 14, 1901, and from *East and West* September, 1903; Madras: National Press.

1904 'Professor Geddes' Report on his inspection and examination of Abbotsholme School, 11 July to 18 July', typescript, Abbotsholme School, Derbyshire.

1904 *City Development: a Study of Parks, Gardens and Culture Institutes. A Report to the Carnegie Dunfermline Trust*, Bournville: The Saint George Press, and Edinburgh: Geddes & Company, reprinted 1973 in facsimile, Shannon: Irish University Press.

1905 'A Great Geographer: Elisée Reclus, 1830–1905', *Scottish Geographical Magazine* XXI.

1905 *The World Without and the World Within: Sunday Talks with my Children*, pamphlet, London: George Allen.

1905 'Civics: as applied sociology, Part I', *Sociological Papers, 1904* (ed. by Editorial Committee, Soc.Soc), London, Macmillan.

1906 'Civics: as applied sociology, Part II', *Sociological Papers, 1905* ibid.
1907 'A Suggested Plan for a Civic Museum (or Civic Exhibition) and its Associated Studies', *Sociological Papers, 1906* ibid.
1908 'Chelsea, Past and Possible', in D.Hollins (ed.) *Utopian Papers: being addresses to the 'Utopians'*, London: Masters & Co.

The Sociological Review was founded in 1908 and Geddes published many articles and reviews in it between 1908 and 1931. These included:
1908, I, 'The Survey of Cities'.
1910, III, 'The late Mr. J.S.Stuart-Glennie', *ibid* 'Town Planning and City Design'.
1913, VI, 'Mythology and Life: an interpretation of Olympus; with applications to eugenics and civics', *ibid* 'Margaret Noble (Sister Nivedita) – A tribute'.
1915, VIII, 'Wardom and Peacedom: Suggestions Towards an Interpretation'.
1916–1917, IX, with V.V.Branford 'The Making of the Future'.
1919, X, 'Public Health in the Industrial Age' (a chapter from the Indore Report).
1923, XV, 'A Note on Graphic Methods, Ancient and Modern'.
1924, XVI, 'A Proposed Co–ordination of the Social Sciences'.
1926, XVIII, Devoted to Coal Crisis, reprinted in 'Making of the Future series', see below.
1927, XIX, 'The Charting of Life'; 'The Village World: Actual and Possible'.
1929, XXI, 'The Interpretation of current events: a sociological approach to the General Election'; 'Social Evolution: How to advance it'; 'Rural and Urban Thought: a contribution to the theory of Progress and Decay'.
1930, XXII, 'Ways of Transition – Towards Constructive Peace'; 'Scouting and Woodcraft: Present and Possible'.
1931, XXIII, 'Victor Branford – Obituary notice'; 'Talent and Genius'.

Garden Cities and Town Planning was launched in its new form in 1911. Geddes exerted some influence on the early numbers:
1911, I, 'The City Survey: A First Step' (in three parts).
1913, III, 'The City Beautiful– In Theory and Practice'.

The Town Planning Review was founded in 1910 and edited from the Department of Civic Design, University of Liverpool. Geddes did not contribute much to it, apart from two articles:
1912–13, III, 'The two-fold aspect of the Industrial Age: Paleotechnic and Neotechnic'.
1913–14, IV, 'Two steps in Civics: Cities and Town Planning Exhibition' and the 'International Congress of Cities: the Ghent International Exhibition, 1913'.

1911 With F.C.Mears, 'The Civic Survey of Edinburgh' (reprinted from *The Transactions of the Town Planning Conference* for the Civics Department, Outlook Tower, Edinburgh and Crosby Hall, Chelsea).
1911 With F.C.Mears, 'Cities and Town Planning Exhibition, Dublin and Belfast' *Guide Book and Outline Catalogue*, Dublin: Browne and Nolan.

1912   *The Masque of Ancient Learning and its many meanings: A Pageant of Education from Primitive to Celtic times devised and interpreted by Patrick Geddes*, Outlook Tower, Edinburgh: Patrick Geddes and Colleagues. *A Masque of Mediaeval and Modern Learning*, ibid., (Reprinted as *The Dramatisation of History*, London: LePlay House Press.

1915   *Cities in Evolution: an introduction to the town planning movement and to the study of civics*, London: Williams & Norgate.

Volumes in 'The Making of the Future series', edited by P.Geddes and V.V.Branford

1917   1. P.Geddes and Gilbert Slater, *Ideas at War*, London: Williams & Norgate.

1917   2. P.Geddes and V.V. Branford, *The Coming Polity: A Study in Reconstruction*, London: Williams & Norgate.

1919   3. P.Geddes and V.V. Branford, *Our Social Inheritance*, London: Williams & Norgate.

1926   4. P.Geddes, 'Introducton: a National Transition', V.V.Branford (ed.), *The Coal Crisis and the Future: a study of social disorders and their treatment*, London: LePlay House and Williams & Norgate.

During the war, the Cities Committee of the Sociological Society published a number of articles publicising Geddes' ideas which were also published as pamphlets:

1916–17, IX, 'The Banker's Part in Reconstruction'; 'The New Model'.

1918, X, 'A Rustic View of Peace'; 'The Doctrine of Civics'; 'The Drift to Revolution'.

1919, XI, 'The Civic School of Applied Sociology'.

1919 'Beginnings of a survey of Edinburgh' *Scottish Geographical Magazine* XXXV.

1921 'Women, The Census, and the Possibilities of the Future', notes of lectures, pamphlet, Edinburgh: Outlook Tower.

1925 'Talks from My Outlook Tower' *The Survey*, LIII and LIV.

1925 'Huxley as Teacher', *Nature*, 115.

c) Works relating to India and Palestine

1915   *Reports on the Towns in the Madras Presidency visited by Professor Geddes 1914–1915*, Madras: Government Press, (Twelve towns and one suburb of Madras).

1915   *Reports on Re-Planning of six towns in the Bombay Presidency*, reprinted Bombay: Maharashtra State Press, 1965.

1916   *Report on the Development and Expansion of the City of Baroda*, Baroda: 'Lakshmi Vilas' Press.

1916   *Report on Model Colony at Kanchrapara*, Calcutta: Eastern Bengal Railway.

1916   *Town Planning in Lucknow: a Report to the Municipal Council*, Lucknow: Murray's London Printing Press.

1917   *Town Planning in Lucknow: a Second Report to the Municipal Council,*

Lucknow: Murray's London Printing Press.

1917 *Report on Town Planning: Dacca*, Calcutta: Bengal Secretariat Book Depot.

1917 with H.V. Lanchester *Report to the Municipal Committee: Town Planning in Jubbulpore*, Jubbulpore: Hitkarini Press.

1917 *Report on the Collieries of Bihar and Orissa*, Orissa: State Printing Press.

1917 *Cawnpore Expansion Committee: a Report to H.H. the Lt.Governor of the United Provinces*, Allahabad.

1917 *Town Planning in Balrampur. A Report to the Hon'ble the Maharaja Bahadur*, Lucknow: Murray's London Printing Press.

1917 *Town Planning in Kapurthala: a Report to the Maharaja*, Lucknow: Murray's London Printing Press.

1917 *Town Planning in Lahore: a Report to the Municipal Council*, Lahore: Municipal Press.

1917 *Town Planning in Nagpur: a Report to the Municipal Committee*, Nagpur: Municipal Press.

1918 *Town Planning towards City-Development: A Report to the Durbar of Indore*, 2 vols, Indore: Holkar State Printing Press.

1919 *A Report to the Corporation of Calcutta: Barra Bazar Improvement*, Calcutta: Corporation Press.

1920 *Town Planning in Colombo*, Colombo: Municipal Press.

1922 *Town Planning in Patiala State and City: A Report to H.H. The Maharaja of Patiala*, Lucknow: Perry's Printing Press.

1919 *The proposed Hebrew University of Jerusalem: Preliminary Report on University Design*, typescript.

1919 *Jerusalem Actual and Possible: a Preliminary Report to the Chief Administrator of Palestine and the Military Governor of Jerusalem on Town Planning and Improvements*, typescript.

1920 *Town Planning in Haifa: a report to the Governor of Phoenicia*, typescript.

(n.d.) *Additional notes on the Zionist Commission's Carmel Estates*, typescript.

1920 *The Hot Springs of Tiberias*, typescript.

1925 *Town Planning Report — Jaffa and Tel–Aviv*, typescript.

1919 'The Temple Cities', *Modern Review*, (India) 25.

1920 *An Indian Pioneer: The Life and Work of Sir Jagadis Chandra Bose*, London: Longman.

1921 'Palestine in Renewal', *Contemporary Review* 120.

1922 'Essentials of Sociology in Relation to Economics' *Indian Journal of Economics* III (3).

1924 'Education in Return to Life', pamphlet, Bombay: published by Prof. M.M.Gidvani.

C. Works on Geddes

a) *Books* (in chronological order)

1927 A.Defries *The Interpreter: Geddes, the man and his gospel*, London: George Routledge & Sons.

1936 P.Boardman *Esquisse de l'Oeuvre éducatrice de Patrick Geddes*, Montpellier: Pierre-Rouge. Contains a full bibliography of Geddes' works.
1944 P.Boardman *Patrick Geddes: Maker of the Future*, Chapel Hill: University of North Carolina Press.
1947 J.Tyrwhitt (ed.) *Patrick Geddes in India*, London: Lund Humphries.
1957 P.Mairet *A Pioneer of Sociology: Life and Letters of Patrick Geddes*, London: Lund Humphries.
1972 M.Stalley (ed.) *Patrick Geddes, Spokesman for Man and the Environment*, New Brunswick: Rutgers University Press.
1975 P.Kitchen *A Most Unsettling Person: An Introduction to the Life and Ideas of Patrick Geddes*, London: Victor Gollancz.
1978 P.Boardman *The Worlds of Patrick Geddes: Biologist, Town Planner, Re-educator, Peace-Warrior*, London: Routledge and Kegan Paul.
1979 H.Meller (ed.) *The Ideal City*, Leicester: Leicester University Press.

b) *Books containing substantial comment on Geddes* (in alphabetical order)

P.Abercrombie (1933) *Town and Country Planning*, London: Butterworth.
P.Abrams (1968) *The Origins of British Sociology, 1834–1914*, Chicago and London: University of Chicago Press.
F.Alaya (1970) *William Sharp: 'Fiona Macleod', 1855–1905*, Cambridge Mass.: Harvard University Press.
C.R.Ashbee (1923) *A Palestine Notebook, 1918–1923*, London: Heinemann.
M.J.Bannon (ed.) *A Hundred Years of Irish Planning*, 2 vols, Dublin: Turoc Press.
Mabel Barker (1928) *L'Utilisation du Milieu Géographique pour l'Education*, Montpellier: Librairie Nouvelle.
H.E.Barnes (1948) *Introduction to the History of Sociology*, Chicago: Chicago University Press.
G.Bell and J. Tyrwhitt (eds) (1972) *Human Identity in the Urban Environment*, Harmondsworth: Penguin Books.
M.Z.Brooke (1970) *Le Play: Engineer and social scientist*, London: Longman.
J.Burchard and O.Handlin (eds) (1966) *The Historian and the City*, Cambridge Mass.: M.I.T.Press.
J.V.Ferreira and S.S.Jha (eds) (1976) *The Outlook Tower: essays on Urbanization in memory of Patrick Geddes*, Bombay: Popular Prakashan Private Ltd.
T.W.Freeman (1961) *A Hundred Years of Geography*, London: Gerald Duckworth.
A.G.Gardiner (1913) *Pillars of Society*, London: J.M.Dent & Sons.
F.Jackson (1985) *Sir Raymond Unwin: Architect, Planner and Visionary*, London: A.Zwemmer Ltd.
H.V.Lanchester (1924) *Talks on Town Planning*, London: Jonathan Cape. (The Professor in these dialogues is a thinly disguised portrait of Geddes).
L.Mumford (1944) *The Condition of Man*, London: Martin Secker & Warburg.
J.Mavor (1923) *My Windows on the Street of the World*, 2 vols. London: J.M.Dent & Sons.
E.Sharp (1912) *William Sharp: A Memoir*, 2 vols. London: J.M.Dent & Sons.
A.Sutcliffe (1981) *Towards the Planned City: Germany, Britain, the United States and France, 1780–1914*, London: Blackwell.

Select bibliography

P.Uttenhove (1985) *Resurgam: La Reconstruction en Belgique apres 1914*, Bruxelles: Crédit Communal de Belgique.

c) *Articles on Geddes and his work*

Anon. (1968) 'The Valley Plan of Civilisation', *Architect's Yearbook* 12: 65–71.
Anon. (1969) 'Actualités de Patrick Geddes, textes et documents', *Architecture d'aujourdhui* 143: v–vii.
M.J.Bannon (1978) 'The Making of Irish Geography, III: Patrick Geddes and the Emergence of Modern Town Planning in Dublin', *Irish Geography* II: 141–8.
M.M.Barker (1914) 'Dublin School of Civics', *Scottish Geographical Magazine* XXX: 604–6.
V.V.Branford (1917) *Citizen Soldier: A Memoir of Alasdair Geddes*, pamphlet, London: Headley.
R.N.Rudnose Brown (1948) 'Scotland and some trends in Geography: John Murray, Patrick Geddes and Andrew Herbertson', *Geography* XXXIII: 107–16.
H.Carter (1915) 'The Garden of Geddes', *The Forum* 54: 455–71 and 588–95.
C.C.Fagg (1928) 'The History of the Regional Survey Movement', *South-Eastern Naturalist and Antiquary* 32: 71–94.
H.J.Fleure (1953) 'Patrick Geddes, 1854–1932', *Sociological Review* (new series) 1: 5–13.
R. Glass (1955) 'Urban Sociology in Great Britain: a trend report', *Current Sociology* IV (4): 8–35.
A.Gliksen (1954) 'The Planner Geddes', *Journal of the Association of Engineers and Architects in Israel* XII: 7–10.
Park Dixon Goist (1974) 'Patrick Geddes and the City', *Journal of the American Institute of Planners* 40 (1): 31–7.
R.J.Halliday (1968) 'The Sociological Movement, the Sociological Society and the genesis of academic sociology in Britain' *Sociological Review* (new series) 16: 377–98.
J.Hasselgren (1982) 'What is Living and What is dead in the work of Patrick Geddes?', in *Patrick Geddes: A Symposium*, special occasional paper in Town and Regional Planning, Duncan of Jordanstone College of Art, University of Dundee, pp.26–49.
M.Hebbert (1980) 'Patrick Geddes Reconsidered', *Town and Country Planning*, pp.15–17.
M.Hebbert (1982) 'Retrospect on the Outlook Tower', in *Patrick Geddes: A Symposium*, Special Occasional Paper in Town and Regional Planning, Duncan of Jordanstone College of Art/University of Dundee, pp.49–65.
W.Lesser (1974) 'Patrick Geddes: the Practical Visionary', *Town Planning Review*, pp.311–27.
E.McGegan (1935) 'Sir Patrick Geddes', *Scottish Bookman* 1: 99–106.
E.McGegan (1940) (with Arthur Geddes and F.C.Mears) 'The Life and Work of Sir Patrick Geddes', *Journal of the Town Planning Institute* 26: 189–95.
B.Mackaye (1950) 'Growth of a New Science', *The Survey* LXXXVI (10):439–42.
H.E.Meller (1973) 'Patrick Geddes: an analysis of his theory of civics,

1880–1904', *Victorian Studies* XVI (3): 291–315.

H.E.Meller (1977) 'Patrick Geddes and his contribution to the modern town-planning movement in pre-independence India', *Urban and Rural Planning Thought*, Delhi School of Planning and Architecture, XX (1): 1–7.

H.E.Meller (1979) 'Urbanisation and the introduction of Modern Town Planning Ideas in India, 1900–25', in C.Dewey and K.Chaudhuri (eds) *Economy and Society: Essays in Indian Economic and Social History*, Delhi: Oxford University Press. pp.330–50.

H.E.Meller (1979) 'Patrick Geddes, 1854–1932', *Stadtbauwelt* 12: 478–80.

H.E.Meller (1980) 'Cities in Evolution: Patrick Geddes as an International Prophet of Town Planning before 1914' in A.Sutcliffe (ed.) *The Rise of Modern Urban Planning 1890–1914*, London: Mansell Publishing Ltd. pp.199–223.

H.E.Meller (1981) 'Patrick Geddes, 1854–1932', in G.Cherry (ed.) *Pioneers in British Planning*, London: The Architectural Press, pp.46–71.

H.E.Meller (1982) 'Geddes and his Indian Reports', in *Patrick Geddes: A Symposium*, Special Occasional Paper in Town and Regional Planning, Duncan of Jordanstone College of Art/University of Dundee. pp.4–25.

H.E.Meller (1983) 'City Development in the turn of the century Scotland', in C.J.Carter (ed.) *Art, Design and the Quality of Life in Turn of the Century Scotland* Dundee: Department of Town and Regional Planning. pp.25–51.

H.E.Meller (1987) 'Conservation and Evolution: Patrick Geddes' work in Palestine', in *Jerusalem Papers* 1 and 2, The Jerusalem Centre for Planning Historic Cities pp.46–57.

L.Mumford (1929) 'Patrick Geddes, Insurgent', *New Republic* 60: 295–6.

L.Mumford (1950) 'Mumford on Geddes', *The Architectural Review* Aug. pp.80–4.

L.Mumford (1966) 'Men and Ideas: the Disciple's Rebellion; a memoir of Patrick Geddes', *Encounter* 27: 11–21.

(1932) Obituary Tributes to Geddes in the *Sociological Review* XXIV (3): 355–95.

H.W.Odum (1944) 'Patrick Geddes' Heritage to the Making of the Future', *Social Forces* 22.

A.D.Peacock (1954) 'Patrick Geddes: Biologist', commemorative address on the occasion of the celebration of the centenary of Geddes' birth held by the University of St. Andrews at the Queen's College, Dundee. *Alumnus Chronicle*, St.Andrews University. pp.1–14.

B.T.Robson (1981) 'Geography and Social Science: the role of Patrick Geddes', in D.R.Stoddart (ed.) *Geography, Ideology and Social Concern*, London: Blackwell. pp.186–207.

W.I.Stevenson (1975) 'Patrick Geddes and Geography: biobibliographical study', *Occasional Paper*, 27 Department of Geography, University of London.

C.Ward (1973) 'The Outlook Tower, Edinburgh: prototype for an Urban Studies Centre', *Bulletin of Environmental Education* 32.

K.Wheeler (1974) 'Note on the Valley Section of Patrick Geddes' *Bulletin of Environmental Education* 33.

C.Zueblin (1899) 'The World's First Sociological Laboratory', *American Journal of Sociology* 4: 577–92.

Select bibliography

D. Theses on Geddes

P.Boardman (1936) 'Esquisse de l'Oeuvre éducatrice de Patrick Geddes', PhD, University of Montpellier.

P.Green (1970) 'Patrick Geddes', PhD, University of Strathclyde.

J.P.Reilly (1972) 'The Early Social Thought of Patrick Geddes', PhD, University of Columbia, New York.

M.Cuthbert (1987) 'The Concept of the Outlook Tower in the Work of Patrick Geddes', MPhil, University of St.Andrews.

Toshihiko Andoh (1988) 'The Genesis of Environmental Education in Britain: Patrick Geddes and Edinburgh, 1880–1904', PhD, Hitotsubashi University, Tokyo.

Lilian Andersson (1989) 'Beyond Bureaucratic and Economic Rationality: a study in Camillo Sitte's and Patrick Geddes' Town Planning Strategies', PhD, University of Gothenburg, Sweden.

E. Select Bibliography of Secondary sources

Abercrombie, P., Kelly, S. & A. (1922) *Dublin of the Future: The New Town Plan*, London: University Press of Liverpool/Hodder & Stoughton.

Alexander, P. and Gill, R. (1984) *Utopias*, London: Duckworth.

Allison, R. (1964) *Encyclopaedias: their history through the ages*, London: Hafner.

Andrews, C.F. (1912) *The Renaissance in India: its missionary aspect*, London: Church Missionary Society.

Armytage, W.H.G. (1961) *Heaven's Below: utopian experiments in England*, London: Routledge & Kegan Paul.

Armytage, W.H.G. (1955) *Civic Universities: aspects of a British tradition*, London: Benn.

Arnold, M. (1869) *Culture and Anarchy*, Cambridge: Cambridge University Press pb.edn. 1969 reprint.

Ashworth, W. (1954) *The Genesis of Modern British Town Planning: a study in the economic and social history of the nineteenth and twentieth centuries*, London: Routledge & Kegan Paul.

Ashworth, W. (1987) *A Short History of the International Economy since 1850*, London: Longman 4th edn.

Barnett, Rev. & Mrs. S.A. (1888) *Practicable Socialism: essays on social reform*, London: Longman, Green.

Basu, Aparna (1974) *The Growth of Education and Political Development in India 1898–1920*, Delhi: Oxford University Press.

Beevers, R. (1988) *The Garden City Utopia: a critical biography of Ebenezer Howard*, London: Macmillan.

Bibby, C. (1959) *T.H. Huxley: Scientist, Humanist and Educator 1825–95*, London: Watts.

Booth, C. (1889) *Life and Labour of the People of London*, London: Macmillan, 1902 edn.

Bose, A. (1970) *Urbanisation in India*, New Delhi: Academic Books.

Boyd Rayward, W. (1975) *The Universe of Information: The Work of Paul Otlet for*

Documentation and International Organisation, published for International Federation for Documentation by All-Union Institute for Scientific and Technical Information, Moscow.

Brooke, M.Z. (1970) *Le Play: engineer and social scientist: the life and work of Frédéric Le Play*, London: Longman.

Burchard, J.and Handlin, O. (eds) (1966) *The Historian and the City*, Cambridge, Mass; MIT Press.

Burgess, E.W. and Brogue, D.J. (eds) (1964) *Contributions to Urban Sociology*, Chicago: University of Chicago Press.

Burrow, J.W. (1966) *Evolution and Society: a study in Victorian Social Theory*, Cambridge: Cambridge University Press.

Chaudhuri, K.N. and Dewey, C.J. (eds) (1979) *Economy in Society: essays in Indian economic and social history*, Delhi: Oxford University Press.

Checkland, O. (1980) *Philanthropy in Victorian Scotland: social welfare and the voluntary principle*, Edinburgh: John Donald.

Cherry, G.E. (1979) *The Evolution of British Town Planning*, Leighton Buzzard; Leonard Hill.

Cherry, G.E. (1982) *The Politics of Planning*, London: Longman.

Cherry, G.E. (ed.) (1981) *Pioneers in British Planning*, London: The Architectural Press.

Clark, T.N. (1973) *Prophets and Patrons: the French University and the Emergence of the Social Sciences*, Cambridge, Mass: Harvard University Press.

Collingwood, R.G. (1945) *The Idea of Nature*, Oxford: Clarendon Press.

Cranz, G. (1982) *The Politics of Parks Design: A History of Urban Parks in America*, Cambridge, Mass: MIT Press.

Crawford, A. (1966) *C.R. Ashbee: Architect, Designer and Romantic Socialist*, New Haven and London: Yale University Press.

Daston, R., Heidelberger, M. and Kruger, L. (eds) (1986) *The Probabilistic Revolution Vol. I: Ideas in History*, Bradford Books.

Davis, K. (1951) *The Population of India and Pakistan*, Princeton: Princeton University Press.

Defries, A. (1927) *The Interpreter: Geddes, the Man and his Gospel*, London: Routledge.

Demolins, E. (1898) *Anglo-Saxon superiority: to what is it due?*, (A quoi tient la supériorité des Anglo-Saxons?), translated by L.B.Lavigne, London: the Leadenhall Press.

Drake, B. and Cole, M. (eds) (1948) *Our Partnership by Beatrice Webb*, London: Longmans, Green.

Durant, W. (1962) *Outlines of Philosophy: Plato to Russell*, London: Ernest Benn.

Escott, T.H.S. (1897) *Social Transformations of the Victorian Age: a survey of court and country*, London: Seeley.

Evans-Prichard, E.E. (1972) *Social Anthropology*, London, Routledge & Kegan Paul, repr. edn.

Febvre, L. (1925) *A Geographical Introduction to History*,London: Kegan Paul, Trench & Trubner.

Fishman, R. (1977) *Urban Utopias in the Twentieth Century: Ebenezer Howard, Frank Lloyd Wright, Le Corbusier*, New York: Basic Books.

Select bibliography

Fletcher, R. (1971) *The Making of Sociology: a study of sociological theory*, London: Nelson.

Freeman, T.W. (1961) *A Hundred Years of Geography*, London: Gerald Duckworth.

Freeman, T.W. and Pinchemel, Philippe (1977–1984) *Geographers: Bibliogaphical Studies*, vol.III, London: Mansell.

Gilbert, B.B. (1966) *The Evolution of National Inusurance in Great Britain: the origins of the Welfare State*, London: Joseph.

Gillion, K.L. (1968) *Ahmedabad: A Study in Indian Urban History*, Berkeley: University of California Press.

Gordon, George (ed.) (1986) *Regional Cities in the UK 1890–1980*, London: Harper & Row.

Gregory, D. and Urry, J. (eds) (1985) *Social Relations and Spatial Structures*, London: Macmillan.

Grosskurth, P. (1981) *Havelock Ellis: a biography*, London: Quartet Books.

Hall, P. (ed.) (1981) *The inner city in context: the final report of the Social Science Research Council Inner Cities Working Party*, London: Heinemann.

Harrison, B. (1978) *Separate Spheres: the opposition to women's suffrage in Britain*, London: Croom Helm.

Harrison, J.F.C. (1961) *Learning and Living*, London: Routledge & Kegan Paul.

Hartog, M. (1949) *P.J. Hartog: a memoir by his wife*, London: Constable.

Harvey, David (1973) *Social Justice and the City*, London: Edward Arnold.

Harvie, C. (1977) *Scotland and Nationalism: Scottish Society and Politics 1707–1977*, London: George Allen & Unwin.

Hennock, E.P. (1973) *Fit and Proper Persons: Ideal and Reality in Nineteenth Century Urban Government*, London: Edward Arnold.

Herbertson, D. (1946) (eds. V.V. Branford and A. Farquarson) *The Life of Frederic Le Play*, Ledbury: LePlay House.

Hill, O. (1970) *Homes of London Poor*, and Mearns, A. *The Bitter Cry of Outcast London*, London: Frank Cass.

Hobsbawm, E.J. (1987) *The Age of Empire 1875–1914*, London: George Weidenfeld & Nicolson.

Hollins, D. (ed.) (1908) *Utopian Papers: being addresses to 'the Utopians'*, London: Masters.

Horsfall, T.G. (1904) *The Improvement of the Dwellings and Surroundings of the People: The Example of Germany*, supplement to the Report of the Manchester Citizens Association, Manchester: University Press.

Howarth, O.J.R. (1931) *The British Association: A retrospect 1831–1931*, London: British Association.

Hughes, H. Stuart (1959) *Consciousness and Society: the reorientation of European social thought 1890–1930*, London: MacGibbon & Kee.

Jackson, Frank (1985) *Sir Raymond Unwin: Architect, Planner and Visionary*, London: A. Zwemmer.

Jaransch, K.H. (ed.) (1983) *The Transformation of Higher Learning, 1860–1930: Expansion, Diversification, Social Opening and Professionalisation in England, Germany, Russia and the United States*, Chicago: Chicago University Press.

[345]

Jones, D.C. (1934) *The Social Survey of Merseyside*, Liverpool: University Press of Liverpool.

Kent, R.A. (1981) *A History of British Empirical Sociology*, Aldershot: Gower Publishing.

King, A.D. (1976) *Colonial Urban Development: Culture, Social Power and Environment*, London: Routledge & Kegan Paul.

Lanchester, H.V. (1918) *Town Planning in Madras: a review of the conditions and requirements of city improvement and development in the Madras Presidency*, London: printed for the Government of Madras by Taylor, Garnett, Evans and Co.

Lanchester, H.V. (1925) *The Art of Town Planning*, London: Chapman & Hall.

Lees, A. (1985) *Cities Perceived: Urban Society in European and American Thought, 1820–1940*, Manchester: Manchester University Press.

Lewis, W.A. *Growth and Fluctuations 1870–1913*. London, Allen and Unwin: 1978.

Linton Bogle, J.M. (1929) *Town Planning in India*, London, Bombay, Calcutta, Madras: Oxford University Press.

Lloyd George, D. (1938 edn) *War Memoirs* Vol.I , London: Odhams Press.

Lloyd Morgan, C. (1894) *An Introduction to Comparative Psychology*, Contemporary Science Series (ed.) H.H.Ellis, London: Walter Scott.

Lloyd Morgan, C. (1896) *Habit and Instinct*, London: E.Arnold.

Lloyd Morgan, C. (1912) *Instinct and Experience*, London: Methuen.

Lloyd Morgan, C. (1923) *Emergent Evolution*, the Gifford Lectures, London: Williams & Norgate.

MacKenzie, D.A. (1981) *Statistics in Britain 1865–1930: The Social Construction of Scientific Knowledge*,Edinburgh: Edinburgh University Press.

McLennan, G., Held, D. and Hall, S. (eds) (1984) *State and Society in Contemporary Britain: a critical introduction*, Cambridge: Polity Press.

Mawson, T.H. (1927) *The Life and Work of an English Landscape Architect*, privately published.

Meller, H.E. (1976) *Leisure and the Changing City*, London: Routledge & Kegan Paul.

Morgan, K.O. (1981) *Rebirth of a Nation: Wales, 1880–1890*, New York and Oxford: Oxford University Press.

Mowat, C.L. (1961) *The Charity Organisation Society, 1869–1913: Its ideas and work*, London: Methuen.

Mukerjee, R. (1926) *Civics*, London: Longmans, Green.

Mukerjee, R. (1926) *Regional Sociology*, New York: The Century Company.

Mukerjee, R. (ed.) (1952) *A City in Transition: social problems of Lucknow*, J.K. Institute of Sociology, University of Lucknow.

Muller, H.J. (1964) *Science and Criticism: The Humanistic Tradition in Contemporary Thought*, New Haven: Yale University Press.

Mumford, Lewis (1944) *The Condition of Man*, London: Secker & Warburg.

Mumford, Lewis (1945) *City Development*, London: Secker & Warburg.

Mumford, Lewis (1961) *The City in History*, London: Secker & Warburg.

Nanda, B.R. (1965) *The Nehrus. Motilal and Jawaharlal*, London: Allen & Unwin.

Nehru, J. (1936) *An Autobiography*, reprinted London 1958.

Nettlefold, J.S. (1914) *Practical Town Planning*, London: St. Catherine's Press.

Select bibliography

Odum, H.W. (1951) *American Sociology: the story of sociology in the U.S. through 1950*, London and New York: Longmans, Green.

Ouvry, Elinor Southwood (1933) *Extracts from Octavia Hill's 'Letters to Fellow Workers' 1864–1911*, London: The Adelphi Book Shop.

Owen, J.E. (1974) *L.T. Hobhouse, Sociologist*, London: Nelson.

Park, R.E., Burgess, E.W. and McKenzie, R.D. (eds) (1925) *The City*, Chicago: University of Chicago Press.

Parker, W.H. (1982) *Mackinder: Geography as an aid to statecraft*, Oxford: Clarendon Press.

Parsons, Talcott (1968 edn) *The Structure of Social Action: A study in social theory with special reference to a group of recent European writers*, New York: The Free Press.

Peel, J.D.Y. (1971) *Herbert Spencer: the evolution of a sociologist*, London: Heinemann.

Pelling, H. (1968) *A Short History of the Labour Party*, London: Macmillan, third edn.

Pentland, Lady (1928) *The Rt. Hon. John Sinclair, Lord Pentland GCSI: A Memoir*, London: Methuen.

Pollard, S. (ed.) (1970) *The Gold Standard and Employment Policies between the wars, 1919–1939*, London: Methuen.

Prochaska, F. (1988) *The Voluntary Impulse: Philanthropy in Modern Britain*, London: Faber.

Ratzel, F. (1896–98) *History of Mankind*, translated by A.J.Butler, 3 Vols., London, Macmillan.

Rice, S. (1931) *Sayaji Rao III Maharaja of Baroda*, London: Oxford University Press.

Richards, E.P. (1914) *Report on the Request of the Improvement Trust on the condition, improvement and town planning of the city of Calcutta and contiguous areas*, published privately.

Richter, M. (1964) *The Politics of Conscience: T.H. Green and his Age*, London: Weidenfeld & Nicolson.

Rolland, Romain (1929) *I. La Vie de Ramakrishna II. La Vie de Vivekananda et l'Evangile universel*, Paris: Delamain et Bonterleau.

Rowntree, B.S. (1902) *Poverty: a study of town life*, London: Macmillan.

Rubinstein, David (1986) *Before the Suffragettes: Women's Emancipation in the 1890's*, Brighton: Harvester Press.

Sack, R.D. (1980) *Conceptions of Space in Social Thought: a geographic perspective*, London: Macmillan.

Sadler, Michael (1891) *University Extension: Past, Present and Future*, London: Cassell.

Seal, Anil (1968) *The Emergence of Indian Nationalism: Competition and Collaboration in the Later Nineteenth Century*, Cambridge: Cambridge University Press.

Searle, G.R. (1971) *The Quest for National Efficiency: a study of British Politics and Political Thought 1899–1914*, Oxford: Blackwell.

Sergeant, P.W. (1928) *The Ruler of Baroda: An account of the life and work of the Maharaja Gaekwar*, London: John Murray.

Sherrington, C. (1963) *Man in his Nature*, Cambridge: Cambridge University Press.

[347]

Select bibliography

Simey, Margaret B. (1951) *Charitable effort in Liverpool in the Nineteenth Century*, Liverpool: University Press.

Simpson, M. (1985) *Thomas Adams and the Modern Planning Movement, Britain, Canada and the United States, 1900–1940*, London: Mansell.

Soffer, R.N. (1978) *Ethics and Society in England: The Revolution in the Social Sciences 1870–1914*, Berkeley: University of California Press.

Southgate, D. (1982) *University Education in Dundee – a Centenary History*, University of Dundee.

Spencer, H. (1876–96) *The Principles of Sociology*, 3 vols, London: Williams & Norgate.

Springhall, J. (1977) *Youth, Empire and Society: British youth movements, 1883–1940*, London: Croom Helm.

Srinivas, M.N. (1976) *Social Change in Modern India*, Cambridge: Cambridge University Press.

Stephen, K. (1922) *The Misuse of Mind: a study of Bergson's attack on intellectualism*, London: Kegan Paul, Trench Trubner.

Stoddart, D.R. (ed) (1981) *Geography, Ideology and Social Concern*, London: Blackwell.

Storrs, R. (1937) *Orientations*, London: Ivor Nicholson & Watson.

Sutcliffe, Anthony (ed.) (1980) *The Rise of Modern Urban Planning, 1800–1914*, London: Mansell.

Sutcliffe, Anthony (1981) *Towards the Planned City: Germany, Britain, the United States and France, 1780–1914*, Oxford: Blackwell.

Swenarton, M. (1981) *Homes fit for Heroes*, London: Heinemann.

Tagore, Rabindranath: Pioneer in Education Essays and Exchanges between Rabindranath Tagore and Leonard Elmhirst, London: John Murray/Visva–Bharati, 1961.

Thompson, Kenneth (1976) *Auguste Comte: The foundation of sociology*, London: Nelson.

Tinker, H.R. (1954) *The Foundations of Local Self–Government in India, Pakistan and Burma*, London: Athlone Press.

Turner, J.A. (1914) *Sanitation in India*, Bombay: The Times of India.

Unwin, R. (1909) *Town Planning in practice: an introduction to the art of designing cities and suburbs*, London: Ernest Benn.

Vicinus, M. (ed.) (1980) *Suffer and be still: Women in the Victorian Age*, London: Methuen pb. edn.

Villiers–Stuart, C.M. (1913) *Gardens of the Great Mughals*, London: Adam & Charles Black.

Ward, B.M. (1934) *Reddie of Abbotsholme*, London: George Allen & Unwin.

Ward, L.F. (1883) *Dynamic Sociology*, New York: D.Appleton.

Ward, L.F. (1898) *Outlines of Sociology*, New York, London: Macmillan.

Webb, B. (1926) *My Apprenticeship*, London: Longmans.

Webb, B. (eds B. Drake and M.I. Cole) (1948) *Our partnership*, London: Longmans, Green.

Weber, A.F. (1899) *The Growth of Cities in the Nineteenth Century*, Reprinted Ithaca NY: Cornell University Press.

Wells, H.G. (1934) *Experiment in Autobiography: discoveries and conclusions of a very ordinary brain – 1866*, London: Victor Gollancz.

Select bibliography

Wells, H.G. (1905) *A Modern Utopia*, London: Chapman & Hall.

West, T. (1986) *Horace Plunkett, Co-operation and Politics: An Irish biography*, Washington DC: The Catholic University of America Press.

Wilson, E.O. (1976) *Sociobiology: The New Synthesis*, Cambridge, Mass: Harvard University Press.

Wohl, A. (1977) *The Eternal Slum: Housing and Social Policy in Victorian London*, London: Edward Arnold.

Wright Mills, C. (1970) *The Sociological Imagination*, Harmondsworth: Penguin edn.

Index